2

On Reason, Morality

and

the would-be gods

~~~***~~~

"The most admirable Justin rightly declared that the aforesaid demons resembled robbers."
~ Tatian (c.120-c.180 A.D.), *Address to the Greeks*, chap. xviii.

~~~***~~~

By

William Thomas Sherman

2022.

On Reason, Morality and the would-be gods
Copyright 2022, William Thomas Sherman
Text last updated: 5 April 2024.

A New Treatise on Hell
Copyright 2001, William Thomas Sherman, Txu1-024-206, 7/9/01

Wm. Thomas Sherman.
1604 NW 70th St.
Seattle, WA 98117
206-784-1132
wts@gunjones.com and williamthomassherman@gmail.com
http://www.gunjones.com

TABLE OF CONTENTS

PREFACE…or Twenty Years After. 7
I. 11
II. 21
III. 32
IV. 38
V. 48
VI. 56
VII. 60
VIII. 69
IX. 73
X. 81
XI. 90

APPENDIX 1: A NEW TREATISE ON HELL 95
  Preface 95
  Introduction 98
  What is Hell? 104
  "The Evil One" 108
  Some Arguments Hell uses to Influence People 120
  The People of Hell 130
  Who Comes from and Works for Hell 137
  Demons and Sprites 142
  Little Folk and Other Spirit People 145
  Devils, False Gods, and the Damned 152
  Gyro 158
  Simon Magus 161
  Gomez 168
  How Hell Characterizes Judaism and Christianity 173
  "Regular" People who work with Hell and Why 175
  Faustus 180
  Hell at Large 182
  Specific Methods and Devices Hell Attacks With 188
  Ways to Fight and Cope with Hell 196
  In Conclusion 215
  APPENDIX A: Spirit People Primer or "Spirit People for Dummies" 220
  APPENDIX B: Proof for the Existence of Spirit People 225

APPENDIX 2: NARRATIVE 243

6

PREFACE...or Twenty Years After.

In my earlier *A New Treatise on Hell* (2003), and accompanying "Narrative" – both of which are included as appendices to the present volume – the primary intent was to raise, introduce, and address the subject of what I have come to denote as "criminal spirit people," and this from a would-be objective and empirical viewpoint. I made clear at that time the endeavor was rather a preliminary inquiry and exploratory dissertation to serve as an lead-in to the subject, and that, moreover, much of what I had to say was conjectural and on a given point perhaps (as it might turn out) flat out in error. Yet the phenomena of criminal spirit people, and the "regular" (i.e., flesh and blood) persons involved in their machinations, are, without question, real enough; only that which is known is to a large extent incomplete, sketchy and there is much still that needs to be further looked into, analyzed, organized, and clarified. But one must start somewhere, and this is what I tried and set out to do. Of course, I can and do not claim that in every instance my observations or conclusions are all or by any means one hundred percent accurate: I admittedly may be only partly right or even entirely wrong in conclusions based on my own experience, surmise or interpretation. But if so, I can at least acknowledge at the outset the possibility, and my willingness to be corrected by others who know or who can explain the given matter better than I can, and doubtless there are others and more informed who can speak far better and more fully on a particular topic or point than myself. Needless to add, there are many things I simply *don't know*; which if I did, would doubtless enhance and better elucidate an understanding of what is known.[1]

By "*criminal spirit people*" I mean "other-worldly" persons whom we vulgarly and traditionally identify as "demons," "ghosts," and or else "gods;" and which as literal labels, I submit, should be stringently avoided. One could or might address the question of non-criminal spirit people, yet inasmuch as the criminal ones are a serious problem and thus call for their being addressed as a topic, the other category of non-criminal has no compelling or practical mandate just here.

According to my view, a spirit person is by definition criminal if they or those under their orders engage in behavior that can be identified as criminal if done by a "regular" flesh and blood person: for example, murder, torture, robbery, physical assault and battery, vandalism. And just as with "regular" people, a spirit person may be more or less guilty. There may be circumstances to mitigate a given individual's putative guilt. One person, for instance, might be classed as a misdemeanor sort another a hard core offender; this other a dumb slave coerced into wrong-doing; this yet other a veteran and career criminal, etc., with there being varying degrees of culpability in between. In my experience and speaking in very general terms, most criminal spirit people would seem to be males, but

---

[1] It would, for instance, be interesting, were it possible, to be able to apply David Hartley's (1705-1757) "vibratory" theory on the subject, and reflect on where that might lead.

there are female criminal spirit persons, though more rarely encountered, on occasion as well.

It is of course out of hand assumed or asserted by some that criminal spirit people do not exist. It is, by contrast, my contention that they most certainly do, but and while stating emphatically the task – for a number of different reasons, not least of which long standing dogma and childish prejudice on the subject – is *far* from being an easy one. For one thing the contest becomes one of: what will determine what is truth and by what authority is said truth established? Sheer violence and worldly, physical might, *or* honest and right reason and science? For if, at least for the sake of argument, we posit criminal spirit people, and such who would dress themselves on occasion with the mantle of would-be "gods," who then will people listen to as authority? Such "gods" or an honest, rational person? For many, perhaps even most, the answer would seem to be the former. Brute force and intimidation, backed by stupefying and utterly astounding legerdemain, is identified as authority, *not* (or less so) honesty and right reason. That is, if what I claim about there actually being criminal spirit person *is* true, there are or will be those who, given the dishonest and criminal character of such preternatural persons, will do what they can to gainsay and refute the honest rational thinker on this question. And this, let it be understood (at least as I understand and see it), is what someone like myself is up against in attempting to make my case; with scoffing and semi-rational skepticism being time honored, routine devices and the shields against such inquiry as I posit. But whether or not you are prepared to accept the key premise, I respectfully request you might at least entertain it tentatively, and in short keep an open, dispassionate and objective mind; suspending judgment as need be till *something like* the full story can be told and more adequately established and delved into.

People's being taken in *unawares*, baffled, frightened, tricked, or have their living circumstances infested is part and parcel of what criminal spirit people do. Where do they come from? Why do they show up at some time and not at another? When will they have left and departed? Where did they go to? Are just some of the facets to their behavior that makes addressing the topic (at present certainly) no little knotty and problematical. Owing in part to such and other daunting challenges and complications, the work now before you is therefore one addressed especially to uncompromising advocates of honest, rational, scientific truth, and *less so* offered as persuasive and plausible rhetoric to exhort the masses to the validity of what is claimed. More wise, intelligent and courageous people must first take up the question in order to properly ascertain the factual viability and correctness of what is being asserted and proposed. Otherwise, the subject too easily lends itself to distortion, sensationalization, and popular misunderstanding generally; something which, naturally and it goes without saying, we want very much to keep wholly clear of. Not only because we desire to approach the topic more soberly, rationally and objectively, and without any unfounded (or loosely founded) preconceptions and assumptions, but because such distortion, false assumptions, etc. are not (when the topic comes up) infrequently employed by the friends and stooges of criminal spirit people as a tool or weapon to purposely thwart and mislead honest, objective, and impartial investigation.

This said, I have no desire here to start wholly anew what I already sought to do in my *New Treatise*. Instead, what I will attempt at present is to review, expand on, and as need be qualify and make more accurate, what is imparted there and in my "Narrative." Though occasionally I will re-state or reintroduce something from them, a proper and more fully adequate grasp and understanding of "On Reason, Morality and the would-be gods" would be immensely aided by your already being familiar with those two earlier writings, and which contain important and more specific details about criminal spirit people based on what I know about them (and for what *that* is worth.) Details, which given their complexity and sometimes unusually bizarre character, I cannot hope, as a practical matter, to cover and address here with anything like due thoroughness and precision. Additional musings, reflections, personal testimonies of mine on the subject of criminal spirit people can also be found among my "Oracles" or internet writings (2003-2022) available at my website www.gunjones.com ;[2] some gleanings from which, in usually re-edited form, have been incorporated into the present work.

I cannot expect solely by my efforts to conclusively prove to all that are reasonably skeptical of the existence of criminal spirit people. In the course of this and other of my writings I have and will put forth some evidences for what I contend. But, as before, readers are, of course, ultimately left to assess and decide for themselves; and we fully respect this. That said, in continuing I will, notwithstanding and for pragmatic purposes, assume the existence of criminal spirit people here; so that it then becomes possible to attempt to describe them and their motives. Again, it should go without saying that final conclusions are for you and others to arrive at. All I can do in the meantime and otherwise is try to make a plausible and convincing case that might, in part if not in all respects, persuade as to the validity and merit of what is claimed and asserted.

If, when all is said and done, I have not, for some, sufficiently demonstrated and proven the existence of criminal spirit people, or accurately described them, it is hoped that the topic might at least be considered; with others subsequently taking up and more successfully where I have failed or come up short. Furthermore, that there are significant gaps in this introductory review or only scarce coverage of a given aspect of the phenomenon, I hope is not only pardonable, but even and to some extent expected: given the topic's novelty, relative newness, peculiarity, and the vast scope of its aspects and implications, both potential and obvious, on various areas and aspects of both scientific study and practical life concerns.

---

[2] Or see also archive.org (https://archive.org/details/wts-oracles) and Scribd, doing a search for "WTS Oracles," for the same documents.

If, in its essence, what I argue and maintain does turn out to be true (and that I myself know to be true as much as I could know anything to be so), no subject within our comprehension has greater implications on the nature, well being, and quality of our lives. That sentient, calculating persons, not identified as human according to the common acceptance and understanding of that term, can surreptitiously invade, plant themselves in our midst and act as criminals, or as persons of authority to be obeyed no less, has had and does have an enormous impact on and huge ramifications drastically affecting the character and viability of our lives: whether the life of an individual, that of a family, and or of society at large.

That I would write on this subject for fun, career, personal profit, or ambition could only be true in a very small and incidental sense. I only wish someone else or others would take it up in my place, and I could get on with other projects more agreeable and of greater interest to me personally. But circumstances are and have been such, including my having ongoing and direct contact with criminal spirit people for *over three decades*, that this task has been rendered an obligation: an obligation, again, it would be far more easy to pass on to others were it feasible to do so. While it is true I have since my adolescence been interested in and read about "real life" ghosts as a leisurely pastime, say for instance in and by way of books by the likes of Peter Underwood, the subject as I consider it here *came to me* – not me to it – and quite viciously, violently and in a protracted way; as recounted in my "Narrative." That such duress continues even to this day, frankly, *does not* make the task of writing what now follows here any easier or less burdensome.

~~~~~******~~~~~

I.

"But Aristaeus, the foe [Proteus] within his clutch,
Scarce suffering him compose his aged limbs,
With a great cry leapt on him, and ere he rose
Forestalled him with the fetters; he nathless,
All unforgetful of his ancient craft,
Transforms himself to every wondrous thing,
Fire and a fearful beast, and flowing stream.
But when no trickery found a path for flight,
Baffled at length, to his own shape returned..."
~ Virgil, *Georgics*, Book IV.

Looked at from a strictly scientific viewpoint, *from whence* their origins and *exactly* what criminal spirit people are is, not so surprisingly, open to question. Based on tradition, they might be thought of as some kind of "human like" and rationally empowered person that preceded man, such as for example "angels" or in the case of criminal spirit people "fallen" angels," and or they may be "regular" persons who have died but who somehow go on or resume living on a different plane or dimension. My own take on the subject (and it is no more than a surmise) is that the notion of "criminal spirit people" refers to *both* of these types; that is, so-called "primordial" spirits *and* humans who, after dying, passed away from this life into another, working under a common leadership and toward a common goal or end; though it is conceivable that leadership and or goal may change; just as it does with human governments and societies.

English physician and author Thomas Browne (1605-1682), alternatively, expressed this view of quite who criminal spirit people are, stating

"I believe that the whole frame of a beast [i.e., an animal] doth perish, and is left in the same state after death as before it was materialled [sic] unto life: that the souls of men know neither contrary nor corruption; that they subsist beyond the body, and outlive death by the privilege of their proper natures, and without a miracle: that the souls of the faithful, as they leave earth, take possession of heaven; that those apparitions and ghosts of departed persons are not the wandering souls of men, but the unquiet walks of devils, prompting and suggesting us unto mischief, blood, and villainy; instilling and stealing into our hearts that the blessed spirits are not at rest in their graves, but wander, solicitous of the affairs of the world. But that those phantasms appear often, and do frequent cemeteries, charnel-houses, and churches, it is because those are the dormitories of the dead, where the devil, like an insolent champion, beholds with pride the spoils and trophies of his victory over Adam."[3]

---

[3] *Religio Medici* (1643), sect 37.

Perhaps Browne's interpretation may be qualified by saying *some* and *certain* departed are or maybe transformed into and, as a result, become part of the "devil" class that he alludes to.

Although in my *New Treatise* I spend some time in an attempt to describe their physicality, here again there is much that simply isn't known. I refer to them as "spirit people," but does this merely mean a person with a tenuous, albeit nevertheless palpable and real, materiality? Off hand this is my impression. Yet again it is not inconceivable their bodies (or the bodies of some of them) are more substantial than that and in a manner not known to us. Based on my own experience, on one unusual occasion I have seen one of them materialize or seem to become embodied like a regular person, but whether this is just a trick or illusion created for the eye, or whether they can materialize into something more substantial entity (like we think of a regular person as substantial with respect to his or her physical constitution) than a ghost or transparent may be possible, but it would seem to me best for the time being to think of this as unlikely and until more is known.

Otherwise, and for practical purposes we will see them here as being in appearance like what we commonly think of as a spirit or ghost in ordinary human shape, size, and bodily. And yet a spirit or ghost, some of them at any rate, can move physical objects that are relatively light in weight. There are others, however, who possess enough power to be able to hold you down for time, say if you are lying in bed, and or strike you a blow to the head. Yet such powers to do such things are apparently only of a brief duration; perhaps, say, *at most* a few minutes.

Alexander the Great is reputed [4] to have said "Sex and sleep alone make me conscious that I am mortal." Should we have reason to think that so-called "gods" go without sleep? Based on my own experience, one does get the strong impression that regularly active criminal spirit people do not sleep. But more than this, I myself couldn't say and once more this is only one of those "impressions" I have.

When in past writings I will have made reference to criminal spirit people, it may have seemed to be implied that I meant the conventional demons and devils of mythic lore. But such depictions are great distortions, and to some extent are really (though not necessarily or entirely) sorts of "costumes" or "characters" they put on to frighten people. Spirit people, just as with "regular" persons, act or behave as expediency calls for or requires. The same individual who may have or be the friendly face in one situation, may be the utter terror on a different occasion; and vice versa. Not that there are not genuinely "spooky" looking spirit people, but it is very wrong to think they are all like that: much depending on the person in question and circumstances in which you might encounter them. So just as with regular people they can vary in their attractiveness or unattractiveness of

---

[4] The original source is, to my knowledge, otherwise unknown.

their appearance. In my own experience I have never seen an oriental Asian, Middle Eastern *Arab*, or Black negroid or other non-Caucasian spirit person, what exactly, if anything, should be read into this is unclear.

In fact spirit people organized crime has some (at least in real or deceptive appearance) very beautiful girls working for them, and which I was reminded of on seeing this clip from the 1947 film "Down to Earth," with that most eminently likable of Hollywood movie stars, Rita Hayworth. In it she plays the goddess/muse Terpsichore; who comes down to earth. Believe it or not, this, and strange as it may sound, is more what like real "hell" is like – at least as seen from the top of their hierarchy: seemingly pretty in its way above (at least some of them); unattractive on the bottom; not unlike how the very rich often are seen as compare to the poor in our world.[5] While the Mr. Jordan character (the "boss," so to speak) of that film is not in *physical appearance* similar to anything I have seen, the sort of personality depicted in essence rings true. He is matter of fact, business like – not, say, a ranging fiend, with blood spilling out of his mouth making obscene gestures or threats, scaring you to death.[6] Also, as we have remarked elsewhere, though you might see them with your eyes, spirit people speak to you in your head, not vocally as we "mere mortals" do. In general purpose and policy, such people are pre-occupied with making others to be as guilty as themselves. For the peculiar misery such as is theirs (or in being one of *them*) not only loves, but demands company.

If the question is asked where are or would spirit people be found if it were possible to locate them as to a "home" of some sort, I again could not presume to address this sort of question (other then to assume it is a "place" and that has time and space coordinates.) This frankly admitted, they might, for instance, be thought of as occupying what would be to us "unseen" parallel realm, or else dimension, separate from our own; from which they can come and go to enter places where we are. In addition, it might be considered as well if they are not, after all, nomads of a kind. Sometimes they are or have been traditionally spoken of as somehow being from "above" or "down below," that is, up in the sky or down under the earth, but in a manner that is (or else is hitherto) invisible to us. All of which above is offered by me as hypothesis, not as a definitive conclusion. Suffice to say, they can come here, but we (as yet anyway) can't go *there*.

Now it is possible to see spirit people, but (and as of this date) only when they themselves permit it; though there are or may instances where they might be seen by accident or without any willful intention on their part to be seen. At others times, we can perceive criminal spirit people, empirically speaking, mostly from

---

[5] With the understanding that I am speaking here in general and not specifically qualified terms.
[6] Of note, by the way, Terpsichore is told that being immortal she cannot cry anymore. Now I in fact had something like this done to me. I used to be able to cry normally, but at one point in my ordeal they did something to me physically; so that I now cannot weep like I used to, as least not on the outside, but only inside (somewhat) and without the tears (with the magician, at the time this first took place, actually telling me I could not cry any more.)

their physical effects; very much like how some scientific phenomena is known, not by our seeing it directly necessarily and or readily summoning it to be seen, but by the effect it has. An example of this, for instance, is warmth from a machine that produces heat. We don't see the heat, but we feel it, and by feeling it we come up with ways of identifying its source while measuring its intensity. Granted an unusual occurrence, say the sound say of a bell ringing in a house where no one is known to be in that house, does not of course compel the conclusion that a spirit person brought about that event. Yet and even so, and howsoever so unlikely, it will not hurt to consider the possibility; as long, that is, as we don't jump to conclusions; not at least and until we have something like more solid evidence to go on.

If, as I premise here to be the case, there are *criminal* spirit people, they by definition can be detected, discerned and identified by their criminal activity; whether stemming directly or indirectly from them. That is, if there is a crime, and the crime cannot be reasonably ascribed to regular persons or a mere accident (e.g., an "accidental" or seemingly unaccountable death), it is not (according to this view) unreasonable to consider the *by chance* likelihood of criminal spirit people being the cause of that crime or regrettable event. This understood, a crime is by no means the only way a criminal spirit person might be detected, but is one criterion that, again given our definition, stands out most prominently and or else prompts an inquiry as to such persons being in some way or not present and involved with some criminal event that as yet can be adequately explained or accounted for. That we do or might have further evidence of the existence of spirit people outside this context (i.e., of a crime having occurred), and if that evidence is reasonably credible or verifiable (actually or potentially), we naturally would want to avail ourselves of that, of course, as well.

Prior to the 17th century that devils, demons, ghosts and witches actually existed was, more often than not, taken as a given, and the possibility of such preternatural entities was not thought necessary to dispute, at least compared to the attitude of subsequent times. A ready example of this can be found in Shakespeare and who speaks of and describes them with an uncanny liveliness and realism. When for example Macbeth, upon seeing what appears to be Banquo's ghost, cries out "never shake thy gory locks at me..." one is hard put to take something so vividly specific as but the workings of imagination prompted by feelings of guilt or remorse. In Hamlet, we know the ghost is not merely the melancholy Prince's delusion because the sighting has the corroboration of two close friends. In King Henry VI, part 2, we have the most unusual situation where Gloucester's wife is arraigned on charges of witchcraft; with clearly being indicated that there is something palpable underlying such charges. In addition to Macbeth, Hamlet and King Henry VI, part 2, other plays where ghosts, or other spirit persons act conspicuous roles are Julius Caesar; Richard III; A Midsummer's Night Dream; and The Tempest.

Yet with scientific advancement, the rise of the Enlightenment and free masonry in the 17th century, and outside perhaps of religion, there was a radical shift away from common acceptance and assumption actual existence of spirit people; indeed away from religion generally; so that such as Joseph Glanvill and Cotton Mather, for example, not only found it necessary to respond to skeptics but at length.

The prosecution and subsequent hanging of supposed "witches," perhaps most famously in Salem, Massachusetts and elsewhere in New England, in the late 17th century empowered the skeptics and acted a conspicuous role in ridiculing and intellectually marginalizing those who took criminal spirit people to be real entities. Yet what is entirely overlooked is that the witch hunters themselves very probably were in some ways and to some extent under the influence and or else manipulated and prompted by criminal spirit people to act as they did, especially when it came to resorting to the death penalty or other cruel and unusual punishments. Skeptics themselves (as such as Mather argued) might, in a given instance, be taking the position they do not because of their seeming love of disinterested, honest, and objective truth, but exactly the opposite. As for witches or the idea of witches themselves, let us be clear here that there are no such thing. Those who might be understandably thought to be witches, if it have any reasonable or plausible basis to begin with, are only regular persons being used by or under the influence of criminal spirit people, and therefore all can agree that, and in that sense, they do not exist as preternatural or supernatural entities. Such other-worldly powers that are mistakenly attributed to them are only and rather the "magic" or fantastical doings of professional criminal spirit persons, such as a "ghost sorcerer" for example, and who uses the seeming "witch" a pawn, front, or as an employee/assistant. It must be remembered also that historically it was not merely some suspecting others to be so that brought about belief in witches, and the alleged ability of their being able to harm using "supernaturally" related   (what we might refer to as) technologies, but at times certain persons themselves, and for whatever reason, did indeed profess themselves to be witches, and who admitted to have wished or sought to cast spells and curses on others that contributed to the belief in witches. The story of Major Thomas Weir (1599-1760), of Edinburgh, is one especially famous case of this.

By the mid 18th century it had become customary to characterize ideas about ghosts as the result of yarns spun out and imagined over a winter fireside or else products of over heated and childish imaginations that too soon dispensed with reasoning. The only overt exception that stands out from this trend is the writings of Emanuel Swedenborg, yet which are understandably treated more as a work of religion than science; though in fairness Swedenborg is (relatively speaking) often quite scientifically minded and realistically precise in his descriptions and explanations. Later on, William Blake was another anomaly where spirit persons could find berth and harbor in poetry and painting, and yet were, nevertheless, deemed no less literally and palpably real for that.

What more precisely and otherwise induced this turning away from acknowledging the reality of spirit people in the 17th and 18th centuries is perhaps more involved then anyone as yet (to my knowledge anyway) has closely or adequately considered. If, for instance, we assume (as I myself do) that there are criminal spirit people working an agenda, such themselves doubtless played a decisive part in encouraging and enacting such a major shift. Moreover, and given the high expertise and technical sophistication of some or certain types of veteran spirit persons, it is perhaps even possible that some advances in science and technology were contributed to by them – but if so, presumably by way of a "devil's" bargain and at a huge price to honesty and morality.[7] When Goethe took up Faust and Mephistopheles as his theme, was he perhaps, whether consciously or no, portraying poetically what was conceivably the most relevant and central character or type of his and the preceding era; with the German doctor mayhap, and in no little way, best representing Western culture and humanity of and at those times?

Sir Walter Scott's "Letters on Demonology and Witchcraft" (1830) William Godwin's "Lives of the Necromancers" (1834) are remarkable for their adamant and strident stance in denying the reality of the preternatural or supernatural; which attitude, as Glanville or Mather would have told you, was an implicit endorsement, whether intended (Godwin) or not (Scott) of atheism. Rather strangely in retrospect, why, for instance, do not Scott or Godwin not allow the least latitude for considering the empirical possibility of ghosts and spirit persons? Instead, they dogmatically presuppose and assume there can be *absolutely* no such possibility. And yet if one were to guess why such learned men agreed to quench properly open minded discussion on the subject, it perhaps seems plausible to infer that they dreaded the wrath of the very persons whose existence they denied.

The Victim of Magical Delusion (1795) by Cajetan Tschink (1763-1813, a German speaking Hungarian of Austria and a Olmütz University professor of Kantian philosophy) is a most curious and unusual work. It defies casual description, otherwise and suffice to say and here for brevity's sake, it is ostensibly a romance, in part gothic, that tells the story of a protagonist who is led from one adventure to another by two strange conjurers, Heirmansor and Alumbrado, the former seemingly good the latter bad in character. Both describe themselves as acting on behalf of "God." The protagonist, while alternatively admiring and loathing them, is ever in awe and in reverence of their baffling and

---

[7] If we posit that criminal spirit people in some significant measure contributed to advances in science in technology, with perhaps at the same time a mind to glut us with power (in addition to corrupting our appetites perhaps), it may be that our being so empowered was seen as a risk worth taking; inasmuch as such advancement might ultimately be used against them. But of course this latter could only happen if criminal spirit people were known, acknowledged to exist, and understood empirically to begin with. An analogy might be like trading arms to an indigenous people, say to gain land, but with the understand that such arms might one day be used against the invader who so armed and equipped then; i.e., it is seen as risk worth taking.

extraordinary mastery of legerdemain, and their *undeniable* superiority over the rest of humanity. They are spoken of as "superior beings" and are in their authoritarian manner dead-ringers for how certain professional criminal spirit people, like a ghost-magician, like to both take on a pompous air and show off their magic; *unless*, that is, you have the common sense audacity to challenge and or ridicule them.

Like Schiller in *The Ghost-Seer* (1787-1789), which latter work inspired him,[8] Tschink goes to considerable lengths maintaining that "apparitions" (i.e., seemingly real spirit persons) are in truth only elaborate hoaxes concocted by adept tricksters; the purpose of which is to form and manipulate the character of individuals at whom they are specifically targeted. Taken in all, Tschink's would-be alternative and "rational" explanation is even more strange, incredible and harder to swallow than traditional "superstition." Yet perhaps most significant is his vehemence in asserting that other-worldly beings are not to be discussed as appearing or as the cause of anything, or that what we take the supernatural to be is almost invariably only human in origin; or if from "God," something not to be questioned or scrutinized. At one point, the protagonist, (addressing the "Irishman," namely Hiermansor) says:

"'I must confess,' I exclaimed, 'you have executed your plan in a masterly manner, I only think,' added I after a short pause, 'the means too human, and ill-becoming a man who can work miracles.' 'Who told you that I can work *miracles?*' the Irishman replied, 'No one except the great ruler of the world, can interrupt the course of nature and alter her laws; at most, mortals may serve him as instruments to execute the wonders of his omnipotence. I do not deny that I have been appointed several times to be a minister of Providence, but no mortal being can work miracles on his own accord. The whole of the power intrusted to me, consists in the knowledge and application of such powers of nature as are concealed from the short-sighted eye of mortals. At the same time I cannot but confess, that the mysterious deeds which I perform by virtue of that knowledge, appear to men to be wonders, because the spectator is unacquainted with the means by which they are effected. Mark well what I am going to tell you now. Although the higher unknown powers of nature are at my command, yet this power has been intrusted to me, only under the condition never to make an improper use of it, and never to have recourse to it while common human means shall be sufficient to attain my view. And this was the case in the affair of the banker, when I determined to save an unhappy family from ruin. There nothing was wanting, as the event has proved, but art and precaution, and consequently I durst not have recourse to the higher power entrusted in me.'" [vol. I, pp. 239-240 (1795 London edition).]

---

[8] Schiller, for his part, was inspired by the very strange, real life career of magician and reputed charlatan Johann Georg Schrepfer (c.1738–1774), and, perhaps as well, Count de Saint Germain (1712?-1784) and or Cagliostro (1743-1795).

Although there are exceptions, actual ghosts tend to be rather rare in the earliest gothic novels. However, the alternative of having a seeming ghost turn out to be a human imposter (à la "Scooby Doo") before long was seen as too contrived and unsatisfying to many readers, and by 1820's real ghosts and vampires began appearing more regularly; as in tales of the likes of Sir Walter Scott, John Polidori, Ludwig Tieck, Friedrich de la Motte Fouqué, Washington Irving, Nathaniel Hawthorne, and subsequently Dickens, Sheridan Le Fanu, Edward Bulwer-Lytton, Charlotte Riddell, Guy De Maupassant, and Ambrose Bierce – to name some of the more well known and popular. Ironically, simultaneously with this outpouring of the fictional supernatural in literature, the earlier and emphatic insistence that real-life ghosts were all and merely instances of vulgar superstition and disturbed imaginations became pronouncedly more muted.

Yet when later in the 19th century the age of spiritualist séances became the rage, skepticism became reinvigorated in view of many spiritualists having been proven or inferred to be hoaxsters; so that to this day it is ordinarily assumed that 19th century spiritualism, and the summoning of ghosts by way of accompanying mediums at that time, was unquestionably and in all cases fraudulent. And yet is not unlikely that in some cases, such as that of the Fox sisters, where the professed medium admits to fraud, they have actually been coerced into doing so by criminal spirit people. Such, at any rate, is a possibility that ought to be given due consideration.

At and since that time otherwise, alleged accounts of true ghosts appeared in the news every now and then, with probably the two most famous being that of the Bell Witch in Tennessee and Borley Rectory in England. The former concerned alleged attacks by a reported poltergeist in Tennessee, circa 1817-1821, but brought to public light in 1894; with the latter purported hauntings at Borley Rectory taking place on and off from about 1900 to the late 1930s. The account of a poltergeist in Surrency, Georgia from the 1870s, is another intriguing case that deserves looking into; regarding which see *Harper's Weekly*, June 3, 1911, p. 26. The 18th century had had its own two quite famous reported real life encounters with the other-worldly persons by way of Epworth Rectory (Dec. 1716-Jan. 1717) and Cock-Lane Ghost (1762) hauntings; which, in passing, had been preceded in notoriety in 1661 by the account of the "Tedworth Drummer." Of course, one needs only inquire into the writings of Cotton Mather for other real-life supernatural claims from that more flexibly credulous earlier time.

By the time movies came into their own, ghosts once more and generally were played down, and outside the very early silents of Georges Méliès, their appearance in a film story was surprisingly scarce, with Victor Sjostrom's "Körkarlen" (1921) being one of the notable exceptions. The 20s and 30s were replete with houses of mystery reputed to be haunted with a ghost, but invariably by the end of the film the supposed ghost or phantom turns out to be someone pretending to be one.

In the 1940s and later, it became however more possible to have real ghosts in a film story. One interesting example of this is the "Topper" comedy series based on the writings of Thorne Smith. The 1942 film "I Married a Witch," also prompted by a work of Smith's, became a precursor for the 60s tv series "Bewitched." In such comedy supernatural films and later tv shows that while the ghost or witch is admitted to, it is understood and taken as a given that such supernatural persons or manifestations are not to be talked about outside of those who are entrusted with the special *privilege* of knowing them. Examples of such in the way of comedy tv shows that employ this device (and that many are familiar with) are: "Mr. Ed," "My Favorite Martian," "Bewitched," "My Mother the Car," "I Dream of Jeannie," and "The Ghost and Mrs. Muir."

In the series "Night Gallery" (1970-1973) and later "Tales from the Darkside" (1983-1987), tv audiences witness the supernatural being both real *and* taking charge over people's lives. Especially and often conspicuous in episodes of such shows are messages that one must surrender and or else is completely helpless in the hands of intimidating "gods," and other-worldly persons and who are unchallengeable and invariably far superior to us; with it sometimes being suggested that a dualistic manichean-like view [9] offers a good sense compromise to traditional religion, which last is often portrayed as puerile, ineffective or ridiculous.

The preceding is, needless to say, only a very general and brief summary of what have been some of the predominant attitudes and outlooks concerning spirit persons and related as found in secular Western culture in the past few centuries, and of course as well one could expound at much greater length and depth on the subject; particularly, and further, for example, how in recent decades ghosts, devils, demons, witches, and vampires have increasingly been portrayed, especially in movies and books, in a supportive and positive light; rather than as evil doers or malefactors as was the case in the past. In all and it else goes without

---

[9] "'I know you, and I thank you,' he replied, with emotion, 'let us die together; this world is not deserving to contain us. What business have we in a world (he added with a ghastly look) in which vice only triumphs, and good men find nothing but a grave?'

"Reader, do not fancy this language to have originated merely from a transient agitation of mind; alas I it originated from a heart exasperated by the concurrence of the most melancholy misfortunes, and this exasperation was rooted deeper than I had fancied at first. It generated in his soul poisonous shoots which injured his religion. He declared it to be impossible a good God could designedly make good men so unhappy as be had been rendered. He ascribed the origin of his misfortunes to a bad principle, which, having a share in the government of the world had appropriated his understanding merely to the execution of its bad purposes. He maintained that it was contrary to the nature of an infinitely good being to effect even the best purposes by bad means; and if there were in this world as much disorder, imperfection, and misfortune, as harmony, perfection and happiness, this would be an undeniable proof that the world was governed, and had been created jointly by a good and a bad principle. In short, he subscribed entirely to the system of the *Manichees*."
~ Cajetan Tschink, *The Victim of Magical Delusion*, vol. III, pp. 146-147 (1795 London edition.)

saying, the topic is easily worth and would lend itself to an entire book or in-depth dissertation, yet which here we can only speak of as having at least touched on.

II.

"Then he said to them, 'My soul is sorrowful even to death. Remain here and keep watch.'"
~ Mark 14: 34.

"As flies to wanton boys, are we to th' gods;
They kill us for their sport."
~ "King Lear" Act 4, scene 1.

Despite widespread coverage of supernatural entities in film, literature, and song, consideration of them as real, or based in something that is real, continues to be down-played, and honest, rational discussion of their empirical possibility dogmatically eschewed by main stream science and intelligentsia. In fact, the silence, strangely or not so strangely, can be characterized as deafening. Indeed, outside books of 20th century "ghost hunters" like Harry Price and Peter Underwood, the topic is not taken very seriously, and when covered is examined usually more for purposes of entertainment, thrills and putting on a show than serious scientific inquiry and analysis. This is all the more odd when every now and then even the government is openly enjoined to explore and fund research or probing into the possible realities of space aliens, and this by some scientists, reputed journalists, and persons otherwise recognized as wise elders or public authority. And yet by stark contrast with the sober contemplation of space aliens, serious addressing of spirit people is routinely avoided and hushed aside. What? Are we to believe no one is even a little bit curious regarding a question of such profound implications? True, we have an area of purported science related to the supernatural or "paranormal" denoted parapsychology, but this approach in practice seems in retrospect more to have served the ends of obfuscation rather than clarification; which, for *some* of its practitioners at least, is perhaps and even its actual intent and design.

Yet if criminal spirit people are something palpable and, could it be that they themselves, including the very fear of them, play a part in having impartial, rational and objective inquiry about them summarily dismissed and brushed aside? And if we posit a "Faust" or "Fausts," that is persons who, whether to a small or great degree, overtly or implicitly, consciously or unconsciously make deals with such for profit and or hoped for security, could we expect any cooperation or admission of the fact from such as these? The problem becomes even more intractable and difficult when we take into account the possibility that such spirit persons would have the gullible, timid, and unreflecting see and receive them as gods, deities, or superior beings of some kind; including also *well-meaning* religious, but who may be duped and fooled by criminal spirit people acting as heaven sent impersonators and who might (in a given instance at least) be a kind of expert at theological sophistry and religious psychology, while in addition perhaps be able to perform magical wonders and seeming miracles.

In *The City of God* (bk. XIX, ch. 9), Augustine bemoans: "...there is great need of God's mercy to preserve us from making friends of demons in disguise, while we fancy we have good angels for our friends; for the astuteness and deceitfulness of these wicked spirits is equalled by their hurtfulness. And is this not a great misery of human life, that we are involved in such ignorance as, but for God's mercy, makes us a prey to these demons? And it is very certain that the philosophers of the godless city, who have maintained that the gods were their friends, had fallen a prey to the malignant demons who rule that city, and whose eternal punishment is to be shared by it. For the nature of these beings is sufficiently evinced by the sacred or rather sacrilegious observances which form their worship, and by the filthy games in which their crimes are celebrated, and which they themselves originated and exacted from their worshippers as a fit propitiation."

To give a further idea of the kind of confusion and complication that might further arise, imagine if the said "devil" were to mercilessly beat and torment someone. Who is to say that said devil is not, after all, merely carrying out necessary discipline on behalf of God Almighty? Wrath of God or evil maniac, how does one tell the difference?

Then there is this odd case or example. In the course of my own ordeal of being bothered by these people, a pear tree in my neighbor's yard straddles the property line with its branches. Normally, the pears it produces are so few, small and rumpled that the squirrels, possums, and other local animals hardly bother to eat them. Then one year the tree became amazingly overloaded with numbers fruit. So much so, that one branch, with a diameter of something like *seven or eight inches*, was so weighed down with the pears that it snapped and broke; such that it became necessary to have what remained of the branch sawn off by the neighbor.

Now even supposing (at least *for the sake of discussion*) this event was brought about by spirit person machinations, does it indubitably point to God's *direct* hand in what took place? My sense is simply "no;" because it is merely a display of power and ability, and does not necessarily imply moral goodness and character; which we must insist is to be expected of someone we see as or representing God. Or put differently, at the very least, we don't necessarily know it is God who caused the specific, unusual event, and therefore can safely assume it isn't (i.e., unless and until we were somehow in a position to know; were that possible.) What's more, God's specific workings in the overall and grand scheme of things are beyond our comprehension; other than to say that good and that which is truly good comes from or originates with God. At the same time high-power spirit people are capable of all kinds of amazing magic and wonders,[10] at least I know personally this to be the case; so that the manifestation of a seeming miracle (outside of nature) does not necessarily imply God (that is, beyond the

---

[10] With *some* evident proficiency in the mantic arts; at least among the more advanced and professional criminal spirit people.

general and universal case of every and anything that is good); anymore that a wonder produced by technology implies direct and divine intervention. Such, at any rate, is my own take on the question.

Some might argue – but does and would not God use Babylon or Assyria, i.e., ostensible representatives of evil, to chastise and justly punish and teach us for our straying into error? The answer is or may be yes, of course. Yet the greater points I would emphasize are these 1) we are ourselves are hard put to know exactly what God is thinking, and 2) even if we are well deserving of correction or even chastisement, Babylon and Assyria themselves *are not God*; despite their insisting that you take them to be so. Put another way, that devils crucify and martyr people this does not, somehow by implication, confer on them the authority to appoint saints. But if by acting badly they somehow might be seen (or it be argued) that they serve the ends of divine providence, we have no way ourselves of determining such a claim as being valid, and therefore are under no obligation to grant them the benefit of the doubt. Yes, we can and should listen to Jeremiah, but *if and only if* it is a veritable Jeremiah could we begin to take seriously or consider adopting such an interpretation of direct divine interposition. Anyone else than a Jeremiah, and we have more than good ground to be suspicious.

The weighing and enduring of life suffering are deep religious questions which I by no means would make light of, and of course we ought to deal with such patiently, conscientiously and with a mind to bearing our cross, whatever that cross is.[11] However, doing so need not necessarily imply that those who afflict us are somehow direct or authorized by God to do so. Were this not the case, then all acts of crime could be interpreted as the will of God. Not that we can't, if so choose, view what suffering we go through as in some way or measure a fit chastisement for our sins (as some religious would or might); only it is nonetheless unnecessary, even if we do so, to confer *legitimacy* on the criminally motivated abuser or oppressor themselves.

Another critical point in this regard that should be mentioned is that it is typical of high-power criminal spirit people, or their underlings, to have you believe that they act out of necessity; when in actual truth do what they do out of choice, and because they believe such action as they take to be a good idea that will further their own selfish interest; at the same time happily rationalizing to themselves that *their self-interest* is the same as *yours and everyone else's best interest*.

On the other hand, and when it comes to those in the lower ranks, the poorer, slave-like criminal spirit person may be compelled to act under the pressure of threatened violence to afflict or abuse you; so that it is easier to pardon, in their case, that they act out of necessity (or something, which *for them*,

---

[11] While certainly making due allowance for receiving God's discipline in *any* circumstance; such is referred to in Proverbs: 3:11-12, and Hebrews 12: 5-13.

is very like necessity.) Of course and specific real life instances, all of this depends on the crime in question, the level of character and intelligence of person committing it, and the pressures, real or unreal, that they see as influencing their action.

~~~***~~~

Any given person is against what is "bad" (or their idea of what they suppose "bad" to be.)" By "bad" I mean things, persons, actions, or events that bring about excessive pain and suffering, and that mar or threaten our peace, happiness, and ability to function normally, and as opposed to merely inconveniencing us. Obviously and otherwise, something can be characterized as "bad" that clearly does not requite evil to bring it about. So that "bad" *as we use the term here* refers to something harmful or destructive that is or verges on life threatening or seriously debilitating.

Despite this, we are far from all being in agreement about evil being a problem; that is, and by the definition we employ here, the ultimate source or fomenter of that which, according to my definition, is bad (in the sense spoken of.) For some, evil doesn't exist, or so they say. For others, evil exists and is a problem. But their ideas as to quite what is the cause of evil are uncertain, erroneous, or else confused.[12] For yet others, evil is deemed as a positive good;[13] indeed, there are those who practically make a religion out of condoning, supporting and or celebrating it.

If we say real evil does indeed exist and is a problem, do those who complain about what is bad, and yet who deny evil's existence, make any sense? Offhand the answer would seemed to be clearly no. Because, for one, if evil exists clearly it can be the cause of various kinds of bad (i.e., the one thing we are all agreed upon is a problem.) Likewise, if we assume there are persons deliberately disposed to evil, and yet we ignore or else deny such persons exist, then how better able are we to prevent or contend against what is bad? Unless and until then we are prepared to admit to and have a resolve to deal with real evil, our efforts to resist, reduce or eradicate what is bad are severely weakened and impaired (certainly in the long term sense.) So, for instance, if an open sewer is exposing the public to contagious germs, it is of little or at best makeshift help to rely on medicines to counteract the illnesses those germs create while we overlook or ignore their source.

And if we say further that there are persons who *deliberately* and *calculatingly* pursue a policy of evil, we will find them, I would argue, mostly (if not exclusively) among high-power criminal spirit persons. And yet the

---

[12] The novels of Charles Brockden Brown, rather interestingly are replete, indeed arguably obsessed, with the instances of evil arising, unintentionally, out and as a result of benevolent motives.

[13] In my *New Treatise* I referred to such persons as "Orkons" or "Orkonists."

acknowledgement of the scientific and empirical existence of spirit persons is (or so one assumes) fairly unheard of. At the same time, if there are actually such evilly disposed spirit persons, they will, not surprisingly, do what they can to generally encourage the view that *a)* there is no evil, *b)* that spirit persons do not exist, and *c)* that if evil does exist it is to be found anywhere *but* among or coming from spirit people. If then what I have been saying for years now about spirit people has a true basis in fact, then in order to combat both bad and evil, evilly disposed spirit people must be assumed to exist and combated. But if (and given our premise) the existence of criminal spirit people is denied, then what hope is there of our effectively removing or eliminating serious "bad" or evil from our midst in any lasting or meaningful sense?

Some would-be progressives and other well-intentioned yet blind and ignorant folk deny the reality or existence of evil; and rather will attempt to explain what are seen as evil actions as merely the result of psychological, sociological, or political dysfunction. While certainly this approach is not without its value and truth, it deliberately fails to recognizes, at least as a possibility, that their are persons and powers who are evilly disposed not by accident or circumstance, but intentionally so and as matter of policy or arch-philosophy that views evil, paradoxically, as a positive good.

If then evil (in the sense of that which is a primal source or inducement of that which is "bad") does exist, it follows that there is someone who is motivated into causing and bring it about. What then could possibly be a motive or motives for someone wishing to bring about evil; that is, evil in the sense of the conscious and willful desire to do extreme harm or injury, and that most would interpret as unjust?

For some, evil might be seen *as a means* rather than an end; e.g., the thieves hold up the bank and deliberately kill the teller so they can have the money to live what they think is a happy and contented life. Fortunately, and to make our lives easier, the vast majority of people would find this reasoning absurd and self-contradictory.

For other sorts, evil might be promoted *as an end*, and or that evil is somehow a necessary adjunct or corollary to good; such as we might find with manichean-like thinking and moral relativism taken to an extreme. At first hand, such a view seems even more logically incomprehensible. And yet there are or maybe those who see a benefit to be gained by espousing such a strange point of view; in which case evil, or the furtherance of evil as positive good, is really a means. The idea being that if the one who promotes the belief in evil can deceive and get others into believing that evil is a positive good, he can use such persons as to profit himself by their evil deeds; while in the process damage and lessen their strength to resist him and his authority: that is, by getting them to believe nonsense. Such an approach is one used and adopted by high-power criminal spirit people.

One of the biggest problems when dealing with spirit people is the power of some of them – thinking here particularly of such who are professional or veteran criminals – to pass themselves off as a "god" or other supposed superior being who implies that he "knows better" than we do. Why listen to a would-be just, correct, or rational argument when here we have a god or someone somehow superior to decide the matter for us instead? Some people are little able or disposed to be honest and rational in the first place, and are perhaps easily frightened as well. How much less rational and more skittish are they if a professional criminal spirit, armed with all kinds of time proven magic tricks, mind control, sophistry, and scare tactics tells them what to do and or think? When honest, right reason and a seeming "god," are at odds with each other, which of the two will people listen to as authority?

It is commonly found to be the pronounced tendency, especially irrational people, to see great good as originating and stemming from those with the greatest wealth – rather than from God the creator. So that, for instance, the frequently spoken of local billionaire is seen by the unthinking as the one who provides all these good things we have or can buy. This is one reason atheism can be used to favor the very materially wealthy; for then the exorbitantly affluent not only are made or assumed to be key public leaders and the embodiment of the collective good and welfare, but in fact end up taking the place of God in the many people's minds as well. That criminal spirit people would strengthen and enrich certain of the very wealthy and use them as the mantle, mask, or the store- front of their own would be godship is routine and only to be expected.

We can assume God, and by definition, is perfect. But is anyone else? It in no way follows that because a person, whether spirit or otherwise, is in some ways superior that they are also perfect. And yet in dealing with a high-power, con-artist criminal spirit person, most (were they to have that experience) will tend to be so dazzled and intimidated by such that they do not pause to consider what are or might be the failings, drawbacks, and weaknesses of this supposed superior being. As impressive as they can make themselves appear, high-power criminal spirit *will* come up short; that is, if you analyze and sift them more critically. To use and analogy, when did one ever find a tank that was also a submarine, and an airplane, and a sports car? Supposed or so-called "gods" are like that. They can't be everything.

Nor is it in the least likely that a criminal spirit person, who are themselves autocratic minded or else work for someone who is, will allow you room to question or challenge their seeming invincibility and deity-like importance and presumed higher station in the grand order of things. Why should they? When by deceit, intimation and bamboozling they can as much as get away with all kinds of robbery and murder, and even and perhaps (to some greater or lesser degree) transform you into an accommodating dupe or slave?

As much then as the more professional criminal spirit person can astound, awe and frighten with magic (i.e., create incredible and convincing illusions), including even and possibly predicting the future, while displaying other real or seeming extraordinary powers and abilities, it is even so necessary and all-important to bear in mind that there are as many and more things they are *not* capable of. And it is such inability that gives the lie to their claims or implied pretensions of being "gods," let alone God. Gods at times are made to look foolish even in Homer. If you are the least bit intelligent yourself, scrutinize and look carefully for a weak spot or frailty in any one or other of them and you will assuredly find and detect it. We will be considering more specifically some of their characters and personalities as we proceed, however, among the traits, to give one example, that very frequently stand out in these people are an extreme pettiness, hypocrisy, self-pity, the need for attention and general childishness, and which traits one also invariably finds as well in their followers.

On the flip side of things, such of the wealthy who have ties to criminal spirit people, like and perhaps those who run big tech, the mass media, and similar are not so much malicious, anti-morality and against equal opportunity *freedom* because they are themselves are inherently that way; rather it is because some such among them had to make deals with the devil (like Goethe's high-minded and well-meaning Faust) in order to achieve their benevolent, worldly success and, of all things, love. Yet if Faust has love by means of such infernal arrangement, it is like joying over flowers in a sewer. As they see it, by making deals with a "Mephistopheles," they are merely being necessarily practical, not evil. Indeed, and if in a given instance we knew the truth, it is not at all unusual for persons we think or assume to be well meaning and highly respected to make some kind of compromise with criminal spirit people appeasement. We see social and political activists from all sides of the spectrum doing this. Did they not do so, it is not that unlikely that you would not hear of or from them at all. If our opponents use steroids,[14] than why shouldn't we? Our cause is just, is it not? Yet they do this not because they wish to aid and abet evil, but because, and perhaps unbeknownst to others, they are simply too foolish and or cowardly to say no to the demands of peer pressure created and introduced by other-worldly organized crime. Christians and other sincere religious, on the other hand, are *supposed* to be different than these sorts who are mindlessly taken in.

Another and related type of offender we can make note of, whether spirit or regular person, is like the hard-core drug addict. They don't really want to hurt anyone. But it is necessary for them to perpetrate some criminal offense in order to get the money that will make possible the continuation of their habit.

Egotism is commonly such a mania among them that they come to think flattering or inflating someone's pride is a just compensation to repay a victim for the horrendous, perhaps even tragic, damage they did or might have suffered at

---

[14] Criminal spirit people empowerment is indeed like that derived from using steroids, but fraught with even more hazardous life side-effects.

their hands. That is to say, if pride can be measured in denominations or quantity or some kind (say, "ego dollars"), they see it is as perfectly all right to use it as a currency or transferable substitute to be exchanged for honesty, justice, (honest) love, peace, etc.

In any event, criminals, of whatever kind, or for that matter ourselves from first hand experience, can and do always invent excuses which make it sound that they or we do wrong act for some ostensibly good, justifiable or benevolent reason. The argument "if it doesn't kill the person, it is good for them" is one famous example of such thinking.

Yet "Judge not lest ye be judge" applies here also. It of course would be wrong for us to attempt to judge spirit people or their regular person adjuncts (or for that matter anyone) in absolute terms; though admittedly our opinion of one or other of them, quite understandably and forgivably, might be very harsh or pronounced. Even so, what we are at here is that if they (i.e., whoever) are in their actions criminal, we have an inherent, common sense right to protect ourselves from them, most especially when it comes to things like violence, terrorism, and vandalistic harassment and shock bullying. And we will be all the more able to do this if we better know and understand what they are really like, what they are or may be about, and how and why they act as they do.

Compounding the danger posed by age-old misconceptions about, and along with criminal spirit persons' purposeful efforts to mislead and deceive, is the not so unfamiliar tendency of many to think more like or as herd animals, rather than rationally equipped human beings capable of something like critical thinking. The final and ultimate test or criterion of what is real for such is what others or "everybody" thinks; that is, the given person cannot themselves reflect or analyze very rationally or deeply so that they then too readily let others do their thinking for them.

In passing and I would urge, it is important in this respect to remember that whenever something is believed or decided as being true or false, whether the proposition concerns a FACTUAL and or a VALUE judgment, a JUDGE is necessarily implied. Nothing is thought of as real or unreal, as true or false (or in part true or in part false) unless there is a PERSON who makes the given judgment. If then and whenever we say something like A is (or equates to) B, or B is C, etc., what is concluded can never be so unless SOMEONE decides it so. I address this point at greater length and depth in my book *Peithology: The Nature and Origin of Belief* and for further and more proper expounding on this point I would direct the reader to that. But suffice to say, and in short, nothing is real or unreal but that SOMEONE decides it to be so. Things can only be as they are – whether as a supposed fact and or as something having or not having value – because SOMEONE ascertains or concludes it must be that way (i.e., must be so concluded.) So that if then we would avoid being fooled or taken in by a con-artist, it is necessary for us to ask WHO this person is making the call, and then how trustworthy and qualified are they to make such a judgment or decision on a given question. There is in point of fact no faceless truth; since any and each assertion or conclusion implies a judge, and therefore a person. Granted, such a fine, albeit crucial, distinction will probably be lost on most, and even those aware of it may in practice forget its crucial relevance and application; still it is something well worth bearing in mind; hence its mention if we desire to decide and conclude more correctly and in a manner consistent with honest, rational truth.

Both ostensibly and by inference and traditionally, the motive of criminal spirit people might be construed as *a)* challenge and effort to diminish the authority of true God in humanity's eyes (at least so the religious will see it), and *b)* to enrich and aggrandize themselves by setting up their own authority and government over others and ourselves; such usurpation being possible insofar as and to the extent we permit and allow such to happen by our overt surrender or acquiescence to their con-artistry, bullying and both obvious and surreptitious manipulation.

One gets the impression that at times they will cause problems to test and experiment on people; either and perhaps because such criminal spirit persons are idle and it gives them something, to do and or are genuinely curious and interested to find something out.

Just as with regular people, criminal spirit people are individuals with different experiences, life stories, and behaviors peculiar to each. Like us their mood, attitude, enthusiasm, optimism and outlook periodically change. They have their good days and bad days. Not perhaps counting so-denoted "fallen angels," as likely as not they have relatives somewhere who would or might be ashamed or embarrassed by being associated with them (i.e., with their being a unrepentant, hardened criminal.) I had heard at one point, for example, that "Simon the Magician," a notorious criminal spirit person, had/has a daughter. Would, did or does she approve of what he does?

And even as much as they are criminal (by our definition) in their behavior, they are not always disposed to act badly if behaving themselves rightly will secure what they want. Yes, they are invariably liars, tricksters and hypocrites, indeed incorrigibly so, at least depending upon the individual. Yet they are at times not without a certain wisdom and they can be quite right and correct on a given point; just as say a lawyer arguing a bad case might be, and it would be foolish to say they are always wrong, or wrong about everything. This said, it would be no less foolish to trust them; when craftily getting the better of you is after all and in practice their vocation. They for example can, in a given instance, imply that they value love, but love on their dishonest, selfish terms; not in fact unlike, say, how a rapist would say that he was motivated by "love." People will exclaim "all you need is love," and yet what kind of love is it that is being referred to? Honest, fair dealing, reliably consistent love? Doubtless not everyone feels the need to be either so qualifying or specific. Similarly what, albeit and perhaps unstated, is included in their conception of that love? Blind obedience? But blind obedience to what exactly? In short, one needs must be on one guard sometimes when someone throws about the word or praises "love" (including "kindness"); all the more so when it comes to criminals spirit people and their barely honest or rational followers.

It is, as well, frequently found among such people, whether spirit or regular persons of this criminal ilk, that they have little or no sense or idea of fairness of justice, and instead see everything as a matter of self-interest: ideas of justice and fairness, as they see it, being merely pretense and or tools of wily, abashed and self-serving diplomacy.

In broad, working terms, and given that due to lack of more direct information it is not possible for me as yet to be more exact, we can speak of three basic classes of criminal spirit people: 1) the executives or bosses (and to which class we can, if we care to, include a single overriding ruler); 2) the lieutenants of the executive or executives, and who typically are skilled and adept at magic (i.e., as in the creation of illusions), deception, mind control, and or perhaps some other expertise, including what is medically related, calculated to frighten, intimidate or fool people, and 3) vassals, servants and slave people. There may perhaps be posited a fourth category; that is, those who are not "lieutenants" or auxiliaries but are not quite slaves either, and whom we might think of as part-time rather than regular employees or servants, and who, depending on circumstances, possess some amount of free-wheeling independence ("Puck" in "A Midsummer's Night Dream" comes to mind as perhaps an out of hand illustration in literature of this type.) Inasmuch as we have contact with criminal spirit people or they come our way, it would seem highly unlikely that we very seldom, if ever, will be dealing directly with the executive sorts. Nevertheless, such leaders are the ones who make policy, call the shots and wield the far greater sway over what goes on. For which reason, it is necessary therefore to keep in mind the command and policy setting position they occupy, and even though they are or might be absent from our own midst; since, for one thing, being the ones in charge, they naturally are the ones most deserving of blame and therefore the ones that most need to be combated and stopped. True the servant may plague us, but the servant after all, whether coerced or for hire, is only or mostly acting on behalf of the master.

To what extent are the overlords working in tandem or direct cooperation with the lower classes? For this, and to my own knowledge, we have as yet no ready or easy answer. But it would seem reasonable to infer and looked at from a broader perspective that we have here something not unlike a military hierarchy with higher ups giving orders, officers under them transmitting and carrying out the same, and the lowest ranks acting and doing as they are assigned, and yet each potentially, depending on their character and disposition, having the ability to do things on their own initiative, and perhaps not always in a way that is entirely pleasing or acceptable to their peers and or superiors.

Last and just in passing here it can and should be remarked certain kinds of "occult" events, and extremist or lurid crimes events have their hour [see Luke 22:53] and season; so that some kinds of extreme wrong-doing that are or may be permissible or appropriate at one time, may be not so at another.

III.

"Expounding the reason of the incessant plotting of the devil against us, he [St. Justin Martyr] declares: Before the advent of the Lord, the devil did not so plainly know the measure of his own punishment, inasmuch as the divine prophets had but enigmatically announced it; as, for instance, Isaiah, who in the person of the Assyrian tragically revealed the course to be followed against the devil. But when the Lord appeared, and the devil clearly understood that eternal fire was laid up and prepared for him and his angels, he then began to plot without ceasing against the faithful, being desirous to have many companions in his apostasy, that he might not by himself endure the shame of condemnation, comforting himself by this cold and malicious consolation."
~ From the writings of John of Antioch (7th century.)

Returning to the question of motive, does a supposed "evil one" seek to breed, groom and indoctrinate people to be his slaves and victims and thus, as a matter of false pride, mock higher good? And or does he merely seek annihilation of who and what is good and morally virtuous in order to more effectively enslave people for his personal benefit? Or both? Evil or willful, deliberate and serious wrong-doing is or can be both an end and a means, depending on the individual and present circumstances. And perhaps in the extended time frame, a high-power criminal spirit person is living in (i.e., as opposed to time as ordinary, flesh and blood mortals see it), it is not necessarily quite one or the other, and perhaps he oscillates between "end" and "means" depending on his mood and exigencies of the moment. One certainly does in any case get the sense that the battle between good and evil is something like a game; for which reason, motives and policy for some, and again given the sort of individual criminal spirit person is, may actual change over time; depending on how things are going. This possibility of changing and overlapping motives, at any rate, is something to be borne in mind when attempting to identify present purposes and intent.

Ideologically speaking and in practice criminal spirit people are manichean-like or moral dualists/relativists in their thinking. With such, "Good" (as, say, Good as it is commonly thought of) is or can be good and "Bad" is or can be good, or "Good" is or can be bad or "Bad" is or can be bad as suits the occasion. Now many will object that this (in most instances at least) would be highly irrational. But then there are, as we know, people like this, and whom it does not trouble or bother them being irrational. Whitman's dictum "Do I contradict myself? Very well then I contradict myself, (I am large, I contain multitudes)" is (outside perhaps a poet's idle musing) in most instances a piece of arrogant self-indulgence that throws rationality out the window when a person feels like doing so. And while such an attitude or approach might be pardonable or even conscionable when applied to light hearted or trite concerns, it elsewhere potentially, when not obviously, lends itself to a condoning of anarchy and lawlessness. After all, we don't expect a "devil" to concern himself with acting rightly, justly, or with rational consistently. Yet if Whitman, our good fellow, says contradiction is all right, then by implication the devil is not so bad either and is a good fellow too.

Conceivably, or at least in theory, a supposed Orkonist (i.e., "bad" or "evil" is *a good*) only adopts a manichean-like guise but is not actually one. That is, he does not formally want to replace Good with Bad; rather he wants to set up and institute is own idea of what Good is or should be. Moreover or alternatively, his pro-evil stance is something adopted to advantage him in present circumstances; i.e., get others to believe nonsense and he can all the more easily ensnare for purposes of enslaving them. Either way, the end goal of his actions and decisions will be the furtherance of his self-interest (such, at any rate, as he perceives his self-interest to be.) Ordinarily and among more common ranks of criminal spirit people and their adherents, some version of manicheanism or moral dualism is the prevalent as a would-be theology and or life-outlook; that is, nothing or most nothing is necessarily good or bad but that it serves craven and crass self-interest.

This naturally raises the question as to what in terms of overriding policy or arch-philosophy do those in the upper hierarchy of such persons see as constituting higher good? Revenge against God? A bitter hatred and contempt, perhaps jealousy, of humanity (ostensibly inspired or instilled by "fallen angels")? Power and dominance over others? Material wealth? Receiving or acquiring the recognition, esteem, and or admiration of others? Military-like defense for purposes of delaying or forestalling divine judgment? Again, for a given individual, it may be more so one or else some combination of these. But, needless to say, what a given individual sees as higher good depends on that individual. And also presumably, individuals though they are of the same party, it is possible will differ to some extent in their priorities of what they value; which itself, also again, may change depending on circumstances; that is, present priorities might at a later time vary from what was decided upon earlier.

For some criminal spirit people wrong-doing is not only excusable but and somehow cosmological necessity, and thus and on this basis they would have you believe they somehow have a resultant claim on everyone. They are, after all, only doing their duty (as they see it.) Everyone owes them, every one must measure up to their standards or answer to their complaints; so all-important are their selfish interests and the role the universe requires of them. Likewise, there is this mentality that if there is wrong or harm done, then God, as it were, will foot the bill when it comes to blame. And if He won't, well then he must not be much of a God to begin with, and or else there is no God. Implied by this is the belief that God is, at bottom, at fault for any and everything wrong, or as Goethe shared it "Nemo contra Deum nisi Deus ipse." While this may and understandably perhaps challenge some philosophically, even so and not surprisingly, such tenet becomes an all too convenient excuse for a very guilty individual to absolve themselves of blame and responsibility. At the same time and in practice, such who do take this view are themselves hardly ever, if ever, either honest or rational moralists, and instead invariably are little better than conscience-less yahoos or anarchists seeking merely to justify their own amorality and hypocrisy – and are not just and would-be righteous philosophers.

These qualifiers, questions, and caveats aside, it is possible otherwise to make out the general characters and personality types of criminal spirit people based on their behaviors; once again bearing and mind that much of these behaviors, emotional dispositions, and attitudes, and again to some greater or lesser degree depending on the individual, are shared or adopted by their "regular" people hirelings or hench-people.

As alluded to earlier, by getting people generally and at large to be more stupid, irrational, coarse, violent, and profane, it becomes that much easier for them to be lorded over, frightened, manipulated, and controlled. Doing this regularly and on an ongoing basis requires the talents of professionals, experts and technicians (medical and other sorts), and some of those working for criminal spirit people can be quite intelligent in a given field of expertise, even on the lower tiers of command. But these are a relatively rare exception, and even among these some can still in other ways will prove themselves to be rather empty headed, and not very bright when it comes to critical thinking. As seen from the perspective of those actually in charge and calling the shots, much ultimately depends on the personality and moral character of the individual working for them, and how best to effectively utilize such strengths and abilities they posses; insofar as doing so serves the ends and aims of the criminal organization.

The encouragement of bad behaviors is, to a large degree and as administered by higher-ups, an on-going matter of carrying out formal policy and doctrine. Things are done a "certain way" and with a mind to discouraging moral attitudes and thinking; at least insofar as what is possible and present exigency will not suffer certain moral principles being advocated; which might conflict with the organization's present concerns. While, true and for example, a particular individual might or will perhaps have a conscience or qualms about forcing themselves physically on another (i.e., without that's other's consent), nevertheless, the practice of violating or forcing oneself on another or their property is, as we could expect and by definition, part and parcel of being a serious criminal. So even if a given criminal spirit person, in a given instance, is not necessarily so disposed to force themselves on someone else, they would or might do so if paid and or commanded to. Yet others might do so readily and on their own initiative, and not feel the least compunction about being so arrogant and obnoxious. In finding one's self contending with such different sorts criminals then, it might be of the one kind, or the other, or both.

In my *New Treatise,* I mentioned how getting or making others guilty is fundamental to (what I there referred to there as) "Hell" policy. One might say that inasmuch as hard-core criminal spirit people are the *most* guilty people there are, it is as if they see themselves as owning the rights to wrong-doing. So that if you act badly, and the more you act badly, you transgress or infringe on what they see as their domain, and thus, at least if they care to make a case of it, you become answerable to them. This is in addition to the (for them) useful principle that, and generally speaking, the more stupid, and irrational a person, the easier it is for a criminal spirit person taking charge at the moment to control and, as need be, make use of such.

To further empower sustain their criminal rule and better establish themselves as would-authority, they seek to eliminate or remove from leadership and influence those who are most moral and rational. At the same time in doing so and by implication, they send a loud and clear signal to other would-be moral and virtuous *not* to challenge their sway and governance. In sum, by skimming off or eradicating the most morally conscientious people from off the top tiers of society, they indoctrinate, and cultivate for their own benefit the subordination of the larger collective (that is situated below in the hierarchy of things); of and in the varied moral, cultural, and political substrata of society.

Yet being "bad" or guilty is not without its obvious down and regrettable side, even for some hardened criminals. For example, some might after all feel remorse or pang of conscience, say, for massacring innocents. One way they have or try to get around this as a possible problem is to deny or play down free will; the logic being that if a person is forced or compelled into wrong-doing, then they are not guilty or else coercion by others makes them seem less so. Such reasoning is not infrequently welcome by many, as they just as soon not be more guilty than they are. Consequently, the illusion is created and encouraged that they *are forced*, say by some mysterious greater order of the universe, to do wrong; so that given this approach a person feels less compunction acting badly since they have been led to believe they cannot or cannot much (depending on the crime) really help it. Guilt implies punishment which they understandably fear; indeed "Fear has to do with punishment," 1 John 4:18. And thus ways are sought to avoid guilt or the implication of such. Similarly, like anyone, they generally would rather fairly pay for something than steal it, at least and so long as their budget is not unduly inconvenienced.

To make this all the more easy to take in and apply as a principle and in practice, there is this pronounced tendency among criminal spirit people and their followers to be overtly childish; as if they were just and only mischievous children behaving in error, and *not really* bad people as their actions might else suggest or indicate. At Matthew 11: 16-17, Jesus says: "To what shall I compare this generation? It is like children who sit in marketplaces and call to one another: We played the flute for you, but you did not dance, we sang a dirge but you did not mourn." "This generation" is acting as a tempter, a veritable devil, and seeking to draw God out to what they want him to do. And yet they are also seemingly and merely playful children just having fun.

As mentioned, they have different personalities, characters and backgrounds and when we speak of certain tendencies and attitudes such might be true of some and yet not of others. Envy, jealousy, and self-pity are characteristic and not untypical of their leadership, and consequently not a few of their followers as well. Indeed, if one has the least familiarity with how some of these sorts typically think, whether spirit or regular person, you will know one of their most favorite rebuttals or grounds of refusal to any and anything you might *constructively* propose for their benefit – and no matter how genuinely helpful and honest to them your suggestion might be – is: "Absolutely not! I do things *my* way – not *your* way," and this oft times with a sneer implying they don't think you are *so* great as to do anything *you* think (i.e., such that they somehow, to their great insult and dishonor, must or would listen *to you*.)

With respect to darker behavior, some will have an acute and unbelievably high tolerance for cruelty, and make light of things that would makes others aghast, blush at, or be utterly ashamed of. Think of the most horrible crime you have ever heard of and the likely hooded of criminal spirit people involve is enormously high, if not a indubitable certainty. But again, only some criminal spirit people are as bad as all that. Truly, many (who do not already know these people first hand) would be utterly amazed at how persistently and incorrigibly bad, and for *very long periods of time* no less, some these people can be. Then there are those somewhat cruel as these, but not so much so. "Fear is the father of cruelty," said James Kirke Paulding.[15] The more terror rules, the more cruelty can be expected. And in a world ruled by an amoral regime like this and for which terror, along with deception, are the basis and foundation of their power, there can be no wonder so many (whether people or animals) suffer so harshly and or unspeakably in this life. Meantime, there are those in a state of dread who see (their own) being cruel as a way of propitiating and securing for themselves the mercy of the high lords of terror. "Have mercy on us! We promise to be cruel also!"

Another aspect to the high power criminal spirit person kind of personality is that they like to pride themselves on the idea that they know virtually everything, at least that is worth knowing; often imagine themselves to be funny when they rarely are ever actually so; all of which putative omniscience only serves, as they see it, to excuse their, in a given instance, callous or even cold blooded arrogance.

There is, based on my own experience dealing with them, good reason to believe that it is not unusual for some of them to get high on some kind of drugs or intoxicants and or group orchestrated hysteria of the moment, and which might on a particular occasion, disperse any competing scruples and or unwillingness to commit crime; which the targets of manipulation might not else be so inclined to commit. Then there are things like bribes and various sorts of mind control,

---

[15] From "The Little Dutch Sentinel," found in his short story collection *The Book of Saint Nicholas* (1836).

including old-fashioned *head games*, and clever sophistry to persuade the potentially reluctant. Simply put, no one can escape the imperative and necessity of the moral law. Even so there are ways, and depending on the individual, that the moral law can, in some degree, be minimized if not quashed and effaced entirely from people's thinking and calculations.

I can by no means speak as somehow having proven the fact, but I would take it as a given that many if not all of the hideous and unnatural practices of tribal and primitive peoples were prompted and instilled by criminal spirit people. Such, at any rate (though others are of course free to disagree), seems to me the more likely explanation for things like cannibalism, human, child, and animal sacrifice in its various forms, as well as and incidentally practices like making shrunken heads or outlandish bodily disfiguring (as in tattooing) and, as in the case of some African tribes, artificially stretched neck and lip extensions. Based on a related or similar surmise, it was criminal spirit people who conveyed to both the ancient Egyptians and Aztecs that if such and such ritual, no matter how absurd, bloody, and or inhuman, proposed was not followed, the sun would not rise. This understood, the ancient Egyptians and Aztecs were (insofar as we know) sure to do what they were told. So that after all, if a cunning and sophisticated criminal spirit person can get people to agree to make complete fools of themselves and act as badly *as all that*, what else can't he incite, acquire, or get out of them?

Hard-core criminal spirit people not unusually tend to be obsessed with degradation and humiliation, all the more so as they are inebriated with absolute power. And this, in part at least, is ostensibly a product of perhaps their own self-loathing combined with a wish to more effectively mold and accustom subjects of such treatment to increased submission and subservience. One famous example of this sort of behavior is how black slaves were sometimes brutally treated in ante-bellum (and post-bellum) America. One might object and say that it was "regular" people who acted thus. Yet for those who have ever been involved with criminal spirit at length, this preoccupation with degradation and humiliation will be readily recognized; hence it is not entirely unreasonable to infer the possibility of criminal spirit people, on some level, influencing such actions and atrocities. Again and more generally speaking, we cannot, just by the mere assertion, say we have conclusively demonstrated that such was necessarily the case in this or in others instances of unbridled cruelty. Yet it won't hurt to make mention of what I think many will agree (particularly those who are already acquainted with these kinds of people) is the more correct explanation for grotesque and bizarre sadism, and atrocities arising out of the same mentality, whether in our own times or historically.

IV.

In "The Ch'-un Ts-ew, with the Tso Chuen: Duke Ch'aou" [And as found in *Chinese Classics* by James Legge, Vol. V. Part II (1872), p. 618], there is this passage:

"Tsze-t'ae-shuh asked his reason for making these arrangements, and Tsze-ch'an replied, 'When a ghost has a place to go to, it does not become an evil spirit...'" [Book X, Year VII.]

The thought expressed here perhaps helps to explain in part ancestor veneration among such as the Chinese and Japanese, and yes indeed, and outside "fallen angels," criminal spirit people one infers originated largely from social outcasts, such as we contemplate with bands of robbers, thieves and other sorts of criminals; as persons who were displaced from or without a proper home, family, and or other morally minded brethren to find healthy companionship with. For this reason, there is certainly a place among feeling and compassionate persons to pity and sympathize with them or that is, some of them at least. Otherwise, and yes, I for one agree troubled and disturbed people such as these *should* be helped. But further permitting them to carry on as violent criminals, clandestinely or otherwise, obviously is *not* a good way for accomplishing that end. Loving one's enemies cannot mean compromising basic principle that, say, involves becoming a willing accomplice to felony crime. Moreover, such sympathy towards these kinds of people, if not guided by caution and tempered with shrewdness can simply be used against a person; as for instance when one pities and trusts a dangerous snake. Our pity is or might be justified, and yet if we are not careful, it might be used (by themselves or another who uses them) merely as a handle to bite and get the better of us. And even if it were possible to care for or take them in, it is or may be only a matter of time before their owner shows up to claim them.

Adopt which or whatever point of view we will, such alienation these sorts of criminal spirit people feel or live with is expressed and comes out in a variety of ways. This would or might include a penchant, with some a pathological preoccupation, to invade and act as an interloper and busy-body in the lives of others; getting attention, say and for example, by shocking and surprising people, *or* by showing off their skill at being able to perform some unusual trick, *or* and similarly by putting on an very elaborate show. Such a "show," for instance might be something expensively paid for and planned out in advance, say like a "Candid Camera" or "Magic Christian"[16] type prank or stunt intended to mock, harass and

---

[16] Speaking of films, "Out of Towners" (1970) is another that comes to mind; but which in its realism is more painful than funny as intended. Although such distressing and untoward events that occur in it are by no means necessarily the result of criminal spirit person machinations, nevertheless it is not in the least out of character for them to bring such about if and when they have a mind to doing so; and assuming as well they budget for the expense to themselves it involves.

or torment somebody. And yes, it is very typical for some of them to have you think them funny; even if it is extremely rare (if ever) for them to actually be so.

Likewise, to compensate for such inferiority and alienation as they feel, they might put on airs and have you think them superior in some grandiose way; as would of course be the case and similarly when a high-power spirit person would have you see them as god-like. Another not uncommon related behavior of theirs is that they will abuse you while acting as though they are your friend, and in fact be even sincere in desiring, albeit in a warped and demented way, to be a welcome part of your life. Also, they will make a pretense to being fair, unbiased, impartial, even sympathetic, would have you believe them to be so; even while standing by when you are being raped or murdered by their associates

In sum, owing in large part to the often deranged, and horrendous crimes they have committed, bolstered by (what for many is) a kind if interminable idleness, they have made themselves detestable, and thus are understandably shunned. And yet the resultant lack of low self-esteem then only prompts them to continually seek attention by and through being trouble-makers. And, for yet certain others of them, they endeavor as well to raise themselves up to positions of highest worldly importance in the seeming grand order of things. At which and at times, they are famously known to succeed.

As a further aspect of this same mentality, it is routine among many of them to make it their business to cause trouble, whether a little bit or much, simply and solely for the purpose of causing trouble, with no other apparent motive present or required; though, of course, and notwithstanding they will, more likely than not, have what they claim is a just and compelling reason for their acting so. And they do this both to get attention and or because they are told to do so by higher-ups, and who firmly believe, and religiously, that violence and vandalism and or other obnoxious pranks are more than justified if carried out on behalf of motives of envy and or self-pity. From which one is inclined to infer that even if in the end evil can't succeed in ultimately conquering or prevailing, it ever and apparently gives them infinite self-satisfaction to just needle and harass others, and secure the often desperately sought-for attention in doing so.

In passing and with respect to the question of spirits and or sprites (i.e., by this latter we mean spirit persons unusually small or tiny in size) entering and leeching on a regular person's body, as, for example, is associated with demonic "possession," it is unclear whether such are *a)* just entering a person to "ride" them, say like a horse; *b)* find shelter and inhabit the host as in a sort of mobile dwelling, and or *c)* whether they are literally "feeding," somehow, off the regular person's physicality; and or some combination of these possibility. It is worth mentioning that these possessing or inhabiting spirits can sometimes be ejected or cast out by placing the regular person victim nearby to heavy metal machinery, particularly thinking of a jack hammer. The vibrations of the latter will, as likely or not and if the vibrations are sufficiently violent and the exposure is for more than just a brief length of time, shake the spirit from the nest they have created for themselves in the host.

While I have no experience of anything so extravagant as depicted in the film "The Exorcist" (1973), perhaps it is not inappropriate I make a passing comment. How does that film stack up and compare to real life? While I like dramatically how the story is resolved in the priest's sacrificing himself at the end, it is worth noting that you cannot medically prescribe for a fictional situation. So that what in the movie seemed insurmountable, in real life circumstances given the presence of more rational, intelligent and inspired people might be less so. For one thing, a given devil is a person like anyone else – who was born and has their personal history. Consequently, consider what difference there is to seeing them in this light rather than as some religious, mythical, or cosmological abstraction. At the same time, they themselves are sick people, medically speaking, and this is another significant aspect to taken into consideration when it comes to sizing them up. Also, spirit people, like regular people, and as individuals are different. The "ghoulish magician," as rotten a devil as there ever was (or as I heard of or knew), for example, forbids swearing like they have in "The Exorcist;" because he himself, and in his own time, has had to hear so much of it.

Things like the shaking of furniture, lights being turned off inexplicably, being held down physically by "demons," and having living apparitions enter your bedroom I have experienced and seen myself, and again if viewed a certain way can be ridiculed. However, this is not so easy to do if you are half asleep in bed; in which case you are more easily frightened. So that in all, a more medical and level headed approach, and perhaps aided by a little bit of humor, will, in most instances, go a much longer way than mechanically invoking sacred or formal ritual. This is not to dismiss or make light of such, but each thing in its proper place, and just because you respect such things doesn't mean a spirit person has to. And to realize this is merely enjoining common sense.

As mentioned in my *New Treatise*, criminal spirit people that act as "bosses" will invariably have "regular" (i.e., flesh and blood) persons routinely carrying out tasks for them, indeed such as are necessary and vital to carrying out organized criminal operations; since of course regular persons can do things that spirit people can't: such as hold money or carry out more involved and heightened forms of violence and vandalism. With respect to vandalism, these sorts of things can be anything from annoying, pointless monkeying to acts of extreme viciousness and violence. To give you an example of the pranks, I have had these people break into my house and repeatedly do things like repeatedly put holes in my clothing, or steal tableware (all my spoons, for instance.) Other sorts of pranks I have had deal with are massive infestations of ants, or constant influxes of rats over the years, and that initially occurred totally out of the blue (that is, after many years in that same location without their ever having been such a problem previously.) No matter how many rats were already removed or exterminated (and there were over the years in question at least several dozens) using traps, air guns, and such, they would return the next Spring (except the present year[17] where there has been no problem – as yet at least.)

Also as remarked elsewhere, the more sophisticated, skilled and adroit criminal spirit people, such as sorcerer magician types, can employ, and in some instances even direct and control, animals to cause nuisances or strange effects; again, such as rats and even insects, but really and for that matter apparently almost any animal at their immediate disposal. In such cases, the animal is prompted, goaded or provoked (or somehow "mind controlled") into a given behavior; though the *illusion* is created that the animal is knowingly acting on its own.[18]

Likewise the same kind of professional criminal spirit people are sometimes in their way medical experts of a kind, and will or might do things like (for example) give you multiple warts and moles *overnight*;[19] disfigure your skin with weird scars; ostensibly give or inflict diseases; prevent bodily functions from working properly, and engage other kinds of physical harassment or abuse; invariably done or carried out on you when you are asleep. Such things, at any rate, I have many times experienced and suffered in my own dealing with them.

Much of this sort of carrying on is as much playing games, confusing, disorienting, and perhaps intimidating you as much as anything else, and which typically serve as reminders that *they will* (or at least would) have their way with you and preferably as they please; the idea being they have you in their clutches and you cannot escape them, and no one can or will help you (since, for one thing, there is no such thing as spirit people, of course, such that the subject can even be talked about.) In addition, the tendency for criminal spirit people to be often idle serves to prompt such mischief and or worse; and sometimes the kind of trouble making they do (whatever that happens to be) can go on relentlessly for relatively long stretches of time, say months, years, or even decades.

Other behavior we might again mention in passing are their acting in a friendly manner [20] or pretending to be neutral while they are abusing or otherwise taking advantage of you. As is known, "the tender mercies of the wicked are cruel." Part of the idea here is that manichean-like mentality of trying to have it

---

[17] That is 2022. Note, the rat problem only first took place about 2011, before that, and outside of pictures, I had never even seen an actual rat in all my life.

[18] Cases, incidentally, where elephants are shown (say in online videos) painting pictures, while holding the brush with their trunk, I myself would be inclined to suspect are the result of a spirit person manipulating the animal (either before hand or at the time of their painting the picture.) This, at least, strikes me as at least one plausible explanation for what otherwise strikes one as seemingly inexplicable.

[19] I have had this happen to me on, at minimum, a dozen and more occasions.

[20] Spirit person magician "powers" can also be used to entertain, amuse and or employed to serve some every day need; such as carrying out a task or labor you yourself would just as soon not have to bother with: very much like what one might see on "Bewitched" or "I Dream of Jeannie" tv shows. Yet that very same ghost magician acting in an ostensibly light hearted way, on another occasion might be participating in the gruesome torture or murder of someone.

both ways. Some will actually be fooled and taken in by kindness offered by an oppressor; while the more wary and wise will see for it what it is. If they do an unsolicited favor for you, you are in no way obliged to owe them or pay for it, and if you don't know and understand this you can be tricked or beguiled into something you might perhaps later regret. If for example they spent a million dollars just to play a funny practical joke on you, say a "Candid Camera" or "The Magic Christian" type stunt (referred to earlier), they might have or insist that it was all in good fun, or served some alleged good (perhaps religious) purpose, and you should be flattered by all the expense they went to and the trouble they took. This sort of attempt at manipulating or working you over is typical of the arrogance and over-reaching found so frequently in them. And yet as cocky and insufferable as such tactics will seem, there are people who will end up timidly acquiescing and foolishly yielding to them. Such scams, put-ons, or charades, and this is also true with magic tricks generally, when they do not succeed in fooling, the criminal spirit person magician will feign his having succeeded.

Trades and exchanges, whether voluntary or coerced, are an essential part criminal spirit people operations, nor, need I add, is it in their nature to deal fairly. That Goethe's Faust could come out with a happy ending thanks to the intercession of Gretchen is about as incredible a fairy tale ending as was ever proposed; my point being not that God could not be so merciful where there is true repentance, but that it is extremely unlikely that Mephistopheles would not know how to get the better of a willing, deliberate compact such as Faust had made. Such after all is part of his professional business, and on this question Goethe, albeit unintentionally, is leading people astray and playing into the wrong hands into thinking works and the pity of good women will of themselves save a person's soul. In this regard, Maturin's "Melmoth" is, by comparison, more like the real thing. Granted it may have happened historically that someone was so shrewd, wise and fortunate that to some significant extent it was possible to have it "both ways," and yet not seriously suffer for it, or at least not too much, in the after-life. But if this has happened, one cannot help but assume cases of playing with fire and winning are, at best, exceptionally rare; at least when it comes to persons who made out and out and deliberate "pacts."

Despite the often grandiosity of the proffered largess, the generosity of professional criminal spirit people is always and ultimately false and an illusion. For they are not such dim-witted fools, in this regard at least, to have it otherwise where they can help it. What they take will always be more than they give, and often what they give was invariably and originally obtained by theft. And it is an inherent part of their character to be the one who is more clever and cunning in trades and exchanges, and other than showing off with magic tricks, they probably pride themselves on this ability to get the better of a deal more than anything else.

Although I have been emphatic and make clear to them my rejection of them and what they do, nonetheless I have gained and profited from having contact with them in the way of learning things significant or even very important about spirit people, including some points and matters I relate about them in my writings. And this is, to some extent how they would have it. That is, they would normally prefer to do some kind of exchange rather than rob you outright. In my own case, and despite my above stated profiting by learning from them, I still would infinitely rather never have had anything to do with them, under any circumstance (if I could help it.) For it is not in my opinion worth it; as more grief and trouble comes from or originates with these people than anyone or anything else, and despite any benefit you might possibly derive from having contact with them.

As well as the traditional cunning, cleverness, and skill at mind control and behavioral manipulation, professional criminal spirit people have what we might (for lack of other ready term) call "spirit people technology." Where and if or to what extent technology such as we ourselves know overlaps with kinds of technology and expertise available and unique to them, I myself couldn't say. I think, however, it can be reasonably and safely assumed or inferred that they have and have had technical advances overtime; both stemming both from their own endeavors and also benefiting from advances acquired or realized by and or in partnership with regular person friends and accomplices. At shock and infiltration tactics, just incidentally, they are without peer.

This understood, I will attempt a description, though lacking and flawed as it must necessarily be, based on my own years contending with, and some of what I have experienced or seen them do. This recounting is both admittedly sketchy and only partial in its coverage of the more strange devices they have and tricks they are capable of. The same is also somewhat covered in my *New Treatise* and "Narrative," and for more on this topic I would refer you to those writings.

As we've stated, the more professional types among them are experts at mind control, including things like hypnotism, and literal (physical) thought control carried out telepathically. Among some of their tricks are transferring thoughts into a person's head, including words, pictorial visions (static or, more rarely, in motion), and even creating and or manipulating a sleeping dream. As an instance of how sophisticated these things can be done, imagine an elaborately detailed painting like Napoleon's crowning Josephine by Jacques Louis David, and then visualize it coming to life in all its tiniest details as if it were not only like a movie, but an actual even taking place at the moment you (in your head) see it. Such a thing I will have experienced in what I have elsewhere referred as "dream productions;" that is, shows they are able to create that are made into sleeping dreams. These dream productions are well likened to a movie or tv show, in which you the dreamer are carried through a series of events; which, in my experience, typically have you degraded, insulted, made to feel isolated, or humiliated. To give you a working idea of how it is possible to recreate dreams (albeit in a different venue or medium than I am discussing) check out for instance the kitchen scene in "Salem's Lot" (1979) or the film "Blue Velvet" (1986). Now how criminal spirit people, such as a ghost-magician, are able to transfer or insert these "dream productions" of theirs into ones head, I do not know to say. Yet it is worth noting that the literature and lore about spirit people going back to ancient times, whether fictitious or alleged to be true, abound with accounts of people experiencing dreams that take the form of messages being transmitted to the sleeper by some other-worldly person. Yet like some ordinary dreams we might have, it is possible to experience them without quite remembering after we wake up what they were about, and the same can be true of "dream productions." In this way the ghost sorcerer or magician's efforts may, on a given occasion, prove ineffective and futile because we don't remember the dream (other than that there was *a* dream and that it *wasn't* an ordinary dream.)

Speaking more broadly, the significance attributed to dreams in clinical or conventional psychology and aside and outside their being used as mediums for persons external (i.e., such as a "ghost-magician") to impart communications of one kind or another, has in my own view been grossly exaggerated; as for example where profound meanings casually are assigned to a dream's contents; as if such were necessarily symptoms of something significant. My own view rather is that dreams are nothing more than a state where our consciousness, because of our being asleep, is operating on what might be called "low batteries." That is our sleeping consciousness is able to function, but only barely; say like a computer running on low power. The sleeping mind can do some thinking, but not properly and far short of what it is capable of when awake; including things like observing measures of time. For example, the events of dreams ordinarily seem to take place in what would seem to be the day time, and yet during the dream we have no conception of what a day (or night) is; nor are we able to track time in any form. Rather what ensues is a sequence of events that, while we are dreaming, knows no clear measure of time. And I suppose a somewhat similar thing is true of space as well; that is, the significance of distance does not (at least in my own experience) occur to us as a matter or issue of any consequence or relevance in a dream. The significance of all this is perhaps by keeping the awareness of time and space muted or separate from us, people, things and events can be distorted from how we would see them than if we were awake; clearly delineating time and space act as a filter or framework to organize and arrange our thoughts and perceptions, and if turned off then our guard is that much less down, and not working; thus arguably and perhaps making the ghost-magician's job easier in attempting to hoodwink and psychologically steer the thinking of the semi-conscious sleeper.

Even so, I have had them attempt in dreams to deceive me about things that even asleep do not ring true to me, and that upon waking I realize what I dreamt was not me but someone else trying to get me to believe or see something in a way that is fundamentally alien to how I do feel and think, and that I myself do and in no way can relate to. I notice, for instance, in a dream production I am made to take on a certain kind of character that is entirely unlike myself: as if another person's personality and accompanying set of value was being projected on to me. The person in the dream is supposed to be me, and yet so much nothing like that it is clearly based on someone else; as if someone had written a script for me to follow; which then in the dream I am made to follow. And upon thinking about it after I wake up, that person in the dream most clearly is not me at all.

Nevertheless, as the years will have gone by (and it has been years), they would seem (so I would infer, that is) to keep a record of my behavior and behavior patterns, which then they are able to incorporate into the fake dream production version of me. Although the characterization now will ring more true, the character is still not me but and rather a far from perfect recreation: *more* like me, yes perhaps, *but still not* me.

One other and very unusual thing I had happen to me just the other day and that I will mention in passing. I was lying in bed awake just before dawn, and I distinctly heard a dog's bark *in* the room! (I do not have a dog, by the way.) Whether this was some kind of ventriloquism or what, I have no idea, but my guess is that this was a case of their being funny, and yes, efforts to be funny can be very usual or typical for at least some of them.

As well as dream productions, it is possible for a technically adept criminal spirit person, like our "ghost-magician," to project images, say for example a face of someone, and or patterns of images *in ones head* while you are awake (as opposed to asleep), and do this in such a pronounced, lively and prolonged manner that you know it is not merely or somehow your imagination. Such, incidentally, may evince some caliber of artistic talent and ability, and this might be observed in say a dream production; where the story and or characters might be cleverly conceived and constructed. Even so, such expressions, whatever the means or medium will *almost* always be tainted with some amount of unnaturalness, or other quality that some will feel offends or disgusts their own sensibility and or morals.[21]

There is of course the age-old sending a "demon" or sprite into a person. The effect of which or harm done depends on the character and strength of demon/sprite in contrast to the intellectual and moral constitution of the victim. Again, this is one of those phenomena I have attempted to describe in other of my writings[22] and would once more refer you on this topic to those.

Despite the "tin foil hat" type mockery we hear of among some, supposedly independent, yahoo type online activists and news reporters, more sophisticated and well-financed criminal spirit persons and their hench-people do in fact have psychotronic weaponry, including "brain torture radios," that is mundane (i.e., human) technology, now in their arsenal. This is in addition to their more traditional expertise at chemical/medical and biological warfare; and which they will resort to as they see a signal or worthwhile advantage to be gained by it.

---

[21] Note. In weeks since first writing this, they have made an effort to come up with "productions" of a more clean and innocent character with admittedly some success, but only slightly so.

[22] In addition to the *New Treatise*, see also my article on the *Somers* Mutiny of 1842 in *The Continental Army Series*, vol. II.

And why, in any case, should it be made to seem impossible there should be such things as psychotronic weaponry, like brain torture radios; when such technology is and would be ideal both for physically injuring and incapacitating an opponent *and* having them stigmatized as being crazy? Is this after all, not a perfect kind of weapon, not least of which those who make it a point to eschew discussion and debate and yet insist on the superiority of their cause? At the same time, there is a materialist hubris and message here; namely that the mind is simply and merely a machine and that can be controlled like a machine; therefore rational and just argumentation isn't worth much and, indeed, is little better than superfluous froth.

In addition to "brain torture radios," I myself have, in the course of my 30+ year ordeal beginning on 1992, been subject to an assortment of strange sicknesses and medical ills, including yellow jaundice, persistent constipation,[23] kidney stones, over-night appearances of warts, moles, and bodily scars, and advanced covid (I have had three versions of covid, overcoming two and am present suffering a third advanced more stubborn type; notwithstanding (i.e., given my very peculiarly isolated circumstances artificially contrived for me) my having have little or no social contact; as well as having been poisoned, including by needle injection, on multiple occasions. Such occurrences may naturally, in a given instance, have required regular person assistance, as opposed to a spirit person's inflicting them. Nevertheless, there is no doubt in my mind at least that many of these physical problems or disorders I was put through were brought on without such regular person assistance, if not necessarily all of them. For more on this subject as it pertains to my own experience, see my "Narrative."

Although my proof is far from conclusive, and while only at the moment positing what follows now as theory and conjecture, I have come to believe, indeed am myself fairly (say, "95%") convinced, certain professional criminal spirit people can control and manipulate computer programs that are presently running. One example of this is playing PC scrabble and on occasions when there is no question of "their" being present in my immediate vicinity, by way of both "brain torture radios" (a worldly, not other-worldly, technology) and telepathic projection (of spirit person origin.) In a particular game, I am winning, and have won. But when it comes to presenting the final score, the computer opponent receives *my* total of (winning) points and I receive *its* own total. That is to say, at the last turn, our actual and respective scores are made to be reversed. For example, on the last move of the game, I might (using round figures here) have 300 points and the computer 200, but when I enter the last move, my score is switched to the computer player, and the computer player's score is made mine. This kind of cheating has happened innumerable times over the years in this in other computer games I have played. If this inference is correct, and the cause of these most peculiar and unexpected game software anomalies is *not* something else, then it would seem professional criminals spirit people can affect the working of a computer by remote means. In saying such, again, I am not in a position to somehow insist that I am incontrovertibly right in this surmise. But if

---

[23] Note. Prior to their invading my life in 1992, I did not even know what constipation was.

nothing else, it would be remiss of me to not at least make mention of this phenomenon; particularly given the not inconsiderable implications such technological ability suggests, including the potential of manipulating software for purposes of determining political elections, and despite and as I fully grant how else silly my speaking of it must admittedly and understandably sound to some.

Meanwhile who, might we ask, has and has had the time and resources to concoct computer viruses and malware, and that requires several million dollar industries to combat? And if there are persons doggedly occupied with creating computer viruses, why not do medical viruses as well? Yet no one, to my knowledge anyway, remarks on the obvious parallel.

Consider then – *if* it is the case that many physical sicknesses and psychological ailments are the result of purposeful vitiation and infection by professional criminal spirit people, including in this perhaps as well their inventing and or "sponsoring" new diseases, how much in utter darkness and confusion medical science necessarily must be when it categorically denies or blindly ignores the possibility of such insidious and malevolent mischief. Not that there aren't or might not be doctors or researchers, suspending long standing conventional assumptions, who would impartially and objectively consider and explore the likelihood of such, but even if there were, there are inevitably persons of great wealth and influence, in some level of vassalage and or partnership with criminal spirit people, who would stymie, and thwart and proscribe their efforts; not least of which resorting to violent and other crime, if need be, to effect this.

As perhaps you might infer otherwise, it is by no means lost on me how much of what I speak of here sounds pretty crazy, and some might rather conclude rather and perhaps for that matter that I am so (speaking as I do.) Yet if you will take my word on anything when it comes to criminal spirit people, totally bizarre tricks and stunts, seemingly inexplicable, and intended to baffle, probe, test, intimidate, impress and or dismay are an innate and essential part of how they act and operate. And there are in facts things even more strange they do or have done I could relate, but will be satisfied here instead, and at least as an introduction; with having at least presented, albeit in an abrupt and rudimentary manner, *some* sense or notion of the extremely weird and kooky things they do and are capable of.

V.

"Justin well said: Before the advent of the Lord, Satan never ventured to blaspheme God, inasmuch as he was not yet sure of his own damnation, since that was announced concerning him by the prophets only in parables and allegories. But after the advent of the Lord learning plainly from the discourses of Christ and His apostles that eternal fire was prepared for him who voluntarily departed from God and for all who, without repentance, persevere in apostasy, then, by means of a man of this sort, he, as if already condemned, blasphemes that God who inflicts judgment upon him, and imputes the sin of his apostasy to his Maker, instead of to his own will and predilection."

~ Irenæus (c.130–c.202 A.D.), *Against Heresies*, v. 26.

More familiar to most, however, than the "technology" and bizarre physical pranks (such as we have partially enumerated) are the hypocritical dodges, con-artist schemes, and sophistry high-power criminal spirit people and their followers adopt and utilize

For starters, they endeavor to recreate the world in their own image, and a person can be left or right, atheist or religious, or whatever as long as you sign up, register, and be licensed with them first. If you don't, it is then intended and or effected that you be made an outcast otherwise.

Someone who *doesn't* or *won't* so give in as demanded, and instead does the right thing in defiance of them, for example, might have charges or accusations brought against him; whether with respect to something he did or didn't do. Only note, if he was (as it turned out) actually guilty of the said charge, he most certainly would *not* have been brought up on said charge *if he had* originally yielded and given in to them; indeed and moreover would have been touted as a model and exemplar of moral, religious, and or civic virtue.

Likewise, it is typical for criminal spirit people and their followers to be the ones most to blame for some kind of wrong-doing of which they themselves are the *worst* offenders. That is, they will be the most vociferous adversaries of some bad behavior or other, and then arrange to have themselves appointed as the illustrious prosecutors for the same; even though and if in fact they would be the ones most to be faulted for the given charge.

Some individual, to give you an illustration of this kind of game, refuses to be a party to or do some wrong the criminal spirit people, et al., ask him to do. As result, he becomes a target of their malice, and whether on his own or being lured into a trap, he is subsequently caught making stupid passes at women. He then is found out on this and perhaps decried in front page headlines, and his reputation and career completely trashed and ruined; even though some of those who saw to his demise are or might be routinely engaged in gang rape or sado-masochism (behind closed doors of course.) A similar sort of thing was for years our hearing widespread concern about privacy being invaded or violated on the internet. Yet sometimes the most vociferous making the complaint are in fact the worst offenders, and indeed are the most shamelessly intrusive people imaginable and who have no scruple intruding on or meddling with people's private things: straining at a gnat and swallowing a camel when it came to complaining that privacy was not being held inviolate and sacrosanct.

It was at one time (and perhaps still is) seen as fashionable and funny for Don Corleone to make someone an offer "he can't refuse."[24] Here again it is really the "witchcraft" people who act like and do this: attributing, foisting or projecting the practice onto others, including Mafioso (who are only acting as directed.) But what is overlooked is that such "offers" are seen by them as necessary; since persons such as they really have little or nothing worth trading to begin with, and no one *in their right mind* would dream of willingly making a deal with them. Coercion then is seen as a viable and indispensable alternative to being rejected or ignored completely.

If one were to ask these kinds of people why they do something wrong to hurt another (thinking particularly with respect to violent or vandalistic wrong-doing) their response (if they allowed you one) will almost invariably be that someone else has snubbed, robbed, cheated, or done them an injustice some how, no matter how otherwise excusable or trifling the alleged offense. They have been wronged! "Justice" therefore demands that they do wrong in retaliation. Two most famous portrayals of this kind of psychology and personality found in deific history and other lore are Eris indignant at not being invited to the wedding of Peleus and Thetis (and evidently and in her way as or more powerful than Zeus), and Maleficent similarly being left out as a guest at the christening of Princess Aurora. Such obsessively wrathful types as these, rest assured, are no mere invention of story-tellers.

At the same time, high-power criminal spirit people and or their immediate followers will make a virtue of a vice, and, further if possible, turn that vice into something chic or fashionable. This has been commonly observed historically, for example, in the romanticization of pirates, highwaymen, outlaws, and gangsters, and subsequently far worse; in which the criminal is treated as a hero or the good guy. While poetic or literary license can sometimes excuse this practice in the interest of story telling, there is on the other hand such a thing as nauseatingly *relentless* propaganda or over doing it. "Good guys finish last," but perhaps only now will some come to realize what, in practice, is the actual or overriding reason for this.

Infiltrating and then impersonating political movements, social causes, religious faiths, and creeds by criminal spirit people, whether through their followers or, more rarely themselves unseen or behind closed doors, is quite routine; nor is it the least unusual to those who are truth-minded and sincere to be

---

[24] This quip, be it recalled, is an example of these people trying to be funny.

fooled and taken in by such impostures. In fact, every now and then some people, including activists, are so unwary and distracted by trifles that their more shrewd and shifty adversaries can use them to actually serve the cause of the opponents as in, say, controlled opposition. Moreover, it can be or get so bad or ridiculous that in some instances you can have one "cointelpro" person (rightly or wrongly) accusing another of being "cointelpro" without realizing that the one accusing is unknowingly "cointelpro" themselves. As rule of thumb to help separate the real from the fake or the dupe, one should (I recommend) be especially wary of social reformers, activists, of whatever cause or creed, who evince a capacity to hate and destroy more eagerly than they do a capacity to be honestly rational and love more deeply in the promotion of their cause.

Of course, among the purposes or aims of masquerade and infiltration are to both corrupt and make the given cause or group look bad to others. This might be done further by placing incompetents, unqualified and or mediocre persons in positions of importance, while ousting, ostracizing, or even assassinating the better ones of intelligence and integrity. The regrettable appointees then might act to have the group or movement make ill advised decisions, misplace their priorities, or instill in its members the necessity of appeasing or accommodating public stereotypes, prejudices, current myths, and false idols of the day; even though doing so goes directly against what is supposed to be their primary interest or purpose. Examples of this in our own time is how groups of such divergent outlooks will come to somehow concur that exponentially increased bad language and swearing in public,[25] goatees, tattoos, junky music (e.g., hell rock or canned romantic modernism), comic book super-heroes *for adults*, the hallowed prestige of contemporary (i.e., monopolistic) moviedom and the mass media, "net worth," – or whatever is the supposed and necessary fashion or a focus of greatest concern, are something *they all like and agree upon* (whether left, right, religious, non-religious, what have you); even though they are or might be dire enemies and completely at odds with each other on almost every thing else. You are permitted to have and express your view; you are allowed to be heard and reach mass followers, but *only* if you have made the requisite offering to whatever the given false idol or high priest(s) of the day happens to be. Viewing the matter broadly and fed on anarchistic New Age swill,[26] one result of this kind corrupting indoctrination of society and the culture in recent decades has been to turn many adults into over-aged and warped children, and many children into under-age and warped adults.[27]

---

[25] This was *before* the advent of the internet; the practice largely something promoted by movies and television, especially in the late 1980s.
[26] With education, art, literature, and music not infrequently being employed to of accustom us to chaos and or glorify mediocrity.
[27] Desirous of free speech? Then you needs must "Rumble" (a colloquialism for a "street fight") to have it. My point here otherwise is to say alternative media (or alternative media venues) are not always so alternative as they might seem, and when the person who is or would be speaking out for truth is required to offer a sacrifice to a false idol, then we might reasonably suspect the integrity of their motives and or the cogency of what they propound.

Notice in this respect that it is not what is right or correct based on better judgment and right (or more right) reason that prevails or deserves sway. Rather the more important standard becomes what "everybody" thinks; and which permits a wider and gaping chasm for error and absurd conclusions to slip in; with truth as a matter of course taking a back seat to rhetoric and mindless crowd propitiation. As per Thrasymachus "truth is the opinion of the powerful." So often then what is interpreted as the *supposed* choice, trends, or preference of the people is actually something they are systematically and patiently over time behaviorally conditioned into accepting, and this by a cunning and clever wealthy minority that are seasoned experts at propaganda, psychological warfare, and wholesale manipulation of the masses.

Sometimes the effect and influence of criminal spirit people, and their hench-people is laid on so thickly that a subject of such treatment becomes incorrigibly zombie-like and irrational under their influence. Having encountered this kind of things many times in the course of my own ordeal contending with criminal spirit people, how true, in their way, then are those vampire movies; where an otherwise good person, once bitten, is transformed in their moral character and basic sense of decency into something *entirely different* from what they were. It is nothing short of amazing, to put it mildly, how when the heavy-duty pressure of high-powered criminal spirit persons is turned on the extent to which it can dramatically transform and pervert some people, and this perhaps in a stupefyingly brief stretch of time. People who earlier were your cordial and cooperative friends, family members, or acquaintances will or might give you the cold shoulder or even betray you virtually *over-night*.

In practice, regular people and even animals [28] are faulted and punished for wrong-doing, yet, as you already know or could expect, *never* for a moment are criminal spirit made mention of, let alone blamed, let alone punished. Yet when it comes to serious wrong-doing, they, and their knowing accomplices, invariably are the most guilty of all. *Of course* God or religion can be censured, ridiculed, vilified, and regularly are. But to even murmur under your breath that there are criminal spirit people or they may share in some blame in what goes on is impiety, heresy and anathema, yahoo madness to be loudly ignored and obliterated with implacable and scowling silence. ("What? Blame the devil? Blame God first!") With such flagrant injustice that necessarily results, is it any wonder the world knows and experiences so much crudeness, depravity, callousness, madness and tragedy as it does? This, in short, I would submit, is after all is one of the primary reasons why life for many is so difficult, indeed for some unbearable, particularly in our time where criminal spirit people dominance and influence on the culture is (or so I argue) dramatically on the increase; with, as some will have observed, suicide rates being alarmingly higher than they were in past decades.

---

[28] Note in this respect how often some new disease or other in named after some entirely blameless animal.

As many are already aware, criminal spirit people and their followers are not lacking what are purported to be compelling arguments to persuade you into going along and cooperating with them and ways to account for why it is they behave as they do; of which we can now review a few of these.

Before proceeding, however, it is necessary to remark, and as I have maintained elsewhere, that there is or may be *some* truth, albeit in the given instance quite miniscule or minute, to any proposition; especially so if the assertion is consistent with good and right morals (at least such will seem true or correct to religious, and who will, as matter of faith, see God as supporting that belief.) Likewise, given our contingent, finite intellectual natures, any proposition, at least taken the wrong way and or perhaps without refined qualification, contains some falsity, albeit and perhaps to a very small degree. Such seeming contradictions or paradoxes when we encounter them can be resolved, however, when we see or realize that the smaller truth is out-weighed by a greater. So, to look at one example, is teasing someone bad? It depends on the people involved, the circumstances of its being resorted to. Naturally, and with certain people in some instances it is quite harmless, even a good thing; while with other kinds of persons and situations it would be extremely cruel and inexcusable to tease someone. Therefore, and summing this up in simple terms for brevity's sake, a statement like "teasing is bad" in one sense is false and in another it is true. It all depends on exactly what and or whom one is talking about.

With then some of the arguments criminal spirit people or their followers do might employ, what they say may have some actual or else seeming truth to it. But as with any argument, it must be fairly and closely examined to sort true from false, and false from true; weighing the one against the other; including to what extent what seems true only appears to be so because it is based on a glaring false assumption. In other words, if we grant a certain assumption, yes the argument is true. But that's just it: is the assumption correct?

Another epistemological caveat that is well to keep in mind is this, namely: *nothing* is or isn't but that *someone* says it is so or not so. Any assertion or conclusion requires at least one individual or more to arrive at it. Typically and for the vast majority of people, we think of something as true or false without realizing whether it is really our own or somebody else's judgment that first caused us to arrive at that conclusion. Either way, when we come to examine the matter closely, we will see that any conclusion we or someone else arrive at requires a person or persons to make it. Every conclusion therefore has and must have a "face" tied or connected to it, or there is no "faceless" truth. Whenever then we contemplate whether to accept or reject a proposition, particular one that is fraught with grave and weighty consequences, we need to include in our assessment an identification of "who" it is that arrives at the final conclusion; since who the person (or persons) is (including their honesty, character, and credibility) does or may affect the reliability and trustworthiness of that conclusion. So, and using a hypothetical, if a spirit person were to come visit you one day (say when you are home all alone), and they seemed to imply they represent higher authority (even and perhaps very God and heaven), *who* is it that says that their claim is true or false, and what are the arguments pro and con to accept or reject that claim?

With these cautions in mind, the following are some arguments or expressions of principle one might encounter coming from criminal spirit people, and or persons representing them; including some we already or earlier mentioned. Although I will not just here, as such, get into whether these asseverations are just or not, it is sufficient to name or mention them for informational purposes; and to note that these things, in one form or another, I have actually heard from these people, and are not mere surmises or inferences on my part.

* God and or Christ is to blame for the worser evil in the world; he makes life too difficult and is tyrannizing people by insisting they be honest and morally good. Just as the religious sees the beauty of creation as proof of the goodness of God, the devil's advocate will ever and similarly point out instances evil and or hideousness as proof, as they allege, of God's badness. And yet, it may be reasonably objected, is God's responsible for what the devil does?

* It is not a perfect world; so it is foolish not to use bad or wrong-doing to your advantage.

* Unless they are undeniable and pure innocents, crime victims are not perfectly righteous and or else they "have it too good," (i.e., are too fortunate, privileged, or well off in some way) and therefore only suffer what they deserve (at the hands of criminals or such as who somehow have the right to judge, condemn and inflict subsequent injury upon them.)

* Evil is ineluctable, and some amount of it *must* be done in the world (not unlike meeting a quota, and as if this were a necessary or unavoidable cosmological principle.)

* Criminal spirit people are not to be discussed scientifically, and for you to do so will only incur their altogether warranted and justifiable wrath.

As was long ago observed by Joseph Glanvill, Mather, and others: those who deny the existence and or workings of criminal spirit people (the latter term, mine not Glanvill's, et al.) take that position in order to reject the idea that there is an after-life and or the idea that any punishment will follow wrong-doing in this life. Consequently, one of the arguments one regularly hears from their criminal spirit person followers (or they through their followers) is that death is the necessary and final end to life. Therefore, "eat, drink, and be merry," have a good time as you please, as if that were all there were to it; which attitude not so surprisingly lends itself (for many at least) to all kinds of reckless, degrading and deleterious conduct; regarding which it is the very policy of organized crime as practiced by criminal spirit people to incite and encourage (i.e., for purposes of augmenting their power.) The more irrational and morally debased people are, the easier it is to corral and make vassals and slaves of them.

And of course, since they themselves are already so guilty, sure it is easy for them to tell you it is no great loss or disadvantage for you to get or become more guilty as well. Extreme misery, such as theirs, not only desires company, but indeed peremptorily demands it.

In practice, they are against fair competition based on merit and free speech generally; unless, that is (with regard to the latter), speech that is to their own advantage. In these kinds of matters then of seeking honors, status, place of office or position, their regular person followers, and with criminal spirit people in many instances empowering them behind the scenes, are not uncharacteristically shameless and brazen liars, bullies, thieves, cheaters and hypocrites; though granted a given individual may or may not be so much this or that particular way. This is the predictable consequence of small-minded persons who see selfish and narrow self-interest as being of far greater importance than public safety and welfare, and or enhanced quality of life for all (even and including, unbeknownst to them, their own.)

One gets the impression that for some kinds of criminal spirit people, if they don't do wrong they won't be able to escape insufferable living conditions; it then being preferable, as they see it, to "rule in hell than serve in heaven." This being the case they come up with a arch-philosophy, pseudo-theology or other ideology making it seem that doing the wrong thing is pardonable, justified, or even mandated by Divine Providence. That everyone else who is not in their straits has their quality of life ruined or drastically diminished as a result, well that's *their* problem.

Another ostensibly decisive factor motivating high-power criminal spirit people is that after long ages of heinous guilt, foul play and brutal savagery, accompanied by interminable frustration, they find themselves sitting on an accumulated and reservoir ("powder keg" perhaps?) of seething anger, resentment, and aggression; that being too overwhelming to regularly suppress, is let lose in the form of the obligatory (as they see it) victimization of others. Like an alcoholic or drug addict, at some point they cannot help themselves, with untrammeled violence and malicious mayhem being the chaos-creating, destructive, and even tragic result.

Yet another and perhaps likely incentive or purpose behind orchestrated and widespread wrong-doing may be that they wish to hasten the end of the world; so that something conclusive can occur to finally bring a close to their dreary and depressing lives; despite appearances some of them do or might put on of being "heavenly."

And the more people they can get to agree with them on any of these points, the more confident and unchallenged they feel in adopting this or that kind of contentious and strife minded attitude or outlook.

Would "satan," or "the adversary," one wonders, have done things differently if he could turn back the clock and do it over again? (Who knows?) [29]

---

[29] It somewhat makes me also wonder if perhaps the age of dinosaurs was not an era when the arch "fallen angel" had the world formally to himself; as a sort of concession to some importance of place in the hierarchy of things he may have been seen as possessing. But when ages later, he as "ruler of this world" had been "condemned," this new state of things thwarted what perhaps he felt was an unwarranted infringement on what assumed was his rightful entitlement. Yet this is only casual musing on my part, and not to be taken as a formal theory.

VI.

"For everyone who does wicked things hates the light and does not come toward the light, so that his works might not be exposed."
~ John 3:20.

As earlier discussed, so often, and most especially so in recent decades, seemingly "popular" trends and fashions[30] are not infrequently the result of long standing, systematic propaganda, manipulation, bribery, mind control, and censorship of opposition, and not really, as is assumed, people's free and thoughtful choice. Unless and perhaps one is already fortified in wealth and prestige, if one doesn't give the nod or bend the knee to some contrived fad or "icon" of supposed popularity, you will almost certainly be marginalized or ostracized, whether openly or in secret, from being able to do business and participate at all. Those with little or no self-respect or conscience take the tip and, evincing no signs of reluctance or questioning, mindlessly prostitute themselves to this crass and arbitrary state of things.

Indeed, when you consider that autocratic criminal spirit people and their followers are at enmity with and opposed to legitimate humanities, arts, culture and tradition generally and insofar as they can't wholly dominate these, they make a point of endeavoring to purchase, take over, traduce, and corrupt these venues and institutions of thought and expression. In furtherance of this, many are named, promoted and funded by them as "leading" artists, scholars, academics, or other authority or expert, in whatever cultural endeavor, whose purpose, whether wittingly or no, is to *deliberately* downgrade the arts, culture and tradition; that is, by enlisting and employing persons usually too glad to do this if you pay them enough money, or else are too incompetent and tepid in sincerity to have any worthwhile and positive impact. Simultaneously, the view that arts and culture are mere entertainment or leisure pastimes of marginal importance and significance is introduced and inculcated into the public large. Indeed, there even some who are all too ready to have arts, culture and tradition destroyed and tossed aside entirely if you sufficiently subsidize them for endorsing this, yet and while masquerading as supporters and advocates of the arts, etc. And thus such are made or become ardent fans in feigned rapture over junk books, movies and music, and who have little or no real taste in good books, films, music, etc. to start with. We encounter this, to cite just two odd examples of recent decades, in films where canned music is mindlessly edited in as mere filler (little more than background noise) to

---

[30] *Some* of the unusually ubiquitous fetishes and bizarre preoccupations, of the last two decades, illustrating this kind of mentality and that, just off hand, come to mind are: tattoos, ambitious graffiti projects, transgenderism *as a virtue* (respecting which see Hippolytus [c. 170–235 A.D.], *Refutation of All Heresies*, Book VI, chs. 13, 15); the *positive* presentation of witches, vampires and the occult, gluten free foods, red velvet cake, chipotle, flinging pairs of shoes onto power lines, and life as interpreted by Lego toys. That a given person might find any of things attractive is (and perhaps) not so strange; rather what is so is how, whether innocuous or no, they became so suddenly *and* widely adopted. Which prompts the question: whoever, as it were and might seem to be the case, first introduced and decided upon these for the rest of us?

accompany the action, and movie trailers punctuated with a pronouncedly violent rhythm and sensibility (*regardless* of the film's subject matter.) That this is done *automatically* and *so perfunctorily* is either reflective of gross lack of taste and talent, and ability, and or a deliberate slap of thinking and feeling audiences.

So much of spiritual, moral and intellectual decay and debasement we have witnessed in our time then, I would argue, is the result of people selling out their centuries old culture and heritage; this based on the idea that only money and material things are what make life worth living. Everything else is seen mere for show or pretence; or as esoteric and frivolous luxuries, gewgaws and pipe dreams. According to this view, the only arts, culture and heritage worth anything are those that sell and take in money; the latter being the primary, if not only standard, but which everyone and everything is at last to be judged and assessed. Yet so much of what sells or not is not a reflection of fair competition, but and rather the manipulation of public perception by self-appointed despots and propagandists who dominate the mass media in its various incarnations.

True, love of money and material wealth over and in place an appreciation of arts, culture and higher learning is an age-old complaint (say in the United States for instance), and far from anything new. What *is* new is not indifference to moral, intellectual, and spiritual advancement, but a movement to purposefully ruin and destroy culture, mind, and spirit; and which only serves the goal and ambitions of criminal spirit people to make us more heartless and mindless; and thus more slave-like. We see this most unapologetically and blatantly in that neo-Hollywood which so celebrates and allies itself with demonic themes; while ridiculing or making a mockery or parody of anything sacred, traditional, or else suggesting of innocence. On the one hand, though the self-appointed cultural commissars identify themselves with real or purported humanitarian causes, they at the same time favor and promote sick depravity (such as realistic violence in films and video games) and the very worst criminality; that is, in the way and to the extent they would popularize, make acceptable, and portray with sympathy the luridly immoral and brazenly vicious. And yet these (i.e., criminal spirit people) whom such mass media moguls defer to, and honor as friends and authority figures, whether openly or on the sly, end up taking them over. Witness how previously family-oriented Disney, and Hollywood in general, in the last thirty years has been devoured and swallowed up whole by those championing new-age devilment and the occult, even to and including putting out programs and shows for children incorporating motifs and subliminal indoctrination promoted by other-worldly criminals. So that even Mickey Mouse can be made into a literal vampire. Meanwhile, comic book super-heroes, at one time and originally intended to promote basic morals and civil idealism in children, are swept away and replaced entirely with homo-erotic super-heroes *for adults*.

One of the most preposterous myths about culture and media success promulgated in our own time is how the public is led to believe that artists characterized as having high status possess such as a result of fair competition and popularity; when really such status stems much more from party membership and decision making of the monolithic powers that be; which latter, generally speaking, themselves and it turn are enthroned only and insofar as they meet with the approval of criminal spirit people based organized crime. Potential challengers and competitors to this order of things are not allowed to participate, are banned or shadow banned, and in some instances are subject to vandalism, robbery, violent crime, or worse; thus giving the latter's competitors a most absurd and ridiculous advantage. It is often so bad that those who cave in and comply with this state of things don't even need actual talent, ability or popular appeal, but are made and designated celebrities by sheer fiat and decree.

This sort of thing is in no small part a result of a major portion of the public gradually and over time coming to believe that it is only wealthy oligarchs hold highest authority and judgment; while the more just minded, honest and right reasoning among us are viewed and treated as sham or deluded persons seeking to trick and win people over to what is, after all, merely personal prejudice and self-serving minority opinion.

But the practice of looking to "gods" or the agents and representatives of such for guidance of course can take on an even more sinister turn then the degradation of the arts, culture and heritage; so that if the "god" or the "gods" will it so, even things like murder can be justified. The following excerpt from Plutarch's biography of Themistocles (ch. 13), although describing an occurrence from centuries past, gives one instructive illustration of how this is possible and conceived of in practice:

"When Themistocles was about to sacrifice, close to the admiral's galley, there were three prisoners brought to him, fine looking men, and richly dressed in ornamented clothing and gold, said to be the children of Artayctes and Sandauce, sister to Xerxes. As soon as the seer Euphrantides saw them, and observed that at the same time the fire blazed out from the offerings with a more than ordinary flame, and a man sneezed on the right, which was an intimation of a fortunate event, he took Themistocles by the hand, and bade him consecrate the three young men for sacrifice, and offer them up with prayers for victory to Bacchus the Devourer; so should the Greeks not only save themselves, but also obtain victory. Themistocles was much disturbed at this strange and terrible prophecy, but the common people, who in any difficult crisis and great exigency ever look for relief rather to strange and extravagant than to reasonable means, calling upon Bacchus with one voice, led the captives to the altar, and compelled the execution of the sacrifice as the seer had commanded. This is reported by Phanias the Lesbian [i.e., of the island Lesbos], a philosopher well read in history."

And, might we ask, how far or removed from us today are such things, even if camouflaged and disguised?

Needless to say, historically we will find no end of this kind of madness and lunacy in both Christian and pre-Christian eras; that is, where right and just reason, along with the most rudimentary morals and decency, are cast aside in the cause of obeying supposedly "divine" others who, without test or accountability, are presumed to be *more wise than* and *superior to* honest and just moral argument and reasoning.

VII.

"And yet, O you great worshippers and priests of the deities, why, as you assert that those most holy gods are enraged at Christian communities, do you not likewise perceive, do you not see what base feelings, what unseemly frenzies, you attribute to your deities? For, to be angry, what else is it than to be insane, to rave, to be urged to the lust of vengeance, and to revel in the troubles of another's grief, through the madness of a savage disposition? Your great gods, then, know, are subject to and feel that which wild beasts, which monstrous brutes experience, which the deadly plant natrix contains in its poisoned roots."
~ Arnobius (c.284-c.305), *Against the Heathen,* Book I.

While I have already and somewhat addressed and commented on the *history* of "gods" and criminal spirit people in my *New Treatise* as they are known through tradition, scripture, and literature, once again the subject is one of deserving an entire book. I will attempt then to scratch the surface on this topic a little bit further. But that is all what presently follows is meant to be, that is, a further scratching of the surface, and yet still and hopefully enough to provide some useful insight into the minds and characters of *other-worldly* autocrats, and oligarchs, and who make it their business to foment anarchy and strife; for purposes of bringing individuals, families, and communities at large to a state of hopelessness and servile subjugation.

In both primitive societies around the globe and from the dawn of civilization belief in supernatural and or preternatural beings has been a matter of course. Indeed was there ever a tribe or developed community or society without ideas of ghosts, spirits, or gods, on some level, and who interacted directly with their lives? I am not in a position to say there never was or were, but if so, such would have been a most unusually rare exception. And this of itself is telling and strong evidence of the palpable reality of the other-worldly, and is a substantial refutation, as I think many will agree, of the idea that such are merely and only a product of people's imaginations.

Some in our times have seen these entities as space aliens or visitors from a planet somewhere off in outer space; an interpretation (you perhaps will not be surprised to learn) I myself reject. Enumerating my arguments for doing so goes beyond the limits and purpose of this writing; so I will not take it up here. However, I would mention that ghosts or spirit persons as they have been understood historically show decidedly human characteristics in both their appearance and behavior. And if we do presuppose there being space aliens, these might be the supposed "fallen angels": that is, persons whose character and level of sophistication would argue something "more than human." But between "fallen angels" or space aliens, which should or would we find more formidable or intimidating as adversaries? Either way, and unless we ourselves abandon its power as a standard and measure of the truth and what is truthful, both types are, would and should be subject and answerable to right reason; for only by right reason could we possibly begin to comprehend, understand, weigh and assess them. This, at any rate, is the position and perspective I assume. For yes it goes without saying, there are those who will wholly abandon honest, just, and right reasoning, but given what, taken in all, I urge and argue for here, that can and will in no wise be we *who would* remain moral and sane.

Quite how such, similar, and related traditional or mythic notions of purported places like "Hades" and "Olympus," if they have any basis in fact, are to be explained and reconciled with what we will presently address, it is not necessary, or possible, for us to get into at this juncture (other than to merely make passing note of them.) Rather at the moment our focus is more simply with (what we call) other-worldly or spirit persons who commit crimes on humanity and the world, and what might be observed, or reasonably inferred about the nature and character of such culprits culturally and historically.

This said, I am inclined myself to take the view that if we speak of there being high-power spirit people possessing and governing what we might call an empire, at the most puissant level their span and scope is global and not merely regional. As far back as the late 18th century there has been and is this idea that some secret group or other is conspiratorially bent on world control and domination. Some will absurdly attribute such power and wherewithal to effect such end to, say, free masons, oligarchic globalists, the Vatican, "Jews," Zionists, communists, or some banking family. And yet even if we posit conjectures such to be the least bit plausible, what could possibly keep such a group together for so long, and without a clear, obvious and ostensible leader and head who never dies? Are we to understand these alleged conspirators, with multi-national connections, can be run and governed as a friendly and cooperating group without any in-fighting rivalries, and do this over the span of not merely decades but centuries? Only high-power criminal spirit people would have such longevity and domineering enforceability to effect such a result, and those aforenamed accused (whether rightly or wrongly) of such nefarious intent are really and at worst merely the pawns, stooges and puppets of the former.

But if criminal spirit person reach and sway is global, is such a regime divided into subsidiary or regional governments? Who is or might be the leader of such a government? Does he have peers? What do his peers think of or say about him? Is such a government, whether in the short, long, or very long term, subject to regime change? Here again, and while we could speculate on any one and all of these and other things, we of course are not in a position, not yet anyway, to quite say. But again, it is still both worthwhile and sufficient, if only in passing, to at least ask and take into account such questions.

Being a fairly well read and yet admittedly only a mere arm-chair amateur in both anthropology and world religions, I am no way in a position to speak in depth or at length as to the myriad ways and manner of how seeming gods, ghosts and spirit people turn up in the numerous and societies, both primitive and civilized, around the world and down through history. Yet perhaps by making some mention, based on what I do know, it will help in make possible a start in having the topic addressed; specifically as it pertains to criminal spirit people as seen from a more empirical perspective.

Outside of legend, myth, and tradition, the true history of criminal spirit people is unknown to us, and of course it is highly dubious and questionable to what, if any, extent such sources as we do have reflect much or anything like the facts.

Many stories about the "gods" ostensibly, if not presumably, originated in some form and measure from criminal spirit people themselves. For what reason might we think such "gods," as are reported of or described by poets and story-tellers, were or were based on "criminal" spirit people? Being spirit people they could speak on the subject as if they knew, and, as would be expected, regular persons would understandably believe and give credit to the story related. Second, because as often than not, the stories show the "gods" to possess what, applied to any one else, clearly is a criminal disposition: say, punish without a hearing, to abduct, rape, torture, and murder. And third, because such stories effectively equate criminals with gods, and, by implication, gods with criminals; thereby excusing and encouraging criminality; not least of which when practiced by persons possessed of higher standing and or designated as higher authority; or given permission to behave so by such who possessed these "divine" advantages. Also, we know (at least, I can say I do from my own experience) that it is in the character of professional criminal spirit people to be highly equivocating and unreliable in their veracity (they are more careful and shrewd than to simply be lying all the time), and because corrupting, hypnotizing and misleading us is an inherent part of their stock and trade; in addition to its being both amusing and of practical benefit to establish their putative superiority and impose it upon the credulous; not unlike say how a sect of racists, assuming themselves by race to be superior, might attempt to justify their criminal abuse of uneducated, primitive people of another color. That is, it isn't wrong if a "god" does it; though common and moral sense loudly cry otherwise.

Quite how "gods" and spirit people compare with each other, say, in ancient Near East, Chinese, India, European and other lore invites some intriguing speculation, and that might lead to some useful and fruitful discoveries; that would in turn provide us with further clues to aid us in our understanding of criminal spirit people generally; thinking particularly of cases where the accounts of one culture closely match those of another and yet are separated by remote time and or distance from each other. One of the most common features we do encounter, for example, in diverse and widely separated ancient cultures is human and animal sacrifice; a religious custom that on the face of it strikes one as lunatic and insane as anything could possibly be. Despite such being made integral or incorporated into many religions since ancient times, I don't see how one can around the moral conclusion respecting that practice; which of course was one of the points Pythagoras, Heraclitus,[31] Lucretius and similar were making in rejecting the conventional worship of their day and locales.

Just here, I will focus primarily on ancient deity and "gods" as found in Western paganism and the Old Testament; simply because I am more informed about these than such as are found in other ancient cultures. Once more and I emphasize, naturally a further comparison with accounts of gods and spirit people found elsewhere around the globe is welcome and desirable for purposes of arriving at a more full picture of what autocratic spirit people are like and how a given culture, had come to imagine them. Finally, it is no little important when considering a supposed "god," in history and literature, whether the identification is seen *a)* literally; *b)* as a figurative abstraction for a natural force; and or *c)* a poetic personification, representing, by way of abstraction, a personality type or psychology. Such distinctions, needless to say, make a crucial and critical difference when the topic of a supposed "god" is under discussion or examination.

~~~***~~~

"He that sacrifices an ox, is as if he slew a man: he that kills a sheep in sacrifice, as if he should brain a dog: he that offers an oblation, as if he should offer swine's blood; he that remembers incense, as if he should bless an idol. All these things have they chosen in their ways, and their soul is delighted in their abominations."
~ Isaiah 66:3.

"And if you knew what this means: I will have mercy, and not sacrifice: you would never have condemned the innocent."
~ Matthew 12:7.

If what is expressed Biblically denouncing blood sacrifice is right, someone might object, what about Christ's sacrifice as found in Christian teaching? This, to my mind at any rate, can be explained this way. Inasmuch as humanity made a deal or got itself into an arrangement with the "evil one," Christ offers himself as a sacrifice to win over the liberation of we the captives; not unlike how someone gives up their life in or to free or rescue hostages. Moreover, Christ's self-sacrifice might also be viewed as, and in its way, a subtle mockery and ridicule[32] of human and animal sacrifice so dear to the *misled*, falsely pious, mad and foolish. How completely absurd and deranged else to think how any "superior" or "divine" person, let alone God the Father himself, would need,

---

[31] "They vainly purify themselves by defiling themselves with blood, just as if one who had stepped into the mud were to wash his feet in mud. Any man who marked him doing thus, would deem him mad. And they pray to these images, as if one were to talk with a man's house, knowing not what gods or heroes are." ~ Heraclitus (c.495 B.C.), Fragment 5, translator: John Brunet.)

[32] "...I gave Egypt as a ransom for your freedom," Isaiah 43:3, also Deut. 7:8, Psalm 31:5, Mark 10:45, and "Do not think that I have come to bring peace to the earth. I have not come to bring peace, but a sword."~ Matthew 10:34-36.

desire or require the victimization of the innocent in order to placate a disappointment or appease a fit of pique.

Most any intelligent person acquainted with more than a superficial knowledge of military history, knows that fought battles (as opposed to overwhelming over-runs or hopeless impasses that result in inescapable capitulation) are most often won where some soldiers are *necessarily* sacrificed in order to insure victory.[33] The sacrifice then of Christ, soldiers, or such who act that role are *not* sacrifices in the age old barbaric sense where someone else is made a necessary sacrifice *against* their will or wishes. It is this latter form that originates with criminal spirit people, and where such was practiced, say, in the Old Testament, I would argue it was seen *a)* as a temporary stop-gap measure or regrettable concession to buy time and indulgence from criminal spirit people and their regular person adherents (long accustomed to such practices which they associated with "divinity"), and *b)* something it was foreseen would be ultimately phased out as true God's people grew to become more properly moral and rational.

None of the immediately preceding is intended, however and even if granted correct, as anything resembling a complete, let alone exhaustive, interpretation of the meaning of Christ's self-sacrifice and or how it relates or might relate, say, to traditional church teaching. Yet it is one interpretation, and one that, to me, seems far more morally plausible than any other. The greater point and otherwise is how historically, time and again, we find high-power criminal spirit people ask or demand human or animal sacrifice, and their reason for doing this, I would argue, is to make people in their hearts more unreflecting and stupid, and therefore more malleable for the purpose of turning them into unquestioning vassals and slaves. Also, we can add, if someone (i.e., a dupe to be used) is given to believe ritual killing or torture of someone is acceptable and all right, if authorized, then he will have that much less reason to object if come the day the real or proverbial axe ever happens to fall on him; and which only further helps to establish and reinforce the strength of a social order that lives in sheer blind terror of unaccountable and arbitrary authority.

We see a manichean-like element or trend in both paganism and the old testament, and where the deity acts morally and there is a concern for justice and basic fairness and decency; and yet at other times not so, or not so much: obedience to the deity, in those days, being all or the most that mattered; irregardless of what would otherwise seem to us so obviously just or unjust, fair or unfair.

If such events had the possibility of occurring today in the present time, should Abraham take the life of his son Isaac? Should Agamemnon sacrifice his daughter Iphigenia? Without hesitation, of course, we would say no absolutely not. Permitting them do that would be flat-out crazy. And yet in each instance

---

[33] "...unum pro multis dabitur caput." *Aeneid*, book V, 814-815.

Abraham and Agamemnon would be doing what *was then* the right thing: that is, unthinkingly obeying the divine command. And yet even back then, how did they know the given someone commanding them was divine and not an imposter taking on the form, say, of "an angel of light?" Both Joseph (Jacob's son) and Achilles (in the Iliad) believed dreams came from God. Yet never did either speak of the devil or implied they even knew of his existence. How, in retrospect, then could their judgments possibly be trusted on this point?

Many now will understandably deplore the cruelty enjoined by "God," whether in judgment or in deed, and that is sometimes presented in the old testament (as we have it.) And naturally quite what the meaning that ought to be read into the account is far from very clear or evident. One spirit person actually told me, in effect, that the bad behavior we find in the old testament Hebrews, at least such as we today would view as bad, cruel or unjust behavior, was a way of weaning them off from earlier, worse behaviors. Observe in addition, that *both* the religion of Israel, and Greek and Roman paganism as well, had brutal practices like animal sacrifices and, as well, the death penalty for what would seem to us relatively minor offenses.[34]

It was also routine in both Greek myths and in the old testament for the someone to be punished without even knowing, or at least being told, that they had done anything wrong; such as Actaeon coming accidentally upon Diana bathing; or Moses striking the rock at Kadesh twice [Numbers 20:11]. This sort of thing is frequent in ancient stories involving the god or gods. If the god sees you do wrong, *that's it*, you are guilty and they can inflict punishment on you without any hearing, or chance to ask or have explained to you what is going on.

Leaving aside how ancient stories or old testament scripture should or might now be viewed, this sort of cold, peremptory arrogance and judgment without a hearing is in any case very typical of autocratic, criminal spirit potentates. And where people listen or defer to the judgment of spirit persons (or those who act to represent them) utter tragedy or calamity can result. It is worth noting in this respect that virtually all the events found in the famous Greek *tragedies* are instigated as a result of the machinations of a "god" or "gods" seeking to manipulate, subdue, chastise or otherwise control people.

Whatever the circumstances of ancient peoples, we today ought not to respond with blind obedience to those in power, but *first* have it clearly understood and established that such authority is and or reflects the moral law, primarily as it is expressed in the Golden Rule and or its corollary (i.e., "Do not

---

[34] And to the extent you were not formally a Hebrew, or not a formal citizen of the pagan state, you could ordinarily expect the punishment you received as an offender could or would be dramatically worse depending on the crime) than that which a citizen would receive. That is, when it came to punishment, a foreigner is going to be treated a lot more cruelly. Better justice then went to those who were entitlement, and entitlement was established by having ties that which was divine; meaning that those outside the formal social order were, by definition, separated from the divine as well.

do unto others...") Only then should obedience of such putative authority be deemed obligatory. Such, at any rate and according to our view, is the approach any morally and rationally conscientious person will take up and adopt.

We might say the fine forms of reasoning such as were cultivated and developed among the ancient Greeks was, aside from some of the more penetratingly shrewd Hebrew prophets and post-Babylonian exile wisdom teachers, absent in old testament times and in the days before Greek philosophy; so that if an ancient person thinking "God" spoke to them can be excused or pardoned when in a given instance they may have been fooled, we who millennia later have the now far greater power available to discern should be able to "test the spirits" [1 John 4:1-6], or at least such should be demanded and expected of persons who think of themselves as more intelligent than a mere child or imbecile.

It might seem that witchcraft and the "black arts" are medieval in origin, or at least many would seem to think that was the case. But of course magicians, witches, sorcerers, and false prophets are spoken of in the Bible. Further we have witches or practitioners of the diabolical, as per:

Homer: Circe in the *Odyssey*
Euripides and Apollonius of Rhodes: Medea
Virgil: Alecto in the *Aeneid*
Lucan: Erichto in *Pharsalia*
Apuleius has fun telling stories of ostensibly *real life* witches in his *Metamorphoses*.[35]

The casting of spells and curses was practiced in ancient Rome, perhaps most famously as related with respect to the death of Germanicus as told by Tacitus. I mention all this in part to correct the false idea some might have that witchcraft and the black arts only came on the scene after the arrival of Christianity; as if such things were only or merely a subsequent response to the latter.

This naturally raises the question: to what extent are those practices known as witchcraft or the black arts separate from or else co-joined to formal paganism (such as, for instance, the Greeks and Romans followed.) My own view is that the two were part and parcel of the same system of criminals spirit people governance, and not separate; except insofar as an organization or regime has discrete departments, but which nonetheless are overseen by one ruler or single group of rulers; so that bright, shining Apollo and the Muses for example were, after all, kinds of cousins of Harpies or the Furies.

---

[35] These anecdotes of his I would imagine or surmise as originating from "dream productions," at least in some measure. Also, by the way, the word "witch" itself is related to "wit," and implies "one who *knows*."

The Greeks and Romans had a deities like Hecate (first mentioned in Hesiod), and who was seen as benevolent, indeed in some cults as the nurse of children, and yet was also the goddess or witchcraft and sorcery. This kind of evil blemish the sometimes pitiless Israelites, even so, were not contaminated with. Witches and sorceries were somewhat tolerated as common in some places of the Roman empire, but certainly not with the Israelites, at least not formally or openly; since Saul was able to secure the services of one without too much trouble.

Yet both the Israelites and, as time went on, some non-Israelites, such as Plato for instance, condemned the gods or earlier notions of the gods handed down as being fallacious. It even got so among non-Jews that by the second century A.D., someone like Lucian of Samosata saw fit to ridicule traditional Olympian deities.

But what then are we to make of those traditional gods like Venus, Apollo, Dionysus, Neptune, et al.? These personalities are in one sense mere poetic personifications; albeit such that *may* have some parallel (unknown to us except in the abstract) in the natural order of things. Alternatively, and in the context of a given story or specifically reported encounter, they were or might have been criminal spirit people masquerading or taking on the particular deific role to further some private scheme; bearing in mind that criminal spirit people, while selfishly motivated, would and do at times like to play the role of the "good" god in order to obtain or effect what they want; like a con-artist, say, who while he would like to rob you has nothing against you personally. An example of this might be spirit people speaking through sybils or priestesses at some oracle. They act as presiders over great events, would have you think them divine, are perhaps possessed of amazing (to us) powers of foresight, and yet they are ultimately actuated for purposes of their own self-aggrandizement and or amusement.

The gospels are most unique and extraordinary for portraying demonic possession as a real life happening. These reports are not related simply to tell a story with a moral, but are recounted as having actually taken place. This I personally find very significant, and in my opinion helps to argue for Christianity's truthfulness. Even if what are alleged as instances of demonic possession were only cases of mental infirmity, why is it we do not or else very rarely (if that) hear similar accounts of purported possessed persons elsewhere in other ancient or pre-medieval cultures?

Of passing note before bringing this section to a close In Book 6 of the *Aeneid*, where Aeneas visits the underworld, Virgil has shades or ghosts see things through the lens or filter of their past lives. For instance, ghosts of Greek warriors see Trojan Aeneas in full armor and run away at the sight. This I would take to be merely an invention, but make note of it as a curious notion possibly that might have *some* basis of truth worth considering.

Tales of ghosts coming back to this world to impart some message are fairly common in both literature and history. Two of the most famous being that told by Pliny the Younger (book 7, letter 27) and the account found in Joseph Glanvill of a murdered pregnant woman (*Saducismus Triumphatus* [1681], pp 1-4). And yet this sort of story, if it has any basis in fact, is, by comparison, told much less frequently in modern times. One possible explanation (and I offer this merely as a hypothesis, and again, for the sake of argument, granting the premise) may be that troubled "ghosts" with a story to tell have less opportunity now than they did in the past to appear among us and make a report of some hitherto unknown injustice done them; either because there are too many such deserving spirit persons to allow them to do this (i.e., it is perhaps too expensive and not cost effective), and or else it is now seen as an unnecessary or futile gesture.[36]

Even more frequently related are claims of "ghosts" visiting places they lived in or where some dramatic event transpired to them when they were alive. The vast majority of these, in my opinion, are more the result of people's wishful thinking or imagining than actual occurrences; partly because it is natural to think this way, and also because ghost actually showing up in our midst (as mentioned in my *New Treatise*) involves some amount of monetary or other (to us unknown) expense that relatively few can afford and or else feel the need for. These sorts of ghostly visitations should be viewed with caution and skepticism; bearing in mind further that such could not take place without the approbation and say so of professional criminal spirit people; who, in a manner of speaking, police them. This, at any rate is (and granted) my own professed take or speculation on the matter.

---

[36] For ghost and related as viewed in the Middle Ages, see *Medieval Ghost Stories* (2001) by Andrew Joynes.

VIII.

"Behold I am against the prophets that have lying dreams, saith the Lord: and tell them, and cause my people to err by their lying, and by their wonders: when I sent them not, nor commanded them, who have not profited this people at all, saith the Lord."
~ Jeremiah 23:32

Before addressing the question of how criminal spirit people should be viewed from the perspective of the church, there are several points to be borne in mind.

We tend, I think, to forget the great difficulties and challenges the church faces, such as:

1) It reaches out to and would communicate to and persuade a wide and great variety of different people; only a portion of which (and that probably a relatively small portion) that is disposed to be both honest and rational.

2) Because God is everything, people mistakenly assume the church will or must be also, and to all people. But this is asking the impossible. So that the church regularly finds itself in the unenviable position of having to please and not offend any and everyone that means well.

3) The church, on behalf of God, sees to it that the sins of the repentant are forgiven, and for this reason welcomes sinners. And yet there are some who interpret grace as license. Not surprisingly, professional criminal spirit people are the sort who would encourage this view; with sometimes confusion and misunderstanding predictably arising as a result of well-intentioned, yet misguided, forbearance and tolerance.

4) While there are of course more pure and stalwart exceptions, church members whether, Protestant or Catholic or other, not infrequently are susceptible to "group think" or the herd mentality of the secular world, and sometimes a parish or diocese (to some greater or lesser degree depending) must compromise with such people in order to win them over or stay in their good graces. Now a more harmless example of this is following and cheering on the local sports team. And yet at other times, this might take the form of current secular mores or political appeasement on a given issue; with perhaps the church in consequence dangerously compromising its core principles. Here as before, professional criminal spirit people, clever con-artists as they are, either themselves or through their agents, do, could or might use this frailty to corrupt and mislead such persons as to what the church needs and or should represent.

One of the reasons the church went out of favor during the enlightenment period was owing to its being seen as a pawn or tool of kings, worldly-minded clerics or the very wealthy, and not without some justification. Similarly in modern times, many will have rejected the church or even the faith, because the church is viewed as less than moral, including not being duly fair, rational, and honest in its views and dealings; too given to appease political expediency of the hour; at other times perhaps acting as authority in a way that, rightly or wrongly, is or seems overly dogmatical; adverse to honest and objective dialogue, and or simply unreasonable. In part this cannot be helped; both because priests, ministers and such are only human and can make mistakes, but also because positioned politically as the frequently are, they are bound to irritate and are perhaps seen as treating some social cause or other unfairly. Add to this criminal spirit persons, in a given instance or circumstance, infiltrating positions in the church, and it is not hard to see the bad reputation the church ends up acquiring in the minds of many, then and since.

Some therefore, ironically, do or have disliked the church because they think, of all things, it is, at least and too often in practice, immoral. In some respects this is or can be true, for the church will sometimes favor worldly power and or a majority to win more widespread support, and this at the expense of more candid truth and justice. In addition, even the church, like any good cause can be misused, warped and twisted to serve bad ends by duplicitous sorts who would insinuate or infiltrate themselves into its ranks for worldly purposes; even thinking themselves well-meaning for and in doing so. Not surprisingly, it is the business of professional criminal spirit to encourage and effect this, *even if* doing so means in some way overtly benefiting the cause in some loud and conspicuous way. And yet nonetheless and at the same time in a more subtle, sly and underhanded way, using the welcome they receive as an opportunity to subsequently do more harm than good.

Now I am not in the least however suggesting that just because someone disagrees with the church that they are necessarily or more likely right or more moral, but and merely that in an individual instance they (that is, the critic or naysayer) *may be*. But then what is the standard and measure for deciding a given question (and that is not core doctrine and dogma, and which are, after all, matters of faith not human disquisition), if not honest and right reason? We cannot in practice say that the religious or secular moralist, in the given instance, is always the more moral and more wise; much of course depends on the individual and the circumstances; and depending on the individual and circumstances that person could or might be hypocritical, irrational, and or glaringly inconsistent in their reasoning. God is perfect; people, however, are not.

Just and serious philosophers, as opposed to merely crafty or disingenuous rhetoricians, are individuals in pursuit of honest rational and moral truth, not crowd pleasers, politicians, or rabble-rousers. And when or how often was a just and serious philosopher ever seen as a popular celebrity? Fairly never, of course, since, historically speaking, it is not in most people's nature to view them so; or to concern themselves with what the more honest, properly rational, and just are about. Even Christ, if viewed a certain way, might be characterized by some as a kind of philosopher but that he possessed the power of performing miracles; otherwise would the more ordinary masses of people, who rarely reflect long or much on the subtlety of spoken wisdom or deeper reasoning, have so taken to or embraced his cause and mission?

But this is where professional criminal spirit people can come in and pose a major danger; since the more high-powered among them can also perform deeds and wonders that will seem like miracles; yet without requiring the Holy Spirit, and concomitant wisdom, that necessarily is present in Christ. Reports are given of saints receiving visits from, say, Jesus or the Virgin Mary, and yet how can we always be necessarily sure that that was who it was who came to them? Why, for instance was the Holy Spirit not sufficient for someone like Catherine of Siena, but that she records lengthy dialogues with Jesus himself? This is not to conclude she was necessarily deceived and or insincere, yet it can't hurt to at least think over the question. For even if and in her case what she relates is, as it turns out, as true as she makes it to be; with others such other-worldly visitations might very well not be what they seem.

For by being "supernaturally" impressive and compellingly awesome, spirit persons acting and commanding with seeming "divine" authorization, can fool well-meaning religious;[37] which latter then can even become unwitting accomplices of criminal spirit people. And of course such practice is not restricted to Christianity, but other faiths and creeds can be "had" or infiltrated in this way as well. So that historically some crime, outrage, injustice or piece of folly is attributed to, say, some Christians, Muslims, or Jews, it is not Christianity, Islam, or Judaism that is at fault, but certain members of that faith, taken in by professional criminal spirit people; acting badly, yet believing they are nonetheless carrying out the will of "God."[38]

Although in ordinary practice it often tends to do so ineptly, Christianity at least has the maturity to deal with hard-core, real and serious evil. Modern science on the other hand, most usually and as reported of, is so craven and cowardly it abandons any attempt to even attempt to do so in a way that isn't short sighted, piecemeal at best, and childish. For example, who or what else empirically (as it were) has then or since even tried to raise the subject of demonic possession in an objective and credible way?

---

[37] One gets the strong impression that Cowper's "disapproving" and conversing "God" was a spirit person; if such was the case, that apparent "madness" would arise from such a relationship and circumstance should come as no surprise whatsoever.

[38] "Not everyone who says to me, 'Lord, Lord,' will enter the kingdom of heaven, but the one who does the will of my Father who is in heaven. On that day many will say to me, 'Lord, Lord, did we not prophesy in your name, and cast out demons in your name, and do many mighty works in your name?' And then will I declare to them, 'I never knew you; depart from me, you workers of lawlessness.'" ~Matthew 7:21-23, and "They will put you out of the synagogue; in fact, a time is coming when anyone who kills you will think he is offering a service to God. They will do such things because they have not known the Father or me." ~ John 16:2-3.

And yet it was no little shock to me that in the course of my 30 year ordeal (and counting) I several times went to or contacted Catholic priests both when in Los Angeles and in Seattle seeking help, I was uniformly turned away and in some instances treated with unexpected rudeness in the process. Typically I would say to them that I understood that they themselves, given their other responsibilities and duties to attend to, might not be able to aid me, but could, would they on the other hand direct me to someone that could?

The response was negative or else there was no response at all. But that they were priests, they might just as well have told me to "Buzz off!" (only in and using the vernacular.)

To this day, I don't know quite why they reacted this way, however, at least one priest (like a friend of Job) seems to imply or suggest that my circumstance was somehow my own fault, and that therefore I was, in effect, on my own when it came to dealing with it. Although I don't know such to be the case, my sense is that they likely had been "talked to" by someone in advance of my contacting them; whether a spirit person or someone in their employ.

But this same kind of gelid response was also what I most usually encountered (though not always) when I contacted police, lawyers, doctors, university professors, government representatives and social activists. For whatever reason, they refused or had no time to bother with my story and complaint. The point being that when it comes to hard-core criminal spirit people one should not assume that church people will necessarily act differently than everybody else. At first, I frankly was amazed and appalled by this, but over time I have come to accept the simple and unavoidable fact that church people, as often as not, are only human and generally are no different than everybody else when it comes to this acutely nasty and, what is for most, a fairly impossible kind of problem; most especially when it manifests itself in its more relentlessly volatile and malignant form. In a word, where professional criminal spirit people make it their business to step in and interfere, the difficulty of whatever you are trying to do or achieve can or will more than likely increase by unforeseen leaps and bounds.

IX.

> "Bliss! sublunary bliss! – proud words, and vain!
> Implicit treason to divine decree!
> A bold invasion of the rights of Heav'n!
> I clasp'd the phantoms, and I found them air..."

In this passage from Edward Young's "The Complaint; or Night Thoughts" (1742), the poet vehemently protests the hoping or expectation of lasting happiness in *this* life. While it is doubtless correct, whether as a practical or religious matter, that it is unwise to seek unqualified and enduring happiness in this life; nevertheless if we, and those we share our life with, are sufficiently charitable, patient, considerate and forbearing of others; act responsibly toward our moral and civic duties, and the world we live in *has not* been reduced to utter anarchic injustice, it is still and even so possible to be *more* happy in the here and now, and more so, certainly, than many would or will allow. In the case of some religious, like Young, they see the quest for happiness in this life as somehow being defiant of God. And yet high-power criminal spirit people, from time immemorial, hold essentially the same view. How then no little strange it is, and based on certain individuals' interpretation of higher truth, that both "God" and the devil should *supposedly* concur on this point.

How then should felicity in this life and in this world be seen? The following are *some* possible thoughts and responses on that question:

A. Life here can in no way be *really* happy here; since it falls short of heavenly perfection, and therefore, at best, can only be a kind of mediocre or tawdry felicity.

B. If this world is tainted with devilish hell, there can be no possible hope of anything like quality or lasting happiness here.

C. Though this world is and will always be imperfect, something *like* or approaching real happiness is possible to the degree we do and will better morally behave ourselves; both towards our fellow man and the animals and environment as well.

Now it goes without saying (I believe), that the "way of the world" cannot possibly lead to lasting happiness if only for the simple reason that in the vast majority of instances, if not necessarily all, that those who rule the world are not happy people in any truly worthwhile and meaningful sense; since those who rule the world possess and or require phenomenal material wealth; for without such they could not rule. Yet who can hold onto vast material wealth for all that long who does not, in some measure, significantly compromise with, find themselves tethered to, having to answer to, and or else (despite their perhaps own benevolent intentions) is done in at last by those who are unapologetically corrupt or evil?

Moreover, every would-be, future, new and unanticipated good or discovery, including true happiness, must rely on the recollection of some past good in order that a greater good can be suggested. Yet the origin of such recollection comes from God and Nature; which in their character are not worldly. Greatest joy in its innate vitality derives from innocence and what is natural (i.e., nature in its humble form or sense.) And innocence and what is natural is obviously not the way of the world; without such joy as the source fount, true and lasting happiness is impossible. On the other hand, if we persistently make the effort to be conscientiously moral, just, and honestly rational, we can at least serve and protect the innocent and that which is natural, and by doing so mitigate and lessen the effects of worldliness both on others and ourselves; at least as a matter of degree.

Though not claiming to speak as authority (religious or otherwise), for myself I nevertheless therefore tend toward the third view expressed above **C)**, and believe very strongly that if high-powered criminal spirit people can be intelligently and humanely combated, and as a result either removed from our midst or at least kept in check, then the chances of happiness in this life, God willing, can, if pursued morally and rationally, be dramatically augmented and enhanced; not to a point of perfection of course, but to an extent far greater than some well-meaning but (in my opinion) misguided religionists, misinterpreting traditional understanding of doctrine, and those who are quite frankly unabashed devil worshipers, as a result of envy and covetousness, would hitherto allow or permit. And I believe this is *all the more true*, since high-power criminal spirit people, who are the worst offenders and trouble-makers and therefore the greatest destroyers or ruiners of happiness, have not yet been adequately addressed scientifically and rationally. Could these other-worldly criminals then be seriously dealt with, contested, combated, contained and or removed with on these terms, it stands to reason that increased and more secure felicity in this life, and for everyone, is, at the very least, that much more feasible. But if this view has any basis, it naturally requires us, at minimum, to know and understand that autocratic and high-power criminal spirit people are the major source and cause of most of our worst ills, and that their age-old regime holds far too much unchallenged sway; whether for our own, their own or anyone's good. For their part, these frown on the idea of our possessing something like real and enduring happiness and peace; because such notion threatens to undermine their governing power and authority; that is and in addition to resenting it out of envy and jealousy.

What sane, rational and moral person does not wish to inhabit a more calm and orderly run world where strife is relegated to civilized games, contests and certain entertainment venues (for those in need of such); where things are done right, with a mind toward honest fairness and basic justice, and where, in short, people better behave themselves, even if that world is necessarily less than perfect? This is not to say that were we to question and take on criminal spirit people power, influence and long standing hegemony, the task would be easy, cost free, and success achieved overnight or even in a lifetime or a few lifetimes. Still we can at least start by raising questions, including what is possible? For obviously, "Life, Liberty, and the pursuit (morally and rationally based) of happiness" are hardly possible if professional criminal spirit people, and their regular person accomplices and collaborators, are granted the power of governance and privileges (including legal exemptions, say, from investigation and prosecution) exceeding those and or disallowed of *everyone else*.

To the extent a people are at peace, extend justice equally and impartially to all, and are relatively happy is in one sense a reflection of how much high-power criminal spirit persons *do not* have power over us. Of course, such measure is not absolute or unqualified; appearances may be deceiving; since in practice professional criminal spirit people operate by stealth, subterfuge, lying in wait, and have the ability to do so patiently over long periods of time. So that presumably regular and constant vigilance is always and obviously necessary; not least of which when it might seem the other-worldly gangsters, not untypically posing as divinity, have (like say, a virus in remission) *seemingly* disappeared, and somehow *appear* to have gone away to plague us no more.

In my own view, the settling of *earthly affairs*, God will, in the end, go with the group that is all around most truthful, just and virtuous; regardless of orthodoxy and religious affiliation; though granted worldly prosperity and seeming contentment by no means necessarily implies God's approval. Consequently, it is only mete that we apply ourselves with due humility in this wise. Else and otherwise, let people and animals be happy if they can be; as long as this does not involve either disrespect to religion; disrespect of others basic rights, including what ought to be seen as the fundamental right of everyone to be free of cruelty, outrage, slavery and bondage not strictly necessary (such as would be the case where bondage is employed to prevent violence.)

Fallen and imperfect human nature in its various forms and complexities, needless to add, makes this much easier said than done. And yet if the foreign and unnatural strife, strain and duress imposed by high-power criminal spirit people is eradicated or significantly reduced, what then might be possible? It is this much we are considering. And while it is well and far beyond the scope of the present work to address more specifically the question of what the general happiness, both for individuals and society at large, should be or may consist of, whether now or in the future, I think we all can agree that *less* violent and felony crime is a most natural and obvious starting point. Notwithstanding time honored cynics and dogmatic pessimists, that when things are brought back to something like their original condition, they *can* to a large degree, perhaps even entirely, be restored to what and how they once were. This becomes even more obviously true, and as such can be more effectively achieved, to extent we can *dramatically diminish*, if not remove completely, the impact and influence of professional criminal spirit people. Despite what has been assumed by some, is it *really so impossible* to "go home again?" Simply put, *not* if high-power criminal spirit can be and are ousted from having power over us, *and* if we do our part to clean up our act (that is, when it comes to behaving more honestly, rationally, justly and responsibly.)

As could be expected or else goes without saying, there are people who would out of hand scoff at or dismiss the kinds of criticisms or suggestions I am making. Not because they don't or would refuse to believe that criminal spirit people exist, but because they don't understand the real or potential power of logic and truthful understanding, or comprehend that spirit people themselves use, and are or can be made answerable to that same power, and, in short, unthinkingly and superstitiously believe that high-power spirit people are somehow or necessarily all wise or invincible: that is to say, somehow above and superior to logic, right reason and morals.

"Be on guard! Be alert! You do not know when that time will come."
~ Mark 13:33.

Quite frequently, and rightly, in the New Testament do we encounter admonitions of this kind and with good reason. Here, and for my purposes, I would interpret "that time" as occasions when we are or might be put to the test when it comes to our faith and moral resolve; with sophisticated criminal spirit people having no intention of making things any less difficult for us; calculatingly taking ruthless and full advantage of our dullness, lack of caution, or indolence when it comes to fulfilling quintessential duties and obligations. Moreover, it is by no means unusual for them to appoint themselves as God's own agents; ready to take advantage of our moral and spiritual lapses; not for the purpose of serving what is right, but to employ such phony pretense in pursuit of their own selfish interest. Fortress Troy, be it ever remembered, fell as a result of napping and stupidity, not the lack of courage or the intestinal fortitude to fight.

Before proceeding on this score, it will do well to review further why it is then so many foolishly acquiesce and give in to the thralldom of other-worldly criminals, including, in part, the psychology underlying such passivity and capitulation.

Although in speaking about autocratic and professional criminal spirit people, I have not been at a loss, and usually quite rightly, to characterize them as merely ridiculous and contemptible, it is nevertheless important to emphasize that in addition to taking on the guise of high authority, even deity, and perhaps, in the case of some of them, perform miracle like wonders, they are in some ways, as well, cunningly clever, subtle, extremely knowledgeable and sophisticated, and can be the most frightening and unbearable terrors, that is when and if they have a mind to it.[39] They more or as much as anyone else know the power of fear, and

---

[39] If I were asked in what film I though the terror aroused by real evil is, *emotionally speaking*, most effectively depicted, I think one of the first choices that comes to mind is "Count Yorga" (1970) with Robert Quarry; particularly in the scene where the would-be vampire hunter is caught off guard and is laughed at and mocked by Yorga. Yet observe, what makes the scene work so

know how to induce, arouse and make use of it. Think of the most brutal, heartless and incomprehensibly nightmarish crimes imaginable and there can be no question they were in some way or measure invariably in on it. Deception, at which they are also professionally adept, along with instilling fear, are, after all, their most powerful tools and weapons. And if they possess enduring sway over multitudes and spans of generations, it is fear and deception as forms of psychological warfare, in conjunction with feigning deity, that serve as their ultimate and winning cards when all other efforts at persuasion fail.

They are hopeless busy-bodies and intruders of common sense, rightful privacy and freedom from physical assault; spend much time spying on others and gossiping, and it is characteristic of them to persistently insinuate themselves into whatever you are doing or whatever you are interested in, whether at home, at work, or in public. In the process, they would have you believe that they are a necessary part of your life activities and concerns. One common form of this is assiduously maneuvering their way into individual lives, families or businesses; seeking willing recruits and adherents, and perhaps at the same time methodically sowing enmity and strife between children and parents, employees and employer (at least if they see there is something to be gained by their doing so.) Of course, a similar thing is or might be done in the political realm. Where power worthy of their interest is or may be up for grabs, they will, as likely as not, be there putting in their bid for control of the same.

With respect to fear, those who have experienced their scare tactics first hand, will readily understand why it is that so many see no alternative but submission and surrender in the face or potential threat of such. Just imagine, for one example, what *medical terrorism* (which some of the more hard-core sorts are not above resorting to) is or might be like in its various forms or manifestations. This capacity for methodical and horrific sadism is presumably *one* of the reasons why the act of Christ taking up and enduring the cross is intended to impart the message that it is far better to have faith and suffer such, even the very most incomprehensible and abominable cruelties, than to yield and give in to what might be the tyranny and false munificence of "god-like" criminal spirit people: no matter how bad or violent the treatment. A similar view is also found in philosophy, Stoicism, for example, and other religions, like Buddhism.

Some not so bright people will, and yet perhaps understandably, reason that if these high-power criminal spirit people are the most dreadful when it comes to dealing out violence, and punishment, then it only makes sense to acquiesce and cave in to them; that is, that by yielding, they will be spared or treated more leniently. One problem with this argument is that one cannot always trust or rely on the word of criminal spirit people. And of course, such a one may, as time gets on, find themselves having to pay more or more imposed upon than at

---

dramatically well is the resultant helplessness of Yorga's adversary. Someone else, in that position, might have thrown mockery right back at him; instead of being themselves drowned by it.

first they were initially led to believe or understand. But whether friend or foe, professional criminal spirit people will seek to take or get it out of you: *one way or another*. Really, when it comes down to it, their friends ultimately suffer worse than their enemies; though just as and with respect to suffering the cross, relatively few will grasp and understand the truth of this.

Granted, as a matter of policy and in an effort to be more consistent, they would prefer to make a good or virtuous person suffer (especially one who is a potential political or other threat to them) than a bad person. Even so, it is the height of madness to assume *anyone's* safety and true well being when such as these are put in charge and or govern. Indeed, if excited or aroused in the heat of the moment, they will as likely molest, rape or otherwise violate an acquiescing follower as they would a physically vulnerable opponent. For some of these high-power criminal spirit people and regular employees are so preoccupied with degradation, humiliation and sundry forms of perversion and sadism that the fact of the matter is they cannot always help themselves. So while a follower might, for instance, be formally guaranteed protection from theft, murder or formal torture, they still very well might be abused in a manner that is deemed as playing, or else is seen as only being "friendly." And there are other points worth making to refute this kind of false belief, i.e., that there is safety in surrender to them, but and which we will touch on later.

If one were ask such "goomer" people how do they protect themselves against the threat of unknown and implacable terror, and if you could get them to speak honestly, many would tell you that they seek shelter in the safety of the herd, and this way of thinking not surprisingly is encouraged by criminal spirit people; it being an easier and more cost effective way for the latter to corral masses of persons than protracted violence and terror. This kind of timidity and accommodating to bullying and or bamboozlement, when pervasive in society at large, is often construed as actually a virtue; while those who do not adopt that attitude are seen as selfish and anti-social; indeed might be dismissed or diagnosed as being a "sociopath."

Although we can have spoken of professional criminal spirit people as being the very worst kinds of people, not everyone by any means see them this way at all. To such as these, they are seen as friendly people, gods, or other-worldly persons possessed of consummate intelligence and life wisdom. And if it is suggested that God, that is, God the creator and as founded in right, rational and just morals, is against such sociable sorts as these posing as seemingly sympathetic "deities," or perhaps (in a given instance) "angels," then God evidently doesn't really love or care, and God therefore must be a bad, and those who believe in Him elitists or snobs. Even some who and otherwise essentially understand what I am warning about can or might still be caught unawares and deceived; insofar that depending on your alertness, level of intelligence and reasoning powers, they might fool you into thinking they came from actual "heaven." Be that as it may, one can safely assume that in this life, evil will never cease impersonating good, availing themselves of whatever costume, or disguise that will take in the careless and unwary; this being then one reason for the admonitions in the gospels to be ever constant in one's vigilance, and to not judge by appearances.

Such who give in then typically become docile and obedient vassals. And yet others might go on further to be enlisted as career or semi-career criminals, and this kind of move might be based in part on the principle (mention earlier) that acting badly or immorally is or can be seen as a form of being sociable, and or a necessary aspect or trait required of worldly success.

From time immemorial secret or privileged knowledge has ever been promoted as something of highest value among persons given to believing criminal spirit people. And they see such secret knowledge, no matter how flagrantly irrational in character, as implying deeper wisdom and understanding. The history of gnosticism that followed on the heals of Christianity is replete with illustrations of such. The person "knows" things in a way you (who are not such a possessor) don't. One spoken phrase, for instance I have encountered is that the given person "knows about these things (i.e., such and matters) *in a certain way.*"

But this idea that secret knowledge or gnostic learning or intelligence – even though it does or might contain some grain or measure of actual truth – necessarily translates into real wisdom or deeper understanding is by no means true. A "modern," for example, might know about rockets and men traveling in space, while a primitive tribesman in some far off bushland is ignorant of such things. And yet despite the great disparity, the "modern" could still very well qualify as a rather childish and stupid person; that is despite possessing that knowledge. And not only is this the case with "goomer" people, but the same as well be true of some spirit people themselves, even the professional sorts; namely because *they* possess the (given) secret knowledge, it supposedly follows that they are wise or understand more deeply. All of which is to say that high-power or professional criminal spirit people, for all their often real knowledge and expertise, can be as delusional and swollen with false pride, and end up believing nonsense as well as anyone; at least if and when their point of view is scrutinized more closely. This is a point that cannot be emphasized too much; *since and insofar as one does or might have dealing with them*, there will be a tendency, and this probably with the vast majority, to assume that the spirit person's judgment can be relied on as correct; that they *necessarily* know what they are doing or what they are talking about; when in fact the very opposite may be the case; and which does not even occur to some as being possible; so unthinking are most people when it comes to necessarily giving spirit people the benefit of the doubt.

As stated previously, it is all too common for many to think that what others believe, particularly when those others are a seeming majority, is *the* truth, and for some indeed *the only truth there is*; when and of course (as I am sure more intelligent and reflective persons will agree), it is not at all necessarily a matter of what people or even most people think that *properly* decides the validity or invalidity of an assertion or belief, but and rather what an honest, objective, informed, and rationally consistent person is in a position to make of and on a given question. Despite this, there are ever those who go or, alternatively, flee in panic with the herd; seeking security, wealth and safety in and with the same, and it is with such group-think that professionals skilled at mind control can best succeeded at fooling and taking in an entire society, and or, perhaps in the present highly confused and traumatized day and age, the entire world.

Whether as a result of short-sightedness, being fooled, or both, most people, it would seem, do tend to see great worldly wealth and power as a (or even "the") sign and measure of ultimate good, and in a manner that replaces (true) God the creator. As could be expected, criminal spirit people, and their hench-persons, as a matter of course will doggedly endorse and promote this view. Yet to the degree that great worldly wealth and power are based on lying, cruelty, and blatant unfairness disallowing competition based on actual merit, the cost or penalty of these offenses is passed on and borne by those who are the willing servants or accessories to such manipulation and coercion. While we are all sinners, and any one of us falls far short of perfect, yet there are others who are way and beyond ridiculousness when it comes to be being blind, callous, and in a daze when it comes to gross immorality and unapologetic criminality, and the more a person reduces what matters in life to worldly wealth and power, the more susceptible and prone they will be such destructiveness and excess; while in the process, knowingly or not, succumbing as vassals or slaves to other-worldly stooge-oligarchs.

X.

"The sun, moon that gives light, bow not down before the glowworm. He who has issued from a wise and virtuous family, who is himself full of virtue, bows not down before the most powerful gods."
~ *Lalita-vistara* (Buddhist) Sūtra

In what I have so far written, there has been some effort on my part to be as comprehensive as present circumstances allow, and yet we have only, *if that*, touched on the myriad surface aspects and complexities of the subject: historical, psychological, medical, legal, sociological, religious, etc.; bearing in mind also that the problems arising from criminal spirit people, and their followers, occur at radically divergent levels of intensity and areas of experience, and which in turn, whether subtly or overtly, may or may not be made apparent to depending our level of knowledge, alertness, and powers of perspicuity. To that extent, such problems, in their various forms, can be likened to a sickness that in one situation may be no harmful than a common cold, and yet in other cases might be as bad or worse than a malignant cancer; with varying degrees of risk and danger lying in between; requiring less or greater care in dealing with. Also like some diseases, a person might go on for years being ill with something; without, perhaps ever, even knowing it.

And yet when it comes to the problem of spirit people organized crime, whether it manifests itself as benign, malignant, or something in between these, all such sickness are, ultimately and in their origin, incarnations, outgrowths of, or in some way tied into or connected with criminal spirit people as the main or source disease, as well of course of our own sins. So that whatever the degree of intensity, and allowing for human sin, error and mistakes otherwise, certainly (and so I would argue) the centrality of these problems arises from and will be found to lie with high-power criminal spirit people who make it their business to subjugate and afflict us. These, and with a reach, whether real or merely apparent, that, practically speaking as least, is almost universal, seeks to infest and hold ultimate power over life in this world; in all of its variety and aspects as is deemed by them feasible. While it is ordinary and proper to hold people accountable for their own mistakes and errors, it is completely blind, irresponsible and flagrantly unjust to ignore the role high-power criminal spirit people play in influencing and planting bad behavior; in inciting and stirring up trouble. Indeed, often times we are led to think people act badly or do something wrong because they themselves desire it. Yet for many and not infrequently the real or overriding reason they act badly, etc., including things like unnatural and prurient proclivities, is rather to appease and win over the good graces of criminal spirit people. But for that as a major factor, the individual, in a given circumstances, might not need or desire to engage in the foolish behavior and or perpetrate the given offense at all.

And yet in practice, criminal spirit people are never even mentioned or considered, at least not seriously, as acting a role in the misdeed or crime. Yes, we can and should blame Othello. But how sane and or honest are we, with respect to judging what happened, if we completely ignore Iago's role in what took place? That high-power criminal spirit people and their followers do and would insist that the behind-the-scenes role they play be overlooked and disregarded goes without saying and is to be assumed as legitimate given. Of course, some smile at the devil: "a friend of the devil is the friend of mine." But what such people forget is that the same "devil" who, say and for example, approves of loose sexual morals or hard core recreational drug taking, also approves of mass murder and torture, and or is inescapable league and business partnership with *someone else* who does.

It is my contention here then that what lies at the heart of indeed the far worst challenges and difficulties which confront humanity faces are persons, whether high-power criminal spirit people or regular person followers of theirs, who bully, deceive, cheat, and wage incessant and violent war for purposes of conquest, and a second group that, either out of stupidity, pusillanimity, or indifference permits the first sorts to do so. Such conquest is in no small part sought by means of by forming or framing our characters, judgments, morals, and morals to be in accordance with theirs; that is, in *getting us to be more like them.* The more selfish, irrational, and depraved we are made or become, the easier it is separate, divide, and thus enslave us. In practice, they operate as a de facto clandestine government, the very deepest of all so-called "deep states;" using what amounts to a kind of secret police and military, armed with both worldly and other-worldly advanced technology, not to mention more than ample financial resources to set-up enforce their rule. At the same time, they recognize as well that a society's culture is its soul;[40] so that to control the culture is to control the soul of that society. "Do not be afraid of those who kill the body but cannot kill the soul. Instead, fear the One who can destroy both soul and body in hell." [Matthew 10:28.]

Acting as a kind of self-appoint secret police, and amoral and hypocritical Censors (in the Roman sense of that title), they restrict or prevent the happiness and success of such who don't conform willingly or who don't, in some manner, subscribe to their protection racket. Theirs is a carrot and stick approach. Make a deal and enmesh yourself in some sort of voluntary contract with them, receive the bone they throw you, or else risk incurring their wrath. They act as if they are above right reason, and treat and view *the Word, honest* reasoning, and discussion as being of small importance. And given what has proven to be their long term staying power, that is *well over and beyond our own regular life spans*, in having dominion over a government, business, or a family, they can phase out and displace those more moral and rational, and empower their opposite; unless, that is, these age-old tyrants and oppressors are recognized, rejected, combated and resisted for what are: the most virulent and insatiable of all plagues.

---

[40] This idea I first heard expressed by geo-political and literary specialist and lecturer Cynthia Chung; my thanks then to her for that.

But one can't just or always just simply do surgery and cut out and remove a cancer. The problem is often or at last far more complex and difficult than that. This is why, *at least from a Christian point* of view, nothing substantial or properly thorough can be done till Judgment Day: that is, only when the wheat and tares can be justly identified, sorted, and separated.

While they ostensibly have been powerful since time immemorial, it needs be noted that the sway and say of professional criminal spirit people has increased dramatically in the last three or so decades (i.e., since roughly the late 1980s); so that, for example, there are things they can do easily now that at an earlier time would have been prohibitive or considerably more expensive to do or carry out. Yes, they have *always* been there, and tragically and devastatingly so, but now they are more out in the open and diversified in their carryings on than earlier. We saw or see this in things like serial killer phenomena, Jonestown, Waco, Heaven's Gate cult, 9/11, school and other public shootings, and the completely insane and wholesale celebration of realistic brutal violence by moviedom and the mass media,[41] and the rampant breakdown of family and basic morals generally, the proliferation of homeless camps, stepped-up vandalism, including lavishly subsidized campaigns of graffiti; epidemics gone "viral," etc, etc.— in other words, all pervasive sheer and utter insanity.

Previously it was and we had the "humanities" as the aim and healthy direction of cultural endeavor; only now to be traded for and replaced by the "inhumanities." Who could ever have devised and wished for such a thing? And yes, this switch is quite deliberate, and there is blatant bribery and literal payment involved in all this. Indeed, the call goes out (from some) to "cancel" culture entirely.

Did such extensive, densely occurring, and eerily bizarre problems, in their primal source, spring from Nature? But if not Nature, from whom or whence did they come?

It is to be expected, most anyone of us will have their own personal thoughts and feelings on the question; and I respect that. But for myself, the proliferated demise of culture and the arts, respect for tradition, the character integrity of communities, institutions, and business is almost entirely the direct result of the increased influx and accompanying deployment of criminal spirit people governance and prerogative in our midst, as much or more so than any other cause. Even those who protest that decadence that others celebrate, while quick to attack secondary and third rate culprits for what has gone wrong, are

---

[41] Arguably the principal fomenter of widespread violent tendencies, and which having in recent decades been transformed into a kind of secular church, has become both the promoter and the exemplar of the very worst kinds of viciousness, depravity and corruption. And though "fake news" has come to be decried as a bad thing, the corrupting aspect of post 1990s movies and the mass media, including purposely mindless and brutal video games, is barely raised or addressed at all by anyone.

often themselves still vassals and zombies of criminal spirit people. To again cite one familiar example or canard of this sort, "Jews" are blamed. Yet if there were *ever* any truth to this kind of charge and inference, to this or any other racial, ethnic, or traditional religious group, it could only be because a segment or certain of its members (whether knowing or unknowingly) are hench-people and dupes of criminal spirit people. In fact, there is no major cultural or political group that is wholly free from being contaminated, in some measure, by the influence of criminal spirit people.

So that our greatest adversary, as it turns out and despite longstanding prejudices and stereotypes, is not a country, race, or nationality, but a person and persons who, and perhaps despite surface appearances or formal affiliations, have no country, race, or nationality. Indeed, they are in their behavior and motivations, *inhuman*. And the conflict that most dogs, ails and plagues us is *not really* left versus right, or would-be progressive versus would-be conservative, rich versus poor, capital versus labor, white versus non-white, female versus male, Christian versus Jew, etc. – *whatever*. Rather it is *them* and their other-worldly hegemony, acting as leadership, against all of us: even and including countless spirit people themselves who are victims and slaves of the crime bosses.

Such have persons employed and installed on *both* and *all* sides of *any* given quarrel or controversy; especially where cultural prestige and authority, material wealth and worldly power and interests are at stake or up for grabs; playing off one side against the other that both in the end will be done in and they as winner over both takes all. And while they may not have great credit or prestige say in a given group at one time, *at a later time they very well may*. We might, for example, understandably see the Romans who persecuted the early Christians with murderous violence as "bad" people, and the Christians as the "good" people. And yet some time later (howsoever soon or long) a group of people *claiming* themselves to be Christian might become the "bad" people inflicting murderous violence on some group or other.[42] And yet in each case there will have been professional criminal spirit people, disguised and or acting behind the scenes or through surrogates; who, more than anyone else involved, *most* incited and insisted on carnage and brutality. When all was said and done, and irregardless of how long it takes, they must always and would have themselves seen as partners or allies to the winning argument; changing sides when and if it is necessary and possible to effect this; encouraging and promoting the idea, whether based on fact or fiction, that the prevailing side ended up being successful because it did the wrong thing.

What in no small part makes this all possible are certain regular persons making deals with these high-power criminal spirit people, the former, although almost invariably unwitting victims themselves, acting as crucial conduits and go-betweens for the latter. Together they seek to force their "religion" on us, and

---

[42] If the existence of criminal spirit people is granted, one can safely assume the Inquisition, certainly in it tortuous and murderous form, involved their regular and direct participation.

which views humans as sort of inferior sub-species (i.e., read "niggers") in relation to the seemingly all-knowing and almighty "gods." True, the technological prowess they can display and capabilities they possess are in many respects genuinely "awesome" and intimidating; and the threat they pose is undeniably formidable and in some ways perhaps seemingly impossible to bear; for professional criminal spirit people are long inured and trained at what they do and can be supremely shrewd in what they are about. Yet it is also true, as we said earlier, that no less many other respects they are weak, indeed wholly impotent compared to many of us; including and at times quite, and perhaps surprisingly, in their being (on some matters) stupid. Moreover, it is all important to keep in mind that at last they possess much, if not all, of these phenomenal powers they have, *only because* they traded away *all that was and is truly good, right*: such as honesty, trustworthy love, peace worth the name, and unfeigned, resilient happiness in exchange for said worldly powers.

Though they will understandably make an effort to conceal and camouflage the fact, they are brazenly despotic in character. And any form of governmental or societal totalitarianism that lasts or endures *necessarily* involving criminal spirit people; since at minimum, no one could possess the might and wherewithal to effect and make such possible for very long without the approval, partnership and cooperation of such. And where a system of governance is less, rather than more, dictatorial and authoritarian, it can only be because the high-power criminal spirit people and their followers find themselves up against opponents who do or will not brook and suffer their consummate arrogance and assumed right to dictatorship.

While it perhaps may (to some) sound harsh to say, the truth is that high-power and professional criminal spirit people are quite literally a disease: in fact, the most virulent and lethal of all diseases. For who does more ongoing and lasting harm and injury than they? Yes and of course, humanity at large can be characterized as sinful and destructive. Yet when have we ever had the chance to do or get by without their corrupting and poisonous intrusion? How then bad would things be without *that* being the case? A virus, after all, does what? Pretends, masquerades, and fools; so that it can get in when and where it is not supposed to; exactly the same practice high-power criminal spirit people (and their operatives) utilize for purpose of suborning and embedding themselves into our individual and collective lives.

Just as one could expect, such other-worldly malefactors are, to many, just *too* intimidating and *too* frightening to reject and refuse. And yet for duly rational and courageous persons, on the other hand (though relatively few they are or may be) it *is* fully possible to say no to criminal spirit people gangsters, including those posing as deity, and reject their imperious solicitations or demands. It takes bravery and courage. Yet is possible, however so unlikely perhaps, for people to be brave and courageous. And a sufficiently honest, rational and daring individual has the ability to size up a given spirit person or persons for their weaknesses as well as their strengths; and this goes a long way to taking the wind out of the sails of hoodlums and con-artists generally. The honest, rational and daring can ask questions like "what are this spirit person's motives?" "What is it that prompts him to act as he does?" "What is it exactly he is believing?" "Is this belief of his justified; does it make as much sense as he thinks it does?" And "does it really make any sense for us to think like him?"

The world has no end of joys and beauties; both those which occur naturally, and also in those things that learning, labor and invention produce, enhance, and make possible. Yet the seeming "gods," who by what amounts to usurpation and or adulterated democratic choice, and who, as often as not, end up and to a large degree ruling the place, won't let us have or experience these God-given blessings for very long without their approbation. They *very much* resent and take exception to felicity unfettered by them; and routinely position themselves as a kind of self-appointed middle-man that does or potentially threaten almost all happiness in this life we might know; again this thanks and due in no small part to "regular" persons making deals with them seeking would-be security and or worldly gain. In consequence, and depending on how shamelessly wicked the age, we are alienated from life and nature because of such artificially imposed and unnatural ties and restraints created by this arbitrary middle-man. On the surface, it is not unusual in our own time finding their de facto, behind-the-scenes, regime espousing social equality and political, but with virtually all reduced to the lowest common denominator; that is with respect to level of cultural, intellectual, educational and social *status*; except, that is, for a tiny minority of the most rich and affluent, and or their henchmen, and who hold control and sway far above and beyond all the rest when it comes to privilege, power of voice, and position; with, as could be expected, all manner of chaos, suffering, depredation, and carnage in pursuit of some never to be realized materialistic utopia. It is a society where, and mostly, promotion and place is based not on honest merit and ability, made possible by fair competition and sportsman-like opportunity, but and rather on party membership and standing that ends up taking absurd precedence over real worth and real capability.

While the good, or those who would be allied to the good, do have right reason, the moral law, religious faith, and with, to some extent, science and technology to aid them, truly effective leaders can be scarce, all the more so when targeted by organized criminal opponents; as history so amply demonstrates. After all, it is the honest, rational, morally conscientious and level headed prophets, saints, philosophers, scientists and other social activists who are those targets that most need to be eliminated or neutralized by their opponents. In the struggle with evil then, most necessary and not of least importance is awareness and education, especially when it comes to making ourselves more rational and more moral; that is, that we might be that much more empowered to resist and thwart it. It is such leaders who can best provide these; hence the need for evil's advocates to routinely oust, displace, or (as need be) assassinate such. And when those leaders are absent or gone, people inevitably grow less moral, less rational, and more doubting, and therefore more vulnerable to tyranny, and voluntary and involuntary servitude to the same.

One cannot live one's life assuming and expecting to *necessarily* change and save the world. But we can try, and may, to some degree, even succeed. And we can all the more protect and preserve some of what life here holds dear if we reject, separate and distance ourselves from those who, whether purposely or no, make it their business to taint, contaminate, demolish, take over, and or ruin everything; namely autocratic criminal spirit people and their followers. In speaking of combating them, I am not somehow or in any wise proposing a substitute for religion; in which latter only is true and enduring hope to be found. But inasmuch as high power criminal spirit people have a material presence and are able to bully and violently insinuate themselves into our lives, there are practical measures that can and should be adopted to counteract their criminal efforts; not dissimilar to how it is possible for us to have medicines to thwart disease, or weapons to suppress violent lawlessness. But just as medicine and weapons cannot solve all problems for which they are designed or intended, we can talk about how in using them we can mitigate some of all that ails, enslaves or otherwise hinders us from living creatively, productively, and peaceably.

In the grand scheme of things, we have the (relatively) good and honestly rational *versus* the morally dissipated and the intractably dishonest and irrational; the latter, as often as not, are, knowingly or unknowingly, led, armed and assisted by and with a veritable army of spirit people, which latter, as budgets and circumstances permit, are organized into crime bands or terrorist cells; to then be drawn upon and deployed as need be; with no end of technology (both in the way of regular and spirit person technology) and financial resources to strengthen and sustain them in attacking us. Their "regular" person allies, never dreaming of questioning criminal spirit people authority, are typically cowardly and or mind-controlled persons; even including some who are very powerful and wealthy who enter into agreements with them for purposes of pursuing personal and material advancement.

The battle of good versus evil is ultimately one of honest, rational truth versus lying and deception; (what is essentially and at heart) selfless and appreciative love versus (what is essentially and at heart) envy and self-pity. But evil (or the "morally bad" if you prefer) *can be* most or even extraordinarily clever, and as we earlier stated as a general rule encourages the manicheanism-like belief that good and evil are both necessary. According to which view, evil is itself a kind of "good." And depending on their moral shrewdness or lack of it, many are easily fooled into going along with this idea. The right and moral view, by contrast, is to say that while someone might not be able to help their morally bad behavior, they at least recognize that such behavior ought nevertheless to be regretted; only they can't help doing it. Some will seek to outgrow and overcome the bad behavior; while others give up trying; but both know that the behavior is bad.

The more manichean view, on the other hand, says that bad behavior, or at least *some* bad behavior and at times is actually a positive good. It is on this *key difference* that professional criminal spirit people try to trip people up, and not infrequently with (what is for them) spectacular success; getting them to think that unless they acquiesce or bow down to evil in some way they cannot possibly hope to get ahead in life. Part of what is intended here is to have people believe the illusion that it is *necessary* to accommodate or make deals with evil; that the protection racket and or "road to riches and success" program criminal spirit people have to offer is essential to both life, safety and the realization of ambition. And when such way of thinking is not knowingly and willingly adopted, it is alternatively surreptitiously inculcated or slipped into peoples' thinking; often and for many, without their even being the least conscious or aware of that taking place.

To compound or make the problem more difficult, it is more than common for people to believe that the ultimate standard of truth is what other people or what "everyone" thinks. For some it is so bad, and they are so lacking in the capacity to reason, that they have no idea, beyond immediate sensation and perception, of what is real, or beyond what others tell them is real. For us who know better, what we want and what we are about is truth *over and beyond* merely what others or everyone thinks. It is the honest, rational truth to us that is and has the higher standing over the *mere opinion of others*; most especially the opinions of those who are not disposed to be seriously honest and rational in the first place. And, after all, if a given group of people do discard honest, rational truth, what or how much does it really matter what they think?

As we said elsewhere, nothing is or is determined to be true or real, or to have value or not have value, but that *someone* decides it is or not so. Ever the question then must be asked *who*, in the given circumstance, is this someone making the decision? How credible or reliable are they? When we say, *we* believe such and such, it is really ourselves, or perhaps is it someone else making the decision for us?

By such method and approach are we better able to sort out the true from the false. For professional criminal spirit people have, practically speaking, no end of means of getting you to believe they are the ones to decide what is real; what is true, what is false. And unless a person knows better, is honest, is sufficiently rational, there is a more than likely chance of being fooled and taken in by such tried and experienced other-worldly con-artists. They have magic, and other amazing displays of power (including bribes); they have means of frightening you; can employ clever arguments and which are sometimes not without their truth; indeed and as bad as they are; they can disguise themselves, speak into your head, as or as if God to some. "Do not judge according to appearance," as is said, "but judge with righteous judgment." For true God speaks through truth, and truth comes to us by honest rationality, a clean conscience, and a sincere and charitable heart. The more we are given to see truth this way, the much harder it is for the enemy to get the better, enter, and take us over.

XI.

"And when you look up to the heavens and behold the sun or the moon or the stars, the whole heavenly host, do not be led astray into bowing down to them and serving them."
~ Deuteronomy 4:19.

"What ailed us, O gods, to desert you
For creeds that refuse and restrain?
Come down and redeem us from virtue."
~ Swinburne, "Dolores," xxxv.

In the course of my own personal and peculiar ordeal involving criminal spirit people and their hench-persons, in which I have been subjected to *all kinds* (too many to list or enumerate) of violence and hounding and robbery and vandalism for over 30 years, the most difficult question for me was not, why should I have to suffer so, but rather: why would anyone act like this? And in addition to this: how could anyone perpetrate so much felony crime and wrong-doing *and* continue, *over the span of decades*, to get away with it? Unless criticism and satire directed at adults be seen as a grievous fault, I had done nothing to these people; indeed, I did not even know who these persons were that were making it their righteous cause to so perseveringly hound and violently attack me. And though I have been poisoned, given diseases, physically and psychologically tortured with sundry methods, been subject to attempted murder and more, I could not get anyone to look into and investigate my story. And when there was a person or two who might have done so came along, they shortly thereafter, and for reasons unknown to me, vanished, it having become no longer possible for me to contact or reach them.

Since the 18th century, doing the wrong thing has, at least in Western culture been frequently glamorized and made to seem popular, and for many a necessary adjunct to being rich, fashionable, sexy, and, not least of all, practical. While Don Quixote has long been laughed at as the ludicrous dreamer, now it is Don Corleone who is come to be seen as the realist.

Yet if the former characterization be seen as true and time honored wisdom, the other is perfect nonsense, and continuing to believe the myth that criminality is a prerequisite to wealth, loved and success is it as much the cause of our worst ills as anything. On the original paper back cover of Mario Puzo's book, I recall there being a side sketch of puppet strings being manipulated; it being implied that the Godfather was the one who did this. Yet *in greater reality*, the Godfather, for all his prowess of conniving and powers of intimidation, is not the puppeteer, but rather the puppet. Perhaps another good example we can mention of what wrong doers are really like is the character of Geoffrey Firmin in the 1984 film version of *Under the Volcano* (which film, in fairness *and with respect to the script*, is much different from the novel; which latter has far more to redeem it.)

The character there, occupying center stage in all the drama, lives his life being jealous and feeling sorry for himself, is concerned with no useful calling or endeavor. Instead, he spends most of his time getting drunk, while implicitly blaming "Godot" for either not existing or else not coming to save either the fallen world or himself.

Is it not perhaps then rather the self-centered, the immoralists, the dishonest and irrational, not the moralists, honest and rational, who are the ones most living in a complete fantasy world, and, along with the ignorant and indifferent, are the ones who most (aside from criminal spirit people potentates) contribute to bring about our collective downfall? Call Don Quixote a fool if you like, but, for all his idealistic madness, he is far less the plague to the world than his seemingly more "practical" and material minded opposite.

And yet in both cases, it is the "devil" or, as he have come to denote them, autocratic criminal spirit people, invisible and behind the scenes, that are actually calling the shots, the ones most directing what is going on, and having the greatest laugh of all.

In what I have hitherto written here, I have tried to present at least a rudimentary picture and idea of these other-worldly persons. Yet as I earlier would have it emphasized, there is so much more that can be said and addressed concerning them, and which, in some respects, I can only begin to hint at. But introductions must necessarily take the form of an outline, and I hope that I will at least I can have achieved a reasonably working and accurate one for that purpose.

In closing then, I hope it won't hurt to once more and briefly go over some of the main points, and one or two others, that I believe need to be most kept in mind with regard to this most extremely annoying, knotty, and trying topic.

If, as I myself take and assume to be the case, there are a substantial number among the myriad of humanity who would allow and differ to authoritarian spirit people power, the very least than can be asked is that the existence of the latter, after being probed and attempted honestly and empirically, be admitted, and further that such ruling power they posses be accounted for and made answerable to the human law and public health, wealth and safety.

It worth remarking that inasmuch as spirit people, criminal or otherwise, can potentially (if not strictly speaking always) see all that we do, then nothing that anyone does can ever be assumed to be concealed. So that not even the most clever and masterly spy or criminal can hide from *them* what he is up to. And the more we can be got to realize this, the less prone many will be to indulge in what is obviously inexcusable and bad behavior; since if it is not hidden, it is highly probable that "someone" can use the knowledge of their guilt to gain leverage against them. We can think God sees all. Yet we do not think of God going to inform on us and get us into trouble. Criminal spirit people, on the other hand, if they think there is something to be possibly gained from it, very well may, and if not just now, perhaps many years later.

There are and always have been the well meaning and not well meaning scoffers and debunkers when it comes to the question of "are there spirit persons.?" But whether or not spirit persons exist, there should be no question that basic honesty and right reason are necessary for purposes of objective determining true from false; from the more probable from the less probable. And yet while we may not be so surprised that some people are reluctant to entertain the idea that spirit people exist, it can nevertheless be insisted that fundamental honesty and coherent rationality are necessary if our objective judgments are to be taken as possessing something resembling reliable soundness and credibility. And yet, and *very* often, there are those who reject this premise and principle, and if they do, can we take their views and opinions on the question of spirit people seriously? In other words, if someone insists that there can be no *empirical* reality underlying the claim that spirit people exist, it is only fair to ask, is that person honest; are they duly rational? Since if it turns out they are not these things, on what conceivable basis can the value of their skepticism be confirmed or validated? And if they will not submit to a simple testing of these qualities in them, what then can be inferred about the sincerity or integrity of their views generally?

Spirit people themselves, whether in fact or in lore, are given to secrecy and do everything possible to prevent close and rational scrutiny and disquisition directed at them. Are we to think that this is the case because they are above or superior as authority to right reason? And if so, what could possibly justify the notion that they are superior? Unrelenting terror and violent intimidation? The reality of the matter is autocratic criminal spirit people would not have the great power over us (as I have argued) that they do were it not for the great number of people, who out of cowardice, ignorance and an aversion to honest rationality, unthinkingly acquiesce and surrender to them.

Experience will show that the vast majority (if not strictly speaking all) of those who know of and trust arbitrary spirit people authority are invariably and without exception childish, rascally, easily frightened, semi-rational and casually dishonest and secretive; evincing the character of herd animals, incapable of critical thinking. This of itself, if correct, lends compelling support to the claim that the authority claims of spirit people need to be treated as, at the very least, suspect, if not of readily condemned out of hand.

Moreover, if criminal spirit people do exist, and if that fact is simply ignored and dismissed out of dogmatic presuppositions, then what good are all the efforts to supposedly save the world through science, medicine, psychology, politics, sociology, or what have you? Put another way if the immune system cannot even identify a deadly disease for what it is, how can it hope to possibly remove or destroy it? The logic is inescapable. And the practical result is that we simply would be leaving it up to spirit people themselves to decide life's greater questions; with our having little or know final say in them.

"But they have such incredible, super-worldly power!" some will object. "What is all your reasoning compared to that? "

Yet what power is that which is only held at the necessary trashing and elimination of honest truth, of basic justice, of real happiness, of peace worthy of the name; most especially that happiness and peace that can be guaranteed the innocent? With criminal spirit in power ruling over our lives you can't, or can hardly, have any of these things in any trustworthy and dependable way.

"But what about the importance of obedience?"

Obedience to who? Obedience to what? Because some will think spirit people are God, or God's surrogates, then it follows, so it is reasoned, that we must obey them. Yet what sense does it make to treat spirit people as authority when they *a)* don't clearly identify themselves, *b)* don't hold themselves accountable, *c)* don't allow reasonable discussion, and indeed are the driving force behind mass censorship of polite, reasonable discussion. If any one can be taken as highest authority, and *if* we have any say in the matter, then it must be assumed that they are honest, rational, and moral, If they are not honest, rational and moral, obedience to them on our part is neither obligatory of required of us. Because if obligatory, if required, by what measure outside of rational morals can such be justified? Yes, we may be forced at gun-point to act a certain way, but it makes no sense to view those threatening us this way as legitimate or respectable authority.

Hitherto, it has been the unwritten law for many that one must make deals, appease, come to terms with criminal spirit people in order to live ones life. My view is rather is we should make every effort possible to cut them out entirely;at least inasmuch as we can do so; while maximizing distance between them and ourselves. The less middle-man there is for us to answer to, the more free are we. What they for ages have asserted as *necessity* was and is rather and in truth a matter of their own *choice* to act as they have done. Likewise, and although handicapped by our ignorance, the docility of humanity to let autocratic criminal spirit people have their way is in large measure a result of our own *choice*, and, again, *not necessity*; despite our having been led to believe otherwise.

As far as pagan theocratic history goes, the only children in *dishonest* heaven are cupids, and whose innocence at the very least is suspect and questionable. If high power criminal spirit people respect children and animals (society's and everyone's rightful darlings) it is only out of diplomatic necessity, and there are no gods for or serving the needs of children, animals and the innocent. "Gods" rather are for lustful, ambitious, narcissistic, and (more often than not) cowardly sorts of adults.

The fuel and subsistence of the demonic is arguably human and animal sacrifice: with children and animals usually being both the worst and most voiceless victims. End this practice and it is conceivable that high power criminal spirit people can be starved from autocratic power.

So often I have heard them (or their representatives) say that, and insofar as we are too blessed or too well off, that we *have it too good.* Cannot their own argument be turned against them? That is, cannot we say *they*, rather, are the ones having and who have been having it too good all along?[43]

That we can at last say no to them; that we can at last resist is only possible to the degree we are sufficiently honest, rational, courageous, moral and as just as possible to both people and animals alike. And yet an honest rationality that seeks to go to the depths and reaches of truth possible, and not merely its outer surface, is far more rare than spirit people wonders and seeming miracles. For this reason, it only makes sense that the study of logic and the practice of reasoning honesty and rightly be all the more encouraged and developed in and among all ranks, including making it, along with reading and arithmetic, a standard component of elementary and high school education curricula.

In any event and in the end, will facts, right reason embolden, rouse and persuade people to consider and address the problem of criminal spirit people? I for one am under no delusion, and there is by no means any guarantee that this will be the case. At this juncture we can only hope it may be so. Indeed, the opposite, that is continued ignoring of the problem, may be just as, if not more, likely what can be expected in the short-term view. But even if the greater world at large stubbornly stumbles on to its own ruin in darkness and ignorance, at least those who care to can and may be made aware of what has been and is going on all these many ages, and take and adopt what steps and measures *they can* to guard and protect loved ones and themselves.

~~~~~~*******~~~~~~

---

[43] It has also occurred to me to ask them how and where they first learned and ever got started in the torture business, for example.

APPENDIX 1: A NEW TREATISE ON HELL

Or

How to Fight Hell

Fourth Edition

~~**~~

"'My lord,' replied Solon, 'I know God is envious of human prosperity and likes to trouble us; and you question me about the lot of man…'"
~~~~~~~Herodotus, *Histories*, Book I

"If Wieland had framed juster notions of moral duty and of the divine attributes, or if I had been gifted with ordinary equanimity or foresight, the double-tongued deceiver would have been baffled and repelled."
~~~~~~~Charles Brockden Brown, *Wieland*

Preface

While some will understandably feel inclined to simply ignore this most implacable and disturbing of topics, they can never expect by their mere doing so to escape from it.

If, as is my contention, the direct impact and influence of "Hell" plays such an overwhelmingly role in bringing about mankind's worst problems, we must address Hell as an issue of utmost priority or else give up trying to solve serious problems to begin with. If we don't get at the very root of whatever it is that does or might ail us, cutting off no matter how many branches and stems of a malady avails us little. If we don't cure the cancer, curing the cold means little.

Needless to say, this is *not* an easy subject. It concerns that which one spends a greater part of their life trying to get away from. Initially, I assumed it unthinkable to write about (literal) Hell, as I believed it would only make it more difficult for people to take me seriously in other matters. In addition to being somewhat puzzled as to how exactly to describe what I had seen and heard, it was a subject I myself would just as soon have put aside and forgotten. Yet when I realized what and how much is at stake with respect to this truly age old problem, I thought I should at least have a go at it, and what is before you is the result.

The possibility that there are spirit people has profound and wide ranging implications for religion, philosophy, law, government, medicine, history, and other various areas of science, implications so great that there is little room to begin to try to take them up here. But to illustrate (in the case of science), if, for the sake or argument, we assume a "Satan" (or someone like or similar to this), it is fairly obvious he could or would, to some degree, influence science, by steering inquiry, falsifying data, and undermining or attacking scientific views inimical to his own interests: a very devastating conclusion to put it mildly. Apply this same "Satan" or "Satan, inc." to other areas of life and study and one can see, and further discover through examination, how extensive and deeply entrenched the Hell problem is. But this is only the beginning, and I believe if we look further we will come to see that crucial life-questions such as whether a people are free rests in the final analysis on how successfully, or not, they are able to deal with spirit people (of whom more as we proceed.)

This book merges my two earlier summaries, "How to Fight Hell," and its "Supplement," and by doing so a more (I think) convenient and better filled out single work has been created. At the same time I have taken this opportunity to include further information and analysis not contained in the previous writings. In one sense then, this *New Treatise* is not so "new" a work, because it merely reproduces the older writings, albeit arranged in a somewhat different manner. Yet the addition of more information and analysis gives it something the essays did not have, and to that extent at least it is new. I considered starting again from scratch and composing something entirely original, but decided against it when I saw that what the essays contained was adequate to get an at minimum a basic understanding of the subject; which (at the present stages) is all that I intended. Yet, at the same time, by combining the essays, and including the new information and analysis, I am able to better organize and standardize the study, while considering the subject somewhat more in depth than before. In these ways, I hope to have put together a synthesis that is both more thorough and instructive to readers than the separate essays. At least this is my wish. Some might prefer the essays for their relative brevity. Some might want that I had done a properly "new" work, or that this "new" *Treatise* was more comprehensive than it actually is. For any such possible disappointment, I beg other's pardon, and can only say that for my own purposes and circumstances, I feel the work is at least sufficient. Without question, the topic is deserving of much, much more addressing and examination than I am providing. Yet as far as what I myself can contribute, I am satisfied that what I have put together here is at least still helpful.

Not a few things I speak of here will certainly sound too fantastical to many, and I don't ask anyone to just take my word for what I say as true. You can believe or not believe what I say as you choose. For my part, I am merely giving here the facts as I know them, and much of what this work contains is simply my own testimony, again, take it or leave it as you please. What is being addressed here is not *really* nor technically supernatural phenomena. Rather they are material phenomena, but *material* phenomena whose proper understanding requires more information than we have available to us, and more intelligent inquiry into their nature and character. My attempting to combine theology with empirical scrutiny and analysis is, granted, a strange or at least unusual approach to a scientific study. But its justification, I hope will be borne out after one has surveyed the work as a whole. It is an approach, which, to my understanding of the matter, is as much as the present state of "Hell-ology" allows.

~~**~~

Note to the Second Edition

The major difference between the present and the previous version of this work is my introducing a few new terms and names, including Orkonism and Goomerism. *Orkonism*, according to the simplest definition, refers to the belief in and or practice of doing the "wrong thing" the "right way." Essentially, to an Orkonist, there is good in evil, if the evil is given its respectful "due;" though given Orkonists might well differ how and to what extent evil is good or justified, or what it's prime motive and purpose is and should be.

Goomerism, by comparison, refers to the practice of doing the "wrong thing" the "wrong way." In practical experience, a Goomerist is someone who is in some way or other under the spell or influence of an Orkonist, but who themselves is not prepared to make such an obvious break with goodness and openly embrace evil. A Goomerist would profit selfishly from the Orkonists ways (at least they believe they can) without making the kind of evil commitment expected of an Orkonist. Also, it is not inappropriate to think of a Goomerist as an insincere bumbling person who attempts to disguise his guilt with his otherwise very real ignorance and stupidity. We might well speak of there being subcategories and classifications, and a given person may be a combination of Orkonist and Goomerist, but these approximations of the two main decidedly amoral outlooks or dispositions are I would think self-evident and clear enough in their basic conception.

One reason for adopting the terms Orkonist or Goomerist, is to avoid having to use such a highly charged, more ambiguous, and value-loaded term like "Satanist." A Satanist may well be an Orkonist or Goomerist, but it doesn't necessarily follow that an Orkonist or Goomerist is a Satanist: the difference, I suppose, being between serving a principle or way of thinking on the one hand, versus serving a specific person ("Satan" or a "Satan") on the other.

~~**~~

Note to the Fourth Edition

In preparing this fourth edition of *New Treatise*, it is somewhat odd looking back now on what I wrote a couple years ago. What is odd is that while what I wrote is (as seems) true or mostly true, with the passage of time I now might see the given topic somewhat differently. I suppose therefore that if I were to write this all over again, among other things, I would probably take a somewhat different approach in terms of presentation and organization, perhaps providing greater emphasis on the nature and role of (so called) "fallen angels." or ""gods," rather than their human servants. This said, I think what I have written holds up well enough, and is still amply adequate to its purpose. Otherwise, for this edition, I have, as best as able, tried to correct or make what I had written earlier more clear, developed previous material, while in other instances added a few new remarks and observations.

The question well might be asked: does this topic insist on a dualistic, that is (great) Good versus (great) Evil, approach to understanding the universe? Must we or is it advisable to see life as primarily a battleground between Good and Evil? Although I address this question more closely in my *Christ and Truth*, the simple answer here is no. We need not nor is it advisable to see the world as one divided, simply, between Good and Evil. Now it is true, under certain circumstances our more immediate circumstances may be fairly and usefully characterized as conflicting between Good and Evil. But though this might make sense for specific circumstances and occasions, we have no needs or compelling grounds to take for granted that the universe and all of life itself is centered and focused around the conflict of Good versus Evil. In other words, (and based on the view I take), the world or universe may be said to contain evil, indeed very powerful evil. But this is of itself no reason to assume that the world is somehow evenly divided as a theatre of war between good and evil, or that the main purpose of our lives is to face and combat evil (or were you disposed otherwise, to battle good.) Great evil, as opposed to small or natural evil, we may learn one day is only accidental to the order of things, and not at all, or never was, necessary as such. While then a great battle between Good and Evil may be said to take place in one part of the cosmos, it is not strictly necessary to assume that that the particular conflict in question has great or any implications for any other part, let alone the rest of the entire universe. On the other hand, it may well have such implication, so that we should not be blind to that possibility also. Yet if we see the conflict between great Good and Evil extending beyond immediate circumstances, we should judge from and based on a rational assessment of specific facts and circumstances, and not overly rely on abstract and speculative generalities

~~~~~******~~~~~

Introduction

"In the senses is deception [in the understanding is the source of truth.]"
~~~~~~~Xenophanes of Colophon.

For many the idea of Hell, devils, spirit people, and ("working") witchcraft and sorcery exist exclusively in the realm of the imaginative, and are either the result of psychological abnormality or else instances of misinterpreted phenomena. At the same time, there are others who think there is truth to these things, and that they are more than just people's imaginings or conjectural surmising. Even so, such persons are usually at a loss to know exactly what such things are about. For the greater part of my own life, I would consider myself to have been in the latter category, while taking great care to be skeptical and guard against superstition and hoaxes. Historically, religiously, culturally and behaviorally there seemed to have been persons and events which bespoke something beyond more ordinarily accepted interpretations of behavior and phenomena. Surely, there has at least been enough evidence over the course of centuries, found and arising from all over the globe, to warrant such suspicions and our being open minded about the existence of "sprit persons" and their possible presence in our midst.

Some might think me a crank or fraud, or someone penning a satire – which, simply given the topic, is understandable. Yet while I have been accused of being crazy, no one as yet to call me liar. I do not expect anyone at my mere saying so, to believe what I claim here is true. My purpose has been primarily to offer an introduction to the subject and also some help and explanation to people for whom this kind of problem might come to them in a drastic way. For the rest, people are free to make of what I write here as they please (as they most assuredly will do in any case.) It is not at all impossible that some who brush these things aside now, will find later what I have to say of value and of use to them.

Without wanting to make silly drama over the matter, over the past decade, I have experienced such persons, events and phenomena that make it undeniable that Hell and spirit persons there from are something palpable, and empirically real. In addressing the topic of "Hell" (and spirit people representative of) I am not (usually) speaking about Hell allegorically or metaphorically, but literally. Leaving aside the question of angels, *spirit people* (normally speaking) are not transcendental beings, demonic energies, abstract forces, but simply "souls," personalities and intelligences with spirit bodies of otherwise dead and departed human beings who, depending on their circumstance, have become, manifest themselves as, or live their lives as persons with rarefied bodies (and of a substance as yet unclear.) As I use the term, spirit person ordinarily, or unless and when I introduce the possibility of other kinds of sentient human-like personages, it refers to people who were born, lived their life, died, and passed on into a future life in the form of what, for practical purposes, we can call a "spirit person."

The quantity of things I have seen and felt in the course my personal and immediate dealing with these people are more than staggering, and the story behind all these events is more than what I am prepared to write at this time. What basically I would like to do instead is try to explain, as best I am in a position to know, what Hell is, how it works, and what steps one can take to protect one's self against its attacks. In raising these issues and formulating the problem here, I prefer to think of myself primarily as a medical researcher examining a horrible disease, with a view toward understanding, and considering what measures might be taken to combat and counteract it. This is certainly not a subject I would choose to write about for interest, curiosity, or amusement's sake. Please believe me, writing this was something I'd rather not have had to do. It is actually very painful to go over many of these experiences. Hell, after all, is "Hell." This work therefore could not, would not, have been written but that its extremely serious and urgent nature demanded it.

In addressing this topic, I don't by any means pretend to try to exhaust the subject, or know everything about it. What I relate here is based essentially on my own particular experience, analysis and study, and doesn't begin to pretend to cover all possible experiences or explanations. The problems with Hell go back as long as the history of humankind. It permeates all aspects of life, and each one of these aspects (that is with respect to Hell's influence on it) is worth a complete study all its own. Some points of morals and religion, with respect to the matter of Hell, will be a matter of honest disputes between some, and it would be impossible for me to go into all of those. What I have endeavored to do, however, is present something like a practical overview of the essential problem with some suggestions as to possible solutions. I am fully aware that there are some points people will have a different interpretation or remedy. But in the case of error on my part, or failure to account for an alternative explanation or interpretation, it will be understood, I hope, that if what I have offered is not strictly speaking true, or even not true at all, it will at least, for the time being, have served as one working hypothesis or warranted expedient for how to view a particular topic raised. Similarly, my pronounced subjectivity, expression of opinion, and perhaps too free generalizing on some points, I hope also will be indulged on the grounds of their being simply and easily identified for what they are. If some of what I discuss sounds very strange and bizarre, do know that many of these things are *very* strange and bizarre; indeed, are intended to be so by criminal spirit people. Realizing this, one will help one better understand why for ages these kinds of things have frequently been unknown, unacknowledged, unbelieved, or dismissed as unreal.

Perhaps, with the passing of time what is put down here will read like something written about a newly discovered land written in the 16th or 17th century, i.e., that is possibly mistaken in detail but which presents (in most if not all instances) essentially correct ideas. God willing, it will, *if nothing else*, encourage and facilitate a more rational and scientific examination of the subject from both the religious and scientific viewpoints.

~ * ~

If literal Hell and criminal spirit people in our midst is actual fact, the question will arise, "why aren't these things already known about scientifically then?" The truth of the matter is, many of these things are already very well known to some, and have been for ages, but where these things are known they are most usually kept secret or denied. For starters, the nature of the subject is *very* distressing and hard to deal with. Encounters with literal Hell can be unspeakably sad, agonizing, maddening, and not surprisingly those who have directly dealt with Hell are not interested in reminiscing about it. They'd rather forget; while at the same time would be at a loss (in any case) to exactly explain the what and whys of what they went through without sounding as though they had some sort of mental problem, or at least much misunderstood. The difficulty is somewhat like describing how magic tricks are done, without being a magician oneself, and this sometimes to people who do not believe there is such a thing as magic to begin with.

All the while, people of or from Hell will routinely threaten "whistle-blowers" that if they talk about what they know, or have seen and heard that "they will have something done to them," and this with the understanding that there is predictably little or no limit to the cruelty the whistle-blower could expect.

Problems stemming from Hell have been with us so long, that for many they aren't even really a problem, but rather a simple matter of "that's the way life is." Such persons, indeed would feel uncomfortable if these problems weren't there – as if there would be something wrong if there was not something wrong. It's like someone who lives with a nuisance for years, such as the loud music of his next door neighbor. After the nuisance ceased, they might feel disconcerted or strange to not hear that noise they were forced to accustom themselves to. Nevertheless, after a longer interval of peace and quiet had transpired, they would come to wonder how they ever got through listening to all that racket in the first place. Others accept Hell because they see it as authority; whether as authority it puts on a malevolent or pretends with a benign form. For yet others, Hell is not merely incidental misfortune and adversity in the course of life, but a terribly painful and tragic predicament or captivity.

One of the fundamental problems when in dealing with Hell is its powers of deception, and not everyone is on the same moral and spiritual, let alone intellectual level, to withstand such. By the same token, the truth can be known only by those disposed to hear it in the first place; regardless of how much facts and arguments one possesses to make one's case. To add to all this, there are some people who do little to resist deception (of which there are not a few); so that attempting to discuss Hell with them is utterly impossible.

Among the obstacles in dealing with this subject is that we tend to think of what is real in terms of the three-dimensional physics world, when in point of fact the existence of spirit people suggest a new dimension (or dimensions) to physics which remains largely unknown, unconsidered, and unexplored.

Another difficulty is that we do not ordinarily think of their being a connection between morals and physics. Yet in light of "spirit person" phenomena, it would seem necessary to ground our understanding of the physics of spiritual phenomena in terms consistent with a healthy and due respect for morals. Spirit persons (it would seem) often are where and what they are for moral or immoral reasons. Furthermore, they are not passive, "unconscious" phenomena such as we commonly view things which science might study, but moral beings, whose physical condition and reason for revealing themselves to someone is to a not insignificant degree a result of their moral character. To illustrate, it would ordinarily seem to be the case that it is going against the "order of things" for spirit people to be making an appearance among "regular" (i.e., flesh-and-blood living) people, and such spirit people who do make appearances, in most cases (though not necessarily all), are law breakers of a kind; such that a given spirit person one might see or encounter are by this definition persons who are disposed (to and in some degree) to act in a criminal manner.[44] In this sense, who one is seeing is the exception rather than the rule; that is to say, most spirit people we will not be able to see, because most are not inclined to break the rules. Now I will not say this is true in all cases of spirits who might appear to us. Some might appear to us without having to break any rules. But these kinds of spirits, if allowed, would be extremely rare. While a scientist studying something will usually want the phenomena to reappear so that it can be studied, for them to want a spirit person to appear is for them to want someone else to commit a crime. Hence there is a possible moral dilemma both for the spirit person who might appear, and the scientist who might in some way study or examine them which can potentially have very dangerous and harmful physiological, psychological and social repercussions to those that might (directly or indirectly) be affected.

While anthropology and psychology allow for the study of people, one should always be aware that even with such conventional sciences that people are souls, not "things," and one is getting off to a false start in viewing and treating them as such. There is then this decidedly moral dimension to anthropology and psychology that perhaps becomes even more pronounced and germane when addressing the subject of spirit people. Such study, if possible at all, requires a radically new kind of scientific approach and outlook grounded in morals. This is not to somehow suggest that morals are not important in other areas of sciences – indeed, morals are very important in science, if for no other reason than the need for honesty. Yet in the case of the study of spirit phenomena, attention to morals becomes even more necessary, both for the purpose of respecting the personhood

---

[44] This might seem like too hasty a conclusion, and I grant that if overall circumstances were different, normal and open interaction with spirit people, might not only be possible but a desirable and good thing. But as it is and has been most of the time and so far as we know, most overt and regular manifestations of spirit people are brought about by certain persons with much power and control, and who desire to manipulate "regular" people, of which more as we proceed. Yet this said, we should note that the reason there might be more good (or bad) spirits around in a given environment (say a city), is for the same reasons there might be good (or bad) "regular" people, which is to sat for moral and cultural reasons. So granted one can't assume circumstances will be all bad or good necessarily such that spirit people can or should be characterized this way. Some people do this, claiming the *entire world*, is half good and half evil for example, which is either obviously false or else scientifically unprovable. Yet even so, and given things how they are and have been as best we know, it is not unwarranted as a practical matter to treat most open manifestations by spirit persons as being the result of malfeasance. I, at any rate, am inclined to take this view as being certainly the far more safe and prudent one.

of who it is you are studying, and as kind of hygiene to protect the scientist from the kind of attacks certain spirit people might make on them. In this way, morals are as important for a scientist studying this subject, as hygiene is for someone studying or engaged in the practice of medicine.

This said, let's not over look the obvious that these matters are not something to be treated merely as a curiosity, but relate to the essence of life and death itself, and that not un-typically regular or spirit people of Hell, who one might be dealing with, are people who are often in some way or other very ill, in great agony, or living amid incomprehensible squalor, and personal tragedy, and not un-rarely have deep rooted mental problems.

More commonly speaking, philosophy is usually ignored or dismissed as an irrelevant and esoteric field of endeavor. Yet in dealing with the machinations of spirit people, applying both idealism (going back to Plato) and more modern scientific philosophy (going back to Francis Bacon), has, in my own experience, proved life-saving and invaluable. A person interested in the subject of spirit person phenomena is therefore most wise to avail themselves of the powers of sound, rational and ethical philosophy as a means of discerning true from false with respect to both fact and morals.

Though it has been a topic with mankind, throughout the world, since time immemorial, serious discussion of spirit persons from a scientific point of view is typically, if not entirely forbidden at the outset, frowned upon.

Arguments like "*We* know...," "Everyone says....," complete silence, or resorting to ridicule are often as much of any kind of argument one will secure from those who categorically deny the reality and possibility of literal spirit persons. While I certainly don't expect that one should out of hand assume the claim that there are spirit persons to be valid, arguments such arguments as those just mentioned can hardly be considered serious, fair, scientific, or objective ones. Indeed, in many instances, and I believe if we are honest with ourselves about it, such attitudes stem more from immaturity and deep seated fear than unbiased, rational reflection.

Among the reasons there has been an understandable resistance to the study of spirit people and spirit phenomena are traditional prejudice, the highly controversial nature of the subject, fear of ridicule, theological confusion and misunderstanding, sabotage of such study by the spirits and or people who cooperate with them, or, if these other reasons don't dissuade, simple fear of Hell spirits themselves or the confusion of them with the benign spirits of spoken of in conventional religion. The Spiritualism of the late 19th century, for instance, consisting as it did of hoaxes, superstition, mixed with what was probably on occasion sound fact and proof, made for such an evidentiary hodge podge that not surprisingly it ended up being discredited altogether.

One sees a related problem arise in the writings of Increase Mather, Cotton Mather, Joseph Glanville, and Emanuel Swedenborg. In the case of the Mathers, the New England Puritans, the approach is not terribly scientific, and is based largely on contemporary hearsay, and scriptural and classical references. To use such as *empirical* evidence, is, not surprisingly a highly dubious proposition. Yet while tradition and theology can only rarely serve as scientific explanation, they do help to provide some kind of initial context and tentative accounting for things we otherwise don't know. Though one need be very careful with the Mathers, it would be equally wrong to dismiss all they have to say in their books as erroneous, inaccurate or without value. For example in *Cases of Conscience Concerning Evil Spirits*, Increase Mather makes note of the fact that "the Devil" can be made to appear as a saint or even Christ himself, which, based on my own personal experience I know to be, for me, an incontrovertible fact.

Swedenborg, the 18th century Swedish scientist and theologian, is more careful. Yet the theological and moral implications of his subject, understandably are given greater weight than the scientific aspects; nor was he always very careful at questioning the nature of who and what he saw, heard and experienced. Yet unlike most others, Swedenborg speaks from first-hand encounters with spirit people. He has some very good things to say on the subject, especially with respect to morals and theology. I would point out, even so, that some of his views, in particular on the ways of people of Heaven, may have, very likely, been deceptions imposed on him by Hell. While the demons he speaks of conversing with are sometimes portrayed as dunderheads when it comes to rational thought (which they often are), he does not (in my opinion) seem to have made due allowance for the often extraordinary power of "Satan's" cunning, including his ability to masquerade as God and His angels.

Perhaps the main obstacles to dealing with Hell scientifically, are both the degree of failure of many religious people to be rational and honestly scientific, and the failure of many scientific people to be courageously honest; while at the same time properly appreciative of what are seen as traditionally religious matters.

~ * ~

What is Hell?

"Ishtar says to her father, Anu: 'If you refuse to make me the Bull of Heaven I will break in the door of hell and smash the bolts. I will let the doors of hell stand wide open and bring up the dead to eat the food with the living; and the hosts of the dead outnumber the living.'"
~~~~~~*Gilgamesh Epic*[45]

---

[45] As translated by N. K. Sandars

For our purposes, Hell can be spoken of in at least three senses (in no special order):

1. As a "geographical" or otherwise physical location
2. The policy, methods, and way of life of "Satan" (for which you may substitute "an autocratic spirit person hoodlum," of extraordinary knowledge, shrewdness and savvy, if you like) and his followers
3. An emphatically immoral and psychologically stunted disposition

Insofar as these definitions are ultimately linked, and in some instances inseparable, it seemed justified for convenience sake, if nothing else, to vicariously use the term as suits us in one of these three senses, or in a sense which combines one, two, or all three definitions. A phrase, for example, like "you can take the ghost out of Hell, but you can't take the Hell out of the ghost," is an example where two quite different senses are used for the same term, and in a manner in which a reader will have little difficulty in getting at the meaning intended.

This said, it is well to narrow down some of these characterizations and definitions.

In many, if not most primitive and ancient cultures from around the world, it was believed that there is an "underworld" which the dead went to, and which, under certain circumstances, they could possibly come back out of. Indeed the idea of an underworld (or something very similar such as the Hebrew *Sheol*) as being the abode for the departed was apparently far more common than that of "Heaven" being such a place. Even in Egyptian theology and cosmology which did recognize a Heaven for the dead, Heaven was for a relative few, and was itself usually spoken of as being located in the underworld. Babylonian, ancient Persian, Hindu[46] and Buddhist thought make reference to a place of darkness and or torment below us inhabited by demons, shades and asuras, The Chinese picked up the idea of Hell from the Buddhists and developed somewhat elaborate ideas on the subject. Interestingly, prior to that they had a more or less common understanding and acceptance of ghosts and spirits, but not "Hell"[47] as such, the world being more peaceably divided between earthly and heavenly beings (though Heaven might be spoken of as both a person and the sky above.).

Mozi (or Mo Tzu), one of China's earliest and most influential thinkers presented a very empirical, if not all that terribly persuasive, argument for the existence of spirits. What is interesting is that such writings of his show that there was doubt on the subject even in his long ago time (the 5th century B.C.); so

---

[46] In the *Bhagavad-Gita* we find: "These [i.e., deceitfulness, arrogance, false pride, anger, harsh speech, and ignorance] be the signs, My Prince! Of him whose birth Is fated for regions of the vile."

[47] Which they came to call *Ti Yü* or "earth prison."

skepticism about these things is not anything new. Some of his arguments are consequential, i.e., it will benefit people if they believe there are ghosts and spirits. Since ghosts and spirits (whether good or bad) witness and see what you do, you have that much less reason to act improperly. Similar ideas and arguments (though – as far as we know – not so specifically addressed) were found among the Greeks. Now if someone argues they are using these arguments to fool people into being moral, then we would have to conclude that they were duplicitous in doing so – an arguably unjust charge to make of someone like Mozi .

In my own dealing with spirit people, the phrase "down below" is very frequently used to designate where criminal or overtly witchcraft connected spirit people (as well as others not so malevolent) come from or "live." Whether this is meant down below relative to us physically, dimensionally, metaphorically, spiritually (or some combination these) is not entirely clear. Some have expressed the view of "down below" being literally beneath the earth's surface. "The earth hath bubbles and these are of them." Though it is uncertain exactly what "down below" means, the most likely correct interpretation is that it exists in what we might call another dimension, rather than down below the earth's surface in this (the third) dimension – though this speculated dimension and our third dimension may in some way be parallel (if, that is, they are not otherwise and after all linked or the same thing.) According to Dante, Hell, or in this case "down below," has a bottom, and on one occasion a spirit person communicated to me the idea that there is a kind of bottom to down below, a desert of some sort, that was alternatively insufferably cold or excruciatingly hot, like a sort of combined Equator and Antarctic. Whether "down below" is restricted to the earth, or whether there are other spirit realms in other solar systems is not known. It can, however, be noted that Swedenborg held the belief that there are a plurality of worlds, aside from our own, which are inhabited by people. If this is so, it naturally raises the question of whether these alleged separate worlds have their own distinct spirit realms, share one in combination by all, or some combination. On this point, Swedenborg, is not to my knowledge entirely clear; though, on the other hand, this may simply be a result of my ignorance of his full body of work rather than any omission on his part.

"Down below," however, is not itself (at least according to some people's views) necessarily "Hell." Hell for some of the people who come from "down below" is something like "down below-down below." So that apparently what actually constitutes Hell depends, to some degree, on how "high up" one is looking from. *Geographical* Hell, as such then, is a place where the worser people of humanity have ended up after this life, yet with the understanding there may perhaps be those who are of lesser guilt in its immediate (if not long term) grips. The people who make up this society are basically the rottenest criminals of various kinds, while keeping in mind points of comparison and degree. As a *general* rule, the worse they were in life, the lower down they go. There are certain means, however, to summon them back up, and this, in effect, is what some kinds of sorcery and witchcraft do. There is reportedly purgatory (or something like it), and it is said that there have been people from down below who made it to purgatory; though this, as seen on the surface, is not all that common. Yet perhaps what actually constitutes Hell down below as such may simply be a matter (as it is "up here") of what kind of people a given person might be with. There is a sense that people down below can live tolerably well; at least some spirit persons don't seem to ordinarily be in such great distress, but for the trouble caused by the Hell people. Hell then, after all, and speaking spatially, can easily be above as well as below. Indeed, on a given occasion there may easily be more Hell going on among ourselves than "down below."[48] It is also well worthwhile to consider that Hell's physical location (and as the worst of all places may have changed, as well there are or may have been more than one such hub or center. As t stands, we just don't know.

One might also think of Hell as a place where people fell; so weighted down by their sins and wrong doing, and thus created an insufferable world by their being together. Oftentimes the way we act toward getting what we want, if immoral, actually frustrates all the more our obtaining what we so desire. If we don't aspire after something the Right and moral way, usually what we want will only become all the more farther off. Quite simply, Hell exists, and its distance from happiness and well-being, partly because of the failure of people to live according to this understanding. Yet bear in mind, how low one goes is not always something that is determined by physical direction, but is also a determination of one's mind, spirit, heart and character. Of interest in this regard, the name "Gehenna," synonymous with Hell, and used by Christ, refers to the valley of Hinnom valley (also at an earlier time called Topeth) just outside Jerusalem (of all places.) It was here children were sacrificed to Baal. The connection is a very appropriate one because making sacrifices of others to appease one's own greed or self-interest is a very Hell-like characteristic.

Hell as the *policy*, methods, and way of life of the Evil One, Satan, Satanism, or (alternatively) Orkonism, I will address as we get on.

In addition to Hell as a "geographical" location and a manifestation of the "Evil One" and his policies, Hell can also be used to denote an emphatically immoral and psychologically stunted *disposition*, or a place where such dispositions pre-dominate among a group of people. But this kind of Hell is really only a by product of and sub-definition relating to the "Evil One" and his policies,[49] with the difference that its speaks of Hell as something immediately and practically real, as opposed to Hell in its more far reaching context.

---

[48] Very possibly, those who we call "Satan" and his fallen angels may, and or pagan "gods," in some manner, live in the air or sky, rather than down-below. Though explaining how such is possible is not easy, there are good reasons even so for believing this to be the case, given their oversight of the world at large, as well as instances where their "Powers of the Air" are displayed.
[49] Or "Evil principle" (rather than person) if you prefer.

One other way to view Hell is as the sin enslaved, sickly past of collective humanity, and which continues to haunt and violently attack the human race. It is like our collective shameful yesterdays that cannot be so easily got rid of, with the Hell spirit people being the very physical manifestation of earlier troubled times and lives.

Hell is what it is, I would maintain, not because God created it, but because, some "Evil One" (or "Ones") and our acquiescing to and going along with him have created it. Hell was in fact something people created; perhaps due to the conniving manipulation of certain "aristocratic" and powerful spirit persons or "fallen angels"

~ * ~

"The Evil One"

"My mind was once the true survey
Of all these meadows fresh and gay,
And in the greenness of the grass
Did see its hopes as in a glass;
When Juliana came, and she
What I do to the grass, does to my thoughts and me.

But these, while I with sorrow pine,
Grew more luxuriant still and fine,
That not one blade of grass you spied
But had a flower on either side;
When Juliana came, and she
What I do to the grass, does to my thoughts and me.

Unthankful meadows, could you so
A fellowship so true forgo,
And in your gaudy May-games meet,
While I lay trodden under feet...

But what you in compassion ought,
Shall now by my revenge be wrought:
And flowers, I and grass and all,
Will in one common ruin fall...
~~~~~~~Andrew Marvell, from "The Mower's Song"

"Their processions and their phallic hymns would be disgraceful exhibitions were it not that they are done in honor of Dionysus. But Dionysus, in whose honor they rave and hold revels, is the same as Hades."
~~~~~~~Heraclitus of Ephesus, *fragment*.

If one can speak of evil or unjustified violent antipathy (towards say mankind) as a major force in the universe, in who or what would it consist? Either we could speak of it as say a malignant substance, or else we could speak of a person who intends unjustified or cruel harm to others. Now if the universe were just created 5 minutes ago, and neither a person or substance of this sort were as yet present, could or should we expect either one to appear? Off hand, I don't think anyone could say. But given how things have been through most of known history, there seems good reason to believe that sooner or later there would be; though granted the why-such leaves much to interpretation[50]

Destruction (such as when we eat, or when skin is shed, etc., etc.) one can fairly assume is a natural activity. Yet it may be that someone (or perhaps a group of someones) a very long time ago got the idea that by tapping into this principle he could appropriate the mantle of great destroyer himself. Few or others would never have dreamt of the idea, and he scared many of them to death with this, and thus founded an empire. This at least will serve as one plausible and preliminary hypothesis.

Now the next question is, if evil was is or has its origins in a person, is this one or more persons? And Is there something about the full vocation of evil which lends itself to being one, rather than more than one, person? If so this would explain our "Satan." But if not, should we posit multiple "Satans?" And whether we speak of one or more, do they remain always in that role, or do others possibly take their place? If there are shifts in power among such people it might easily go on unheard of or behind closed doors in such a way that even people who work directly for the "great evil power" might not even know that such transition had taken place, but went on believing the basic structure of power and authority otherwise left intact. Such possibilities are, at any rate, worth posing when attempting to form a picture of Hell's hierarchy, and also how such is viewed by its servants and slaves.

In more developed and traditional Hinduism, Shiva is the power of destruction opposite to Vishnu the power of life. Yet destruction as we noted is obviously to some extent natural, and Shiva is sometimes seen as life affirmative in his destruction (as when trees lose their leaves in autumn.) If we posit say a Satan, this would not then necessarily imply either that Shiva was Satan or Satan was Shiva. It is perhaps in this gray area between natural destruction and unnatural destruction that we might say Satan or a Satan is able to conceal and legitimize himself. The ancient Hindu's had a Satan whom they called "Mara."

---

[50] So far and in what follows Satan, Hell, orkonist have been and will be used fairly loosely and interchangeably. There is, in part, an intended purpose in this, inasmuch as the "Evil One" is not always so easy to pin down, either with respect to immediate evil going on or evil going on in a long term sense. When one of the high priests says to the blind-folded Jesus:" prophesy who struck you" both the blindfold, the blow and the taunt are in the true spirit and practice of this or this sort of character.

The *Dhammapada* speaks of him as the "devil of confusion," and it was he who tried to sorely tempt the Buddha. Of course, he is also known as "the Devil" or else "the Devil" is used to refer to someone very like Satan.

All this raised, it is necessary then to consider who and what Satan is or possibly is,[51] because for we otherwise have very good historical and traditional reason to believe he is the founder of Hell, or ought be thought of as the founder of Hell (as such), and using the name as a practical label. There is significance in the fact that the terms Satan and Hell can be used synonymously. Now could Hell be posited without Satan? Perhaps, but he reflects the complete arrogance and selfishness which are ostensibly at the root of all Hell behavior. At minimum, and even if only a hypothesis or conjecture, he serves as a starting point for conceptualizing such unfathomable arrogance and selfishness.

I was told by a spirit person (for what it is worth, granted and perhaps not much) that he is actually different than the serpent,[52] who, by comparison, is some kind of primordial personage. In a vision shown to me by a spirit person, the serpent looked like something out of William Blake. He is an actual serpent, but with a more knowing and laconic eye to his appearance than an ordinary serpent. Also he more like a lumpy black eel than a snake, but an eel with a more rounded head and without any appendages to his body. He appeared very large, but whether he actually is so I could not say. While his look isn't one of hostility, it is rather wry and sarcastic, as if to say, "you don't know how it really is." At the same time, he is someone in no hurry and respects your right to believe what you want to believe. When Adam and Eve fell, people sometimes forget or overlook the fact that the serpent fell also. Whether he is now active doing anything as such or just keeps to himself, again, I don't know. There must have been some beauty in the serpent, but beauty that was a deception; beauty that was without compassion, beauty that brought with it pain, death, and decomposition.

Again, Satan evidently or else possibly is someone else. It was also related to me that he has been around for about 11,000 years. He has been spoken of by some Church Fathers as having an "aerial" or else some sort of spirit body, but

---

[51] Maimonides: "They said in the Talmud as follows: R. Simeon, son of Lakish, says: "The adversary (satan), evil inclination (yezer ha-ra'), and the angel of death are one in the same being...The Hebrew, satan, is derived from the same root as seteh, 'turn away' (Proverbs. IV. 15); it implies the notion of turning away and moving away from a thing; he undoubtedly turns us away from the way of truth, and leads us astray in the way of error." *Guide for the Perplexed*, III. 22.

[52] The Book of Revelation, 20:2, speaks of Satan as the ancient serpent. But if they are the same this would leave us to determine how the "most cunning of the animals" is at the same time a fallen angel (or else regular person.) Maimonides citing one of the Jewish sages, says Satan (Samaël) rode the serpent (in the Garden of Eden). *Guide*, II.33. Now if Satan is posited as two persons, then perhaps we could speak of a himself *and* the serpent together as "Satan," it being not uncommon among Orkonists to take on multiple or dual personalities, mostly, it would seem, as a stratagem to better deceive people or otherwise make hiding their guilt that much easier.

that he is subject to pride and envy.[53] Exactly who he is and what his relationship to the serpent is, one can only speculate. Religious tradition has him to be a fallen angel. On the other hand, it is not too farfetched to consider the possibility of him as is just a regular human person who (for whatever exact reason) ended up "down below" and founded an empire there. This said, the fallen angel view is perhaps more plausible inasmuch as certain of these spirit people can make it seem as if they really came from Heaven while making reference to things without ostensible parallel in mundane experience. In the parable about the seed dropped on the ground, Christ speaks of the "birds of the air;" who snatch up those who are not planted in the Word. These "birds" would seem to be a reference to fallen angels. On the other hand, in the oft quoted passage in Isaiah 14:12-15, it is not clear if the person spoken of is literally Satan or merely one of his people, nor is it entirely obvious whether a fallen angel or else merely a fallen (human) person is meant. Note in this passage, also from Isaiah (in this case 24:21):

> "On that day the Lord will punish
> the host of heaven,[54] in heaven,
> and the kings of the earth, on the earth"

that there are bad people in "heaven," or what, no doubt, is more specifically meant, the sky.

Naturally on the subject of Satan there is and will bound to be some disagreement. For some Satan is an actual being. For others Satan represents a way of thinking, namely overweening pride, impiety, rebellion against legitimate and just authority, extreme materialistic greed, arrogance and selfishness.

It may be more prudent and advisable to use the term, "Satan, inc.," "Evil, inc.," an "Orkonist," or some such, rather than "Satan" to avoid possible confusions, and because "Satan," like the word "demon" is so emotionally charged. Yet this acknowledged, here we indulge its use.

Some might argue that Satan is the name for the person sitting on the throne of Hell at a given time, and not necessarily one specifically named person as such. Another idea would be to say there are multiple ruling Satans. In any case, for at least convenience sake, I will normally be referring hereafter to Satan as an actual person, with the understanding that who are what he/it is might be construed differently; including the view which sees "him" as a certain kind of state of mind, character, or disposition rather than a literal person. At the same time, since so much of what we know of "Satan" is so vague and questionable, we might instead sometimes use casually made up names to denote the personification, or personifications, of evil. For instance, in my dealing with

---

[53] Yet if this is true, what are pride and envy that they should subject Satan? The answer apparently has something to do with Satan's being born good, and then forsaking his goodness.
[54] See also Acts 7:42.

devils I might refer to their (to me) unknown leader as "Zombo," "Monster Maker," "the Green Goblin," and the group to which they belong "the secret order of Zombo," both to ridicule and demystify them, while by doing so also admit that our knowledge on some specific points is obviously sketchy..

Satan is, says the Bible, the most powerful person in and ruler of "the world," hence he acts like, and many, including some well-meaning but deceived people, actually believe that he is God. And while Satan knows he is not God, he would like to be, and if he can't actually be God, he will do the next best thing – attempt to impersonate him, and get people to believe him to be such. And in this he to some extent tragically succeeds. As opposed to God's, Satan's is a possessing and consuming power, not a creative and generative one, but is more based in the power of destruction. Now as we said destruction is natural. But if we assume a Satan it is understood that the destruction and harm he brings about is unjust, unmoral, and therefore unnecessary. If that is not the case, we would have to dismiss the notion or hypothesis of Satan at the outset.

In the Bible,[55] Satan is described in these various ways, all of which I think are very helpful in understanding him:

> The prince of this world: John 12:31
> A liar, a murderer, the father of lies: John 8:44
> Snatches the Word from hearers: Mathew 13:19, Mark 4:15
> Perverts the Scripture: Matthew 4:4, Luke 4:10
> Fashions himself into an angel of light: 2 Corinthians 4:4
> Blinds the mind of unbelievers: 2 Corinthians 4:4
> Can produce false miracles: 2 Thessalonians., 2:9
> Prince of the power of the air: Ephesians, 2:2
> Gentiles are under his power: Acts 26:18
> Deceiver of the whole world: Rev. 12:9
> God of this world: 2 Corinthians 4:4

It is Satan's purpose, in effect, to set up his own universe while using God's. That God allows such a thing shows the utmost importance he places on individuality and free choice. At the same time, that Satan is as powerful as he is shows his phenomenal cunning and intelligence. Although not actually mentioned in the Cain and Abel story, it is reasonable to infer his presence in it, even though technically he may have found to a way to absolve himself of what happened. That he could have played such a major role in the events of that story, yet found a lawyerly way of removing himself from technical guilt, might be thought of as typical Satan. While he apparently doesn't necessarily need to actually "do" evil himself, he can and will, nonetheless, usually get others to do it for him. Evil, like

---

[55] As given in Henry B. Halley's *Bible Handbook*, pp. 497-498. Halley's reading of the 2nd Thessalonians passage mentioned here, however, might be seen by some as a rather free interpretation.

any extensive and involved enterprise, ultimately requires team work and careful coordination between people; so one can assume Satan is a consummate and methodical organizer.

With Satan and Orkonist people typically things are seen and done in the reverse, and whatever God or goodness is seen as doing or having, Satan almost invariably has his own version. To match innocence, for example, Satan has (or invokes) childishness, for sympathy there is selfishness; for sorrow self-pity; for humility arrogance; for love hate; for admiration envy. etc. While in Heaven, the meek inherit the earth, in Hell it is the most low down criminal and reprobate who is empowered and rewarded. They have their own brands of religion, philosophy and science; which are the reverse of true religion, true philosophy, and true science; his own interpretation of everyone and everything *as seen through himself*. If Heaven has a cathedral, Hell will have one just as big. If Heaven has a Christian congregation, well Hell can have one. If Heaven has a saint, you can be sure Hell will have one of their own, or perhaps might have one of their own, in a given instance, accepted (by some deluded people) as a Christian saint. In this way, it is fairly routine of them try to pass off their version of someone or something as Heaven's; in order to dupe or take someone in, whether or not the person or thing in question openly admits to evil or whether they pretend to good.

What we typically see Satan, a Satan or a high-powered Orkonist doing then is appropriate to himself God's good things and pervert them to serve what he sees as his own interest. Selfishness and pride, for example, have their proper and natural place in life, but, of course, with Satan such things are taken to a point where they are unnatural and unhealthy. The hard core Orkonists takes things to a religious extreme; while the more mild one or Goomerists are prompted to casually believe they can have it God's way *and* Evil's way simultaneously. By patiently encouraging the latter belief over time a Satan or Orkonist can often times draw a fence-sitter, such as a "Goomerist," over to the "dark side" without the latter quite realizing how they have exposed themselves by carelessly or ignorantly letting their guard down.

One version of "Satan" which we might consider is to see him as the consummate miser and tyrant. Deception, fear, violence, and an appeal to the body over mind and heart are the keys to his power. Like the incorrigible miser and tyrant, he is envious, wants love. Yet due to his bottomless selfishness he is incapable of real love. Instead he can only do things to hurt people, and by this means get their attention, awe, and respect – the closest thing to love for someone like himself. He tries to make himself look interesting through astounding sorcery shows, accompanied by displays of tremendous worldly and spiritual power; including the capacity to make himself or those working for him see divine. He reportedly, and there seems good reason to believe he (or again they) can cause earthquakes and change the weather. He also (and certain individuals working for him) can read people's thoughts as easily as you are I might take in a quick newspaper article. These feats are accomplished by means that amount to nothing more than the orchestration and use of numerous slaves and servants; along with powers which one might characterize as very sophisticated kinds of technologies at his disposal. One of his greatest areas of expertise is understanding human behavior, psychology, and physiology and how to take advantage of and manipulate them. Naturally it is well to consider other physical forces, such as various kinds of wave lengths of light and or other energy which presumably he or his people would be familiar with, and probably have no little capacity to manipulating.

People who think evil is somehow noble or attractive, as for example as in the romantic notion the Miltonic Satan, would be better advised in knowing that Hell is a self-absorbed, withered old miser, who doesn't seriously care about anyone else except as they serve his own personal use, least of all children and animals. Evil pretends to benevolence; loves prestige, power, material riches; feels sorry for himself and requires sacrifices of both the guilty and the innocent. His capacity to deceive, on the other hand, can often be awesomely powerful. A true Satanist, if he really believed what he seems to claim, and needed no one but himself, would simply mind is own business and not bother anyone. But there are apparently few or no true Satanists in that sense, and the vast majority of them are preposterous hypocrites vaunting their self-importance, while using and making beyond-ridiculous demands on others; all in a futile and ludicrous effort to remedy appalling evils and difficulties they suffer which (arguably) they recklessly created for themselves.

A "Satan's" character shows up in his those that serve him; so that by knowing know those people, I believe we can get a better idea of who he is and what he is like, or else at least form a useful and reasonably accurate character profile of a given Satanist or more precisely, Orkonist. Typically one will see patterns of repeated behavior among his people, for example, over-weening self-pity, a cynical, overly negative minded philosophy, utter selfishness, hypocrisy, cowardice and greed. In a review, Edgar Allen Poe makes humorous reference to "His Satanic Majesty." Which is often times how he is treated, so accustomed are many people given to his way of doing things. Similarly, William James once referred to the Prince of Darkness as a "gentleman." If we are judge by those who work for him, these kinds of characterizations are either highly misleading or flat out false; though granted for many Satan in costume can seem quite glamorous or even heavenly (though only in a superficial way really.) One is, in my opinion, better off seeing Satan as the eternal old-fogey, Pharisee, and kill-joy: sometimes rabid, pompous, bullying, and insincere. After deception, instilling fear is his greatest power. Some of the cold, brooding, and oppressive architecture and monuments of Imperial Rome, and certainly those of the Nazi regime, are or were very much Satanic in character, and quite deliberately designed with a mind to fostering ubiquitous fear and awe. At the same time, perhaps ironically, childishness, including an insatiable desire for attention, are not at all unusual among an Orkonist's traits.

As is often well known, but perhaps not as widely appreciated, is that he (and some of those who work for him) routinely "assumes a pleasing shape;" thus disguising his true nature, in order to entice more attractive and likable people into working for him in order, in turn, to fool yet others. At the same time consider: it stands as a possibility at a given time, and in theory at least, that someone may be worse than Satan; even if we assume he is worst most or the rest of the time. These are some of the logical yet peculiar sorts of questions that arise when examining this topic, and which one I think should keep on eye out for.

While a Satan's ways can be remarkably clever, intelligent, and quick to adapt to changing circumstances, he certainly is not without his blind spots, and will sometimes evince very human stupidity and shortsightedness. Behind great evil is always an insecure person attributing ludicrous importance to themselves and their own private interests. Not all who serve Satan (or a Satan) like him, indeed many of them hate him; only they are too much under his power not to do him service on occasion. He is both very proud and arrogant, and at other times feels sorry for himself, using whichever approach best suits his purpose. This same combination of proud arrogance and hypocritical self-pity has been frequently found among his agents and attendants.

He and his immediate subordinates (perhaps, and depending on the individual in question, fellow fallen angels if you like) are, to my understanding, like a spoiled, immature rich family that is alienated from everyone, and are proud yet at the same time defensive and sensitive about being rejected. Because of their arrogance and cruelty, they are despised by others. Though they believe themselves to be superior to others, many of them show their insecurity by their need to get attention, and show off. One sees this in much of the reported behavior of the pagan gods of classical antiquity.[56] Curiously Satan or a Satan person will act superior and mock someone perhaps using some extraordinary kind of trick, like a prognostication, following this with an "I told you so!" Yet in this and other ways this would-be superiority is given the lie to by their excessive need for attention (at least if you are dealing with one who is actively engaged in causing trouble.)

Gratitude, kindness, fairness, moral consistency are foreign to them, or employed in conjunction with others tricks, and if they are treated with sympathy or generosity they will just as soon take advantage of a would-be friend as an enemy. The reason for this is that they have this overweening need to assert their superiority over another; that oddly, or not so oddly, arises out of their insecurity about themselves. At times one can't help suspecting that, as horrible as this

---

[56] Deut. 32: 16-17, Psalm 96:5, 1st Corinth. 10:20-21 (and also Justin's 1st Apology, ch. 5) state or imply that the gods of the pagans are demons. Yet it would perhaps be more correct to say that "demons" or criminal spirit people impersonate or pretend to be such gods. At the same time, gods, such as in classical mythology, can also reasonably be said to have their root in poetic imagination, metaphor and personification, and therefore are not always or necessarily a bad thing when viewed merely as poetic or didactic devices.

sounds, Satan and his people are, in varying degrees and so to speak, suicide cases progressively destroying themselves while trying to get others to join them. This latter behavior perhaps arises out of an overwhelming sense of shame and utter disgracefulness they cannot completely deny or rid themselves of. By getting others to be like themselves, they feel less alone in their guilt. Indeed, it seems very likely that it is from this need that they make wrong doing a law and commandment among some people.

A more adept Satan or Orkonist rarely lies as such. Rather what he does is tell half or partial truths, and "make the worse argument appear the better;" makes the smaller truth seem the greater truth, and the greater truth the smaller truth. This is a far more shrewd and effective way of fooling people than simple lying. Not inappropriately, Socrates, in *The Statesman*, likens the Sophists to practitioner of magic, from which we can deduce that the converse is also probably true to some extent.

In a most critical and vital sense, the wars between (more) good and (more) evil are a contest to define reality. Anyone or anything can (to some degree or other certainly) be seen with the eyes of Heaven or the eyes of Hell. In a very real and meaningful way, how one looks is *how* and ultimately *what* one sees.

One regularly finds that those who are evilly disposed are prone to be more irrational, dishonest, immoral, then those disposed to be good. This certainly is one way of distinguishing the one sort from the other. Reality, according to the Orkonist is merely what they say it is. To a person of morals and reason, by contrast, God and or Right Reason will be seen as the ultimate determiner(s) of what is true and what false, and what is bad and what is good. Because God's way is rational and moral, the Orkonist ultimately rejects both morals and reason, but not, of course, without frequently being a hypocrite about it; availing himself of these as occasion suits. Yet rather than reason and morals, Satan will more often get people to look to public opinion, and to those who have the most material wealth, as the ultimate sources of authority. This is obviously desirable seeing that Satan is so talented at manipulating the one, while not un-typically exerting controlling influence over the other.

According to Genesis, there is ongoing war and enmity between the woman and the serpent. This is well to think on because we are otherwise at a lost to explain why there should be hostility leveled at the human race. Probably the most profitable way to see Satan's impact on mankind, both past and present, is in the story of Cain and Abel. "Satan" encourages someone to be jealous or resentful of another, and by this means incites them to murder. It is on the basis of this principle, which I call the Cain and Abel paradigm that practically all evil is ultimately brought about. The specific deception applied, persons involved, and circumstances can vary greatly. Nevertheless, the fundamental principle remains the same. The devils of Hell, for example, are lead on to attack their living human cousins on the basis of it. As well, regular people who wrongly attack someone else are goaded into doing so ultimately on the grounds or influence of the same Satanic or Orkonic impetus, i.e., "murder your brother if he is having it better than you, or else is in some way seen as a threat to your well being." This is why one is right to call a person with allegiance to Satan or Orkonism a coward and a traitor, since (aside from Satan himself and his reported fallen angels associates) such compulsive evil-doers are, as "Children of Adam," our brothers. Satan, by contrast, if interpreted literally and as a fallen angel, is not even human, and, in this sense, might justly be viewed by us as a kind of malevolent space-alien.

As a natural part of creating his own universe, a Satan or an Orkonist's main purpose is to achieve power and dominion over others. To accomplish this he most endeavors to take God's place in people's hearts and minds, and get them to see reality in his terms. Very simply, if he can display sufficient amounts of worldly and spirit-related power in a given person's direction, that person can be quite easily be fooled into thinking that he (Satan) is God or at least an all-powerful god.

In attacking people, it is imperative to understand that a high-powered Satan seeks not merely to destroy or degrade people's lives, but their very souls, since one's soul has a value beyond this life. If all Satan needed to do was to destroy people's lives that would probably be a relatively simple matter for him. But destroying their souls is a far trickier business, and to achieve this requires a great deal of planning and well-calculated strategies.

"Destroying" or ruining people's souls could be said to among other things involve: [57]

1. Making people more guilty, and more thoughtless
2 Drawing people away from truthfulness, into falsehood, deception, dishonesty, dissembling.
3. Making what is "good" look "bad" and making what is "bad" look "good."
4. Getting people to seek a lesser good at the expense of a greater good, such as in encouraging people to concern themselves with petty matters, while making a trifle of grave and serious ones. This might take the form of promoting something good in its way in order to lure people into their confidence, the way a con artist might do favors in order to deceive others. The particular "good" being promoted might be all right in and of itself, only here it is used as bait to disguise or make excuse for some intended or concealed wrong doing.
5. Promoting selfishness and envy.
6. Dividing people in order to better conquer them; setting individuals, groups, and nations against each other. By doing this disharmony and instability is brought about, thus making it easier for the Orkonist to come in and begin taking over. It is all the more desirable to do this to parties which are supposed to be friends or family, since such bonds are usually strongest. There is an old comedy

---

[57] Some of these to some extent overlap.

gag, in which person "A" hurts person "B," making "B" think that it is innocent "C" that hurt him. "B" will then wrongly blame "C," and "C" will be angry at "B" for wrongly accusing him. This device a regularly working devil will use quite regularly. This also includes getting people to wrongly blame and make scapegoats of others.

7. Getting people to be idle, waste their time; squander resources.

8. Destabilizing or destroying morals, religion, law, reason, legitimate government, education, positive cultural traditions, family, and (what William Blake might characterize as) "visions of eternity," (i.e., true art and poetry), in other words mediums of love or else practices or institutions that bring people together in a positive fashion.

9. Promoting idolatry, especially (but not exclusively) the love of money – the God of Hell is invariably a false or specious idol, such as money.

10. Promoting the use of witchcraft and sorcery, two very powerful tools to bring about a person's psychological demise.

11. Promoting irrational atheism, hatred of God, religious schism, gross carnality, degradation, visions of life seen through restrictive time (as opposed to eternity.)

12. Promoting and bestow powers of worldly authority to mediocrity, giving such ignorant, irrational persons to think they know it all, are highly talented, that they have the inside-track view of things.

13. Making it seem as though the (given) devil is serving some religious function, when in reality he is really nothing more than a criminal, though a criminal who cleverly knows how to operate in a legal way if he needs or is forced to.

It is very characteristic of Satan or a Satan to patiently nurture bad habits, beliefs and behaviors over time in an individual and society, rather than by dramatic strokes. Such an approach serves best to disguise his presence and activities, and in his way he is able to achieve more long lasting results. People are never given to doing evil for its own sake. Rather people do wrong thinking that it will confer on them a greater good. This ostensibly is the heart of Satan's strategy.

Perhaps much of Hell's strength is that Satan gets "his" people (by means of manipulation) to believe in what they believe in so deeply – even though what they believe in makes absolutely no sense whatsoever as far as their own real self-interest is concerned, and out of this (false) belief comes great power, the power itself finding its core in the fervid belief.

~ * ~

(Eve:) "Adam, my lord, this fruit is so sweet and delectable in my breast, and this handsome messenger (Satan) is God's good angel: I see by his apparel he is the envoy of our Master, the King of Heaven. If you spoke anything hurtful of him today he will nevertheless forgive it, if we two are willing to pay him deference. What will it avail you, such detestable quarrelling with your Master's messenger? We need his favour. He can interceded for us with the Ruler of all, the King of Heaven. I can see from here where he himself is sitting – it is to the south-east – surrounded with wealth, who shaped the world. I see his angels, moving about him on their wings, the hugest of all throngs, of multitudes of the most joyous. Who could give me such discernment if God, the ruler of Heaven, had not sent it directly to me?"

~~~~~Anonymous, 10th century Anglo-Saxon poetry, retelling of *Genesis*, manuscript Junius 11.

Satan's policy often takes the guise of otherwise legitimate religions; by which device he will attempt to make such negative behaviors justified. This masquerade I sometimes refer to in my dealing with these people as the "crack-brained, stone age religion." As said, Satan has a false version of just about everything, and this includes false ideas about who God is, eternity, hope, and Heaven. Satan can act in a benevolent way, and even help people, but only as a way of gaining their trust and confidence in order to perpetrate a great evil. This is usually how he and his people are able to infiltrate religious denominations.

A Satan can take what is already yours and fool you into thinking it is a present from him. Likewise, he will routinely take what is a general blessing from God and try to trick people into thinking those blessings came from him, and further that he is God, and that he should be obeyed as such. Likewise, many false religious, even well meaning religious people, have fallen into these kinds of traps and mistaken Satan, in disguise, for God. This can be brought about in a quite cunning way. To illustrate, let us imagine one of God's blessings to be some sort of wonderful, glowing light. Satan will put a concealing cover over the light, leave it there for an extended period of time, and when he needs to use it, uncover the light to his intended victim, claiming that that light came from him. The deceived person then in a given instance will believe him, and follow his command to do something wrong, thinking that by doing so they are serving God. In this way, like a thief or embezzler, wrongly holding someone goods, he will then in a miserly fashion, pay them tokens out of their own riches, all the while having the uttermost gall presenting himself as generous and self-sacrificing.

They can pervert one's idea of what's good; all the more so as the person being deceived is corrupt. As well a Hell influenced perhaps will invariably have a extremely compartmentalized view of reality, in which they will deal with things going on as separate, unrelated matters. This of itself, of course, is neither unusual or necessarily a bad thing. It is just that with Hell influenced people it is taken to an irrational and preposterous extreme, whether they be uneducated or highly educated people.

A carrot and stick approach is essential to and ever present in the Orkonist way of doing things. If you do the wrong thing, for example be a regular liar, hypocrite, or someone indifferent to injustice or cruelty, he will let you "prosper." If you don't do things the wrong way to some degree they will make things harder for you. Not surprisingly, such a system tends to produce slaves and martyrs. Hell will at times make it sound as if Jesus wants people to be crucified in order for them to be better Christians. While it is correct, Christ wants his followers to bear their cross willingly, this is simply a recognition of the fact that Satan and Orkonists do crucify those who try to do the right thing. It in no way implies that Christ desires people to be crucified. The cross, among other things, offers, for those who endeavor to live rightly and justly, a noble and righteous way of dealing with Satan's inevitable oppressions and persecutions, and a means to express their love and faith if they so choose. Yet the devil will preach the cross, but from obviously quite different motives than that of a real Christian.

While it is difficult to know how much evil he is willing to do, or how much he is actually capable of, Satan, it goes without saying, is quite the wily operator and will as a rule prefer to tempt others into doing wrong rather than do wrong himself. He will use people to trap, ensnare and ruin their fellows into doing evil. For doing his will, his followers are often empowered and given means to avoid certain pains. These regular persons or devils become addicted to the power and dope Satan gives them, and in the process alienated from what is truly good, such that it can reach a point where they can hardly keep from causing some kind of trouble in order to retain these "benefits" Satan confers on them. Thus they are led to see him as ultimate authority, at least as a practical matter. Though many devils will acknowledge God separately from Satan, typically they will see God as distant or irrelevant to their interests. Alternatively, they might have a false idea of God created for and imposed on them by Satan, by which they are fooled into thinking that God, if not actually Satan, is very like him.

Not least of his great illusions or instances of masterly sleight of hand, Satan can remove himself for a season (at least make it seem so), such that it will appear he is no longer there or else was never there in the first place, irregardless of what extreme ruin and heartache he might recently (or long ago) have caused.

~ * ~

Some Arguments Hell uses to Influence People

"Oftentimes, to win us to our harm,
The instruments of darkness tell us truths;
Win us with honest trifles, to betray's
In deepest consequence."
~~~~~~~*Macbeth*

The arguments Hell or Hell people use to deceive and influence people are critical to its practice of luring and ensnaring people's hearts and minds. Most any of us are familiar with many of these, but below, in any case, are some which one might hear or encounter. Though my addressing of these is far from thorough, it serves at least to identify them, and which is perhaps the more important thing.

a. *Since God created the world, God is to blame for evil.*

From my own experience, it will, I think, help people to know that while God allows evil, it is mistaken to think that He is or ever was the source of it. For one , it has been argued that it is God's absence, brought about through our rejection of him, which constitutes evil. At least, such is one case made. The source of all evil it would seem is bloated selfishness, immorality, irrational exercise of authority. Evil is ultimately an act of an individual's free will, i.e., a person's *willfully* turning away from God, who – far from being the source of any evil – is, by proper definition, all goodness itself. In this sense God cannot be responsible for one of His creatures turning away from him. Yet this said, it can be admitted that the argument that if God is the Creator then he must therefore have created evil is not without its persuasiveness. "My God, My God why have you forsaken me?" reminds us of how difficult it is amidst the pains of life to just dismiss the argument out of hand, and philosophically. In this way, the Orkonist's argument is not entirely unreasonable. Yet, if nothing else, we can fairly say we don't know what the true answer is. Causation on the primordial level may have been something quite different than causation as we normally read into phenomena,[58] and the issue otherwise is one of those things that certainly allows of different interpretation. This said, as a practical matter, we do see, (indeed catch red-handed), Satanists and Orkonists purposely and regularly acting as destroyers. So that to accept their argument that God created evil only furthers their influence and consequently the amount of evil going on. If we reject the argument, on the other hand, as a very practical matter we empower ourselves against evil, while possessing good reasons to assert that *their* claim is at least un-provable.

b. *Evil is necessary to promote good. Life without evil would be flat, uninteresting.*

That this is false is I think proven by the fact that both higher and lower ranking evilly-disposed persons invariably seek the approval and admiration of rightly acting folk, then ever the other way around. Furthermore, since when were we ever without evil to have known the difference? Many have taken the view that the world, as designed or created, is supposed to be separated between good and evil. While this would seem to be true to some extent, it is equally true that the presence of good or evil can be maximized or minimized. There is no reason therefore for saying the division must be equal, or that evil has equal or as great importance as good. Conceivably, people could if they so chose live in circumstances where evil, while present, was, even so, must less pervasive or noticeable. That it is not actually so (in given circumstances) is merely, after all,

---

[58] We might, for instance, adopt Kant's point of view and say that causation (as we know it) does not apply to incomprehensible noumena, or higher levels of spiritual existence, of which God and the angels might be said, (at least with respect to our phenomena based perspective) to have their unique Being and Reality relative to our own.

the result of people's choice. Without evil, or with significantly less evil, we would have growth, creativity, and the singing God's (i.e., truth's, justice's, freedom's and beauty's) praises in infinite and multi-varied ways. A good case can be made that extreme pain is not natural, and but for sin's great influence in the world we would never (or at best rarely) know it. There is manageable evil, as in pain in physical exercise, which is acceptable, but there's no requirement that things like pain in physical exercise or hard work need to be forced upon or compelled of another. Ordinarily, the only pain in life that is natural is that which we are willing or not willing to accept ourselves. For example, if we work we might feel the pain of labor, yet if we do not work we might suffer from not eating or not having money. Such as these are the only pains in life that are truly natural. All truly serious or extreme pains, including death, are, in their ultimate origin, the result of Hell's maliciously bringing them about.

Much that is seen as evil in nature and animals is arguably effects that over the course of millennia have been brought about by repercussions stemming from man's Fall – such at least is a theory it would do well to consider, or a notion similar. According to this view, evil such as the serpent and our Fall have brought about are of such a momentous character as to have mutated nature, both at the time of the Fall itself and subsequently. If animals kill another for food that is as bad as it ever really gets for an animal. An intelligent person who considers the matter will readily see that to speak of an "evil animal" doesn't make sense; since, for one thing, animals haven't the necessary egotistical motive for being evil. Instances where they are seen to act with particular viciousness are almost always, if not always, caused by sprites and demons triggering and inciting aggressive behavior (as I will address later on.) As well, much of cruelty in nature is ultimately preventable by the rule of man on the earth through just reason and compassion, and as much as one might object to this claim, one cannot deny that such a belief could only serve to promote peace and justice, and as a result render the quality of life better for everyone.

c. *Satan has a "Right" to rule the world.*

The truth is, Satan or a Satan is nothing more than a criminal and a tyrant, responsible for his own plight, and should be viewed as such. Yet bear in mind, to many people Satan is God, or at least that is who they know and understand God to be, though for such people equating God with Satan is not something they do because they consciously reject God, but rather they misunderstand and do not really know who and what God is. Those things spoken of as being the devil's are typically things he or his people stole from others, and if Satan has a right to tax others for doing evil, Godly disposed people then have as much right for taxing him and his followers for being happy. Some of the best humor can sometimes be found in challenging the devil on this point.

By someone's showing themselves to be a most heinous arch-criminal, almost universal legal tradition would seem to show that they forfeit their own rights to surplus wealth and extraneous property, that is to say that minimum amount of property which a person is usually understood as requiring. As a matter of law and justice, therefore, the assumption that great worldly property is automatically the devil's is a false one. Great criminals, I think all will agree, should not be accorded special rights and privileges, including that of possessing phenomenal amounts of wealth; least of all when they go on behaving in an egregiously criminal manner.

However, it needs to be made clear, that to the extent one might agree or continue to agree with Satan or a devil will only make things more difficult for them in dealing with Hell's people. And to the extent people go along with him, Satan can, in a sense, claim rights based on the acquiescence of others. Regardless of whether he is technically justified in doing this, such claims (at least in principle) are, both by "good" and "bad" people, commonly accepted as having *some* amount of legitimacy. Understandably then, judgments and determinations of whether Satan's claims on another are justly founded usually need to be decided on an individual case-by-case basis, with just impartiality toward all concerned, including the Satanist or Orkonist.

d. *A certain number, or quota, of people and animals must be harmed, otherwise things will be worse for everyone.*

If there is any truth to this argument, it is that of the terrorist threatening that if people don't cooperate with him and he will cause trouble – which, of course, is little or no real argument, let alone solution, at all.

e. *All sin and evil are the same, and there is no real difference between one sin and another.*

Though there is some truth to this statement, it is obviously false to say one sin is no worse than another. Hence we read "greater is his sin," "Take the beam from your own eye before removing the speck in another," etc. As a practical matter, it is very obvious that it would make a great difference to someone whether they lived next door to an unrepentant murderer versus someone who had only a few speeding tickets offenses. Any of us can think to ourselves any number of acts of wrong doing for which there is another that is worse. It is a point that is self-evident to all in the day to day living of their life. While, true, it is not correct to call doing what is wrong "good," we can, on the other hand, speak of an act being less worse than another or "not so bad." It is very customary for Hell to accuse or blame someone for not being righteous, or, in other words, complain they are not perfect. Yet in truth, they don't really even know what righteousness is, and simply use finding fault in others to excuse their more serious crimes.

f. *People of Heaven are nothing but a bunch of snobs, who don't care about the suffering of others, particularly those in Hell.*

To promote this line, Satan will on occasion have sent around false spirits pretending they are from Heaven who act in such a way to encourage this belief. Similarly, Satan (or someone working for him) will sometimes, and in effect, put on a wig and beard and pretend to people that he is God or Jesus, either in their thoughts, or else (though much more rarely) even do so visibly, in spirit person form, before their eyes. Consequently, people are at times lead to mistakenly think of God as being oppressive, cruel, and mocking. Of course, the truth is exactly opposite – a fact which Satan seeks to hide from us. Under Satanist and Orkonist influence, many people are persistent in having us blame God for life's woes, but due to some remarkable sort of myopia never bother to blame Satan or Orkon people, no matter how directly implicated those people are.

g. *The people of Hell "have nothing to live for."*

This is a very often used catch-phrase by some of Hell's spirit people, the point being that if they have nothing to love for, their evil doing is excusable. Yet if they have nothing to live for, would this not be as much reason for their doing good, or doing nothing, as for their doing evil? They say they do wrong because they suffer so, and have nothing to live for. The greater truth, again I would maintain, is that they suffer so, and have nothing to live for because they are always, and won't stop doing, the wrong thing.

h. *It helps the suffering souls of Hell if people "up here" allow or help them to engage in wrong-doing.*

The truth is the exact opposite, because their wrong doing, in the long run, only makes them more guilty, thus drags or keeps them down lower, while contributing to a general mentality that wrong-doing is somehow necessary.

Now it is true, on the other hand, that demanding moral perfection of others can for some lead to great frustration and their abandonment of any effort to do good. One solution to this problem lies in frowning on worser behavior, while being more lenient and indulgent of lesser wrong doing, with a view to mitigating the overall situation over time. A satan, of course, does all this in reverse, that is by making light of great evils while ridiculously blowing out of proportion minor ones.

i. *Hell has all the money, the riches, people who believe in God, on the other hand, will not have any money.*

On the surface, this is frequently true for the simple reason Orkonists and Goomerists rob everyone. For those who actually are taken in by this belief it sometimes seems to work, and in others eyes it appears to work as well. Even so, it all ultimately turns out to be a kind of illusion; for evil's way is not that of growth and prosperity, but destruction, and the final result is always and inevitably bitter disappointment. In addition, we can also point out that Heaven minded people have riches Hell could never possess; indeed which (most of) Hell could hardly even conceive or dream of. Evil people came to power through crime and murder, not charm and appeal as they will end up pretending. Yet how typical of them to try to get others to think otherwise.[59]

    j. *If people "up here" have it "too good" it will make the people of Hell resentful, be unfair to them, and therefore incite Hell people to cause more trouble. Consequently, it is better if people are not too happy or successful, even if they are not acting wrongly or immorally.*

In point of fact, as a person from down below told me, things are better for people down below if things are well and done rightly up here.

    k. *Because good people of an earlier time had it bad, for example they were martyred, it would be unfair for those coming after them to be happy and have it good.*

A good person does not grudge the happiness of others, on the contrary they welcome it, as long as it is not had as the direct and purposeful result of immediate or very recent evil and wrong doing. It is only a Satanist, Orkonist or Goomerist, or someone under the influence of these, who grudges others' happiness.

    l. *There are so many evil people that Hell will ultimately overflow, engulf the earth, and even take over Heaven.*

This is pure nonsense for a number of reasons, not least of which it costs a great deal of lives and money to get people from down below up here. If there were no healthy life left up here, people of Hell would actually all sink down lower. Granted Hell can and has made great inroads on the well-being of our world, yet there is a point at which they must halt lest they harm their own interests. Unfortunately they don't stop short of their evil doing as much as we would like, yet it is well to know that there are checks, nevertheless, on what they can and will do. To use and analogy, someone, for instance, might want to drive their car over 100 miles an hour. Yet even if they can ignore legal checks, there still may be physical ones they have to recognize and respect.

    m. *The human intellect is supreme (God if not to blame is irrelevant) and will be able, for example, by science and technology to solve all our problems.*

The purpose of this kind of argument is to promote human conceit, selfishness, and arrogance, as well as a too casual materialism. A person of sound

---

[59] Know also that for every someone who thinks they can obtain wealth and prosperity through evil there is invariably someone somewhere else more able and capable of doing evil than them and who could easily have them at their mercy should they cross paths.

mind shows humility about what an intellect can know absolutely, and takes pains to qualify and categorize assertions and beliefs according to criteria of verifiability. The presumption of human infallibility, whether here and now or in the long run, has no standard or measure outside itself to verify its truthfulness and universal validity. In other words, no one short of God Almighty could confirm the absolute supremacy of the human intellect and its conclusions. This conclusion is not intended to discourage our intellectual powers or their betterment, far from it. Instead it cautions us to be more cautiously humble and careful about making major assumptions that are not sufficiently consistent with more honest, down to earth, and thoughtful reasoning.

n. *People who love animals don't love people.*

Nothing could be more false, and indeed the opposite is true, as we find in the lives of St. Francis of Assisi, St. Martin de Porres, and John Chapman (Johnny Appleseed.) It is true, we hear of people who pamper their pets and are completely indifferent to the sufferings of people, but such cases in reality are rare, and often this kind of situation is one not unusually brought deliberately about by Hell people for propaganda purposes. Some well-meaning Christian theologians deny that animals have a soul. My response to such is that if animals have no souls, then why would Christ want himself to be likened to a lamb? To say that lamb has no soul would completely deny meaning to the symbolism, by reducing God to an object. Where the Bible uses objects with respect to presenting God, they are used to symbolize God's attributes, not his actual person.

With respect to a similar matter, Orkonists and Goomerists will often attribute society's major problems to young people, when children one way or another are invariably a reflection of their parents, and of how their parents run society. This is not to absolve a given child or young person from obligations and responsibilities. On the other hand, it frankly is one of the most contemptible things to see adults with money, status and security blaming young people (who usually have none of those things) for society's ills.

o. *If it is done (or it happened) it is justified.*

If a person from Hell ever gives you this as an argument, tell them, if this is so then, then anything and all things that can be done to hurt them in the future will be fully justified. This said, let us courageously welcome that God's will be done and face whatever troubles that confront us, and in a way deserving of His grace and blessing.

p. *A blackmailed or otherwise threatened person is to blame if they do not give into the demands of the blackmailing terrorist, and is responsible for any harm that comes about through the terrorist's acts.*

A person blackmailed or threatened by a terrorist is never responsible for what the terrorist does. It is a prima facie matter that it is the person who commits the crime that is guilty, not some other person who does not cooperate in wrong doing with them.

q. *By doing evil one becomes more free.*

Some Hell people are led to believe that by breaking all rules, there will be no rules and then they will be free. But this is a hypocritical belief because the principle itself is a rule (based on a certain erroneous kind of reasoning.) What's more, in practice, such people still observe rules that suit them, and therefore violate their own doctrine. The plain truth is the more one does the wrong thing, and the more serious that wrong-doing is, the less one is able to do anything worthwhile, most especially those things that require love, creative intelligence, skill, inspiration, or talent. Granted an evil doer might possess extraordinary shrewdness of one sort of another. Yet most of the time such powers can only be used for negative and destructive purposes, for which use the person inevitably ends up being involuntarily employed as a slave of someone else more powerful than himself.

r. *"God made us this way, so it can't be wrong that we do what we do."*

Then God made us the way we are, that is to oppose and convict you.

s. *The fact that Satan is so powerful shows he is right.*

If he is right why is he or his people not happy? And if they are happy what are they bothering us for? I believe, if not Satan himself as such, his people like in a degenerate and hideous kind of Disneyland fantasy world, which he has created out of people's and animals lives, as well as a host of things, he has robbed or misappropriated from others.

t. *In promoting evil in others, Satan is serving God.*

As best as one could surmise, what might be the basis for such an argument is this. Because a person commits any sin, (indeed merely being born of Adam makes one guilty of sin) all that sin are justified in being damned as being an offense to God and unworthy of God. Therefore Satan is justified in attacking any and everyone, and is justified in doing so as a way of defending God's honor and worthiness. Though there may well be some truth to this argument, it leaves open the question of what specific right Satan or his followers have to inflict the amount of pain and suffering they do. In other words, what possible reason is there that this or that Satan or Orkonist should be entitled to such a right? Going back to the Book of Genesis, the Fall of man mandated four things: 1. Loss of paradise (and its blessings), 2. Death, 3. Pain in childbirth, 4. working by sweat of one's brow. At no point is it obvious that Satan or some Orkonist is justified in attacking us, and making us miserable. Satan's response might be, that it is not he who attacks us. He merely tricks us into attacking our fellows. If we unfairly attack our fellows, it is our own (humanity's) fault, not his.

Hell will act as if they are a team with Heaven, but this is merely one of their standard deceptions. The truth is they act *very much* on their own in the courses of action they take. So while Hell will pretend to be a public interest group, it is in reality the most self-seeking private sector enterprise possible – indeed, by definition.[60]

On this same point, it is well to remember evil can in no way bring any good. Only God himself, the good God allows the devil, or goodness from God that devil has somehow stolen from another, can bring about good. No good can ever come from evil. A Satan, Orkonist has basic rights just like anyone else. But respecting these is a far cry from viewing them as spokespersons for God. And, needless to say, it is a more than a grave mistake to grant to Satan or Hell any authority beyond those basic rights which any given individual is otherwise entitled to, even if he should appear you in the form of a 500 foot tall giant spirit (perhaps in the form of kingly "Leviathan" as portrayed in the front piece to Hobbes' book) casting thunderbolts and shaking the earth. Doing so would only be granting authority to a maniac. Of course, a Satan or Orkonist does posses some good since God, after all, created them, but not enough to justify their holding lawful governmental power, or absolve them from just punishment or legal accountability.

If one believes devils are divinely appointed and empowered to cause problems, then one might just as well believe, say, an arsonist is divinely appointed and empowered to cause the problems they do. True God permits such troubles to take place, but only because we, as sinful collective humanity, allow it by way of our immorality and irrationality. At the same time, the phenomenal power and wealth evilly disposed people might have is not had cheaply. They pay Satan atrocious prices to have these, which is to say by becoming his slave. What good a Satan or Orkonist has to offer is invariably something stolen from someone else, and no good as such ever actually comes from them, though people are often led to think quite otherwise. For all their great powers, Hell people, the vast majority of the time, can't really do *anything* except hurt others.

Why do Hell people have as much power as they do have? For as much the same reason some terrorists are able to hold a building full of people as hostages; which is to say through desperation, cunning, weapons. Hell does have

---

[60] Speaking personally, I feel very strongly that nothing in Bible should be read as justifying Satan's religious role, and that he should be seen as what, in relation to us, he really is a criminal. Texts which seem to express otherwise should, in my opinion, be taken with great caution or viewed with a great deal of skepticism, as any such suggestion that he is somehow God's agent only works to empower him.

invisible people working for them and has been at this game for many, many years. Their incomprehensible relentlessness, obstinacy and determination are rightly legendary. As much as they are willing to make us miserable we must, with faith and love in God, put as much effort and more into doing the right thing is order to fight them off. As individuals, Hell people, aside perhaps from Satan, are usually very weak and feeble. It is as a gang that they have any real power, a Satan, or possibly a higher powered Orkonist as well, being the only ones allowed to be true individuals in that organization. Further, for all the often phenomenal powers and wealth they possess or display, they rely on commonplace devices like lying, secrecy and cheating for any and all of their operations to work.

Living in darkness as he does, a Satan or Orkonist may be a fool to his own propaganda, and fools himself as much as anyone else with his own misleading arguments. It may at times be made to seem (by say a sorcerer) that Satan, fallen angels or false gods have it well, live like aristocrats of a sort, unlike regular devils of Hell. This may in a way be correct, but it is safe to assume such is a specious, if glittering, kind of happiness and prosperity. Satan's way is against Gods way, i.e., the rational and moral way. But he and his people still desire to enjoy the good things in life, want to have it good, and typically they want to have it both ways. If Satan consistently did things according to his own doctrine of absolute selfishness he would probably end up in a black hole or something very like.

As far as "Hell," as in Satanist or Orkonist policy and method goes, but what after all, on the every day, practical level, is their central argument? That in the end, viciousness, violence and lying rule all souls and the universe – not love, reason, morals or truth. Violence and subterfuge, not virtue, love and truth are, in the final analysis the top authority. Hell's says, "Our viciousness and violence, conquer love and virtue, so we will always win in the end." The opposite argument says "Life is not worth living if viciousness and violence are ultimate authority, and we would rather lose our life (in martyrdom or battle) rather than suffer such a rule. If, as a matter of physical reality, God allows your violence, then God allows our rejection of your violence. In this lays our meaning, dignity and freedom. In this is final authority." While in Christianity and other faiths God, or else one's self, willingly suffers in order to win or redeem others from their sins. With Satan or in Orkonism, the idea is reversed, that is, others will be made to suffer; so the Satanist or Orkonist doesn't have to suffer for their own sins.

Another way of seeing the overall situation would be this. On the one hand you have a group, the Orkonists (again, who might or might not be Satanists as such), who have more money and worldly power than anyone else (much if not most of the time.) They achieve their success by the most extreme immoral and criminal means both imaginable and unimaginable. Yet because they do the wrong thing, they ultimately are less happy in terms of real happiness than those who try to do the right thing – to say the least. Yet for many who "do the right thing," they, by contrast, have to get by and often do not have the great worldly wealth or power. Now for a time those who try to do the right thing are, in their humility, willing to put up with the arrangement. But some of the Orkonists, and then Goomerists, unhappy as they are, can't leave it at that; being prone to violent envy as they are, and then go an attack these materially less well off people who are otherwise happy. The people who do the right thing might then turn to their religious faith, armed resistance or both, believing that no matter how bad things can be made for them by the seemingly all powerful Orkonists, God has the final say and power, and will continue in believing in doing the right thing. Because they know that in doing what's right is true happiness, and they will use the occasion of their persecution to show their faith and love of Love. The Orkonist, on the other hand, being so selfish, can never love Love so much. Yes, the Orkonists will challenge them, "but where is God? God is not here to help you. We have the power" "The religious reply "God's will be done if He can or will help us or not, we trust in His judgment. Besides you are never really happy. It's all phony with you people." "Well, says the Orkonist, "then neither will you be happy because we will make a special point of making you miserable. Not only that but we can make it appear as if you are one of us, make ourselves appear as one of you, or make you look bad in any number of ways." The religious answers: "The most beautiful and adorable people in the world have been ruined and destroyed by the likes of you, not to mention poor animals as well. I'd rather suffer and die than compromise against principle with their murderers. But I know you are clever, so that I will, with God's grace, arm and empower myself accordingly. Truth will tell us apart when all is said and done. Meanwhile you should give up your devil way of doing things, and learn what it means to actually make sense."

Some people will think they can simply avoid affiliation with either side, and many to some extent are able to do so. Yet it is, as some have argued, like Swedenborg did, only a matter of time before they are thrown one way or the other, that is either in the direction of freedom or slavery.

~ * ~

The People of Hell

"There are demon-haunted worlds, regions of utter darkness. Whoever in life denies the Spirit falls into that darkness of death."
~~~~~~~*Isa Upanishad*

Based on the awesome power they manifest and how well organized they often seem, Hell (as an organization) can fairly be seen as kind of totalitarian society run by a dictator, though as in Kafka's "The Trial" one living under such a regime doesn't necessarily know who's in charge or making the decisions. After a very small number of dictators, crime bosses, and their main henchmen, the vast majority of Hell people are slave living in various degrees of degradation. The punishments and torments of Hell, as commonly thought of, are something its inhabitants or members inflict on each other or else other people. Some of history's most infamous political and corporate regimes would seem to have very clearly been modeled on Hell's system.

I would say the psychology of Hell's leadership is something like this. They are basically tyrants, utterly selfish and who feel sorry for themselves. They don't necessarily want to do evil, indeed, they tire can tire of it; only it is only by doing evil that they obtain or hold on to the power they possess. On the other hand, evil doing can be something they will take delight in. They are extremely egotistical and one of their gratifications in life for some of them is to be someone monolithically important. Sometimes this will be display in great shows they might put one, with themselves as center or as one of the centers of attention. Nero and Hitler are good examples of "regular" people of Hell who liked to indulge and reward themselves in this manner, and as such delighted in being a "great man in a great show."

In "Hell," people will be punished both for doing the right thing or the wrong thing. Whether the one or the other depends on the current political climate, and what suits Satan's policy, and (for lack of a better term) the "public's" opinion. As much as any other, the purpose of Hell is to empower its tyrants. This said, a true Satan is a sole and ultimate ruling tyrant and brooks no real or potential rivals. The other tyrants are in effect his staff, but their power compared to his is minimal. It is often claimed by people from "down below" that the wrong-doing they are engaged in serves a sort of egalitarianism in the grand scheme of things necessary to make up the disparity in happiness between people "above" versus people "below." If we do things the right or moral way, they say, it won't work for dispossessed people like us.[61] Not only people down below, but underprivileged people up here will adopt this kind of thinking. The real truth, however, is that it is Hell's purpose to empower Satan's and better realize his agenda – not help the "little guy," who things turned out badly for, as is frequently pretended.

The tortures traditionally associated with Hell, go on in no small part because of people, who themselves desire to avoid torture turn a blind eye to or else cooperate in the torture of others. To compound all this, crying and weeping are typically forbidden among Hell people, and as a rule are physically prevented.

In one way it is hard, and in another it is easy to try to figure out why Satan and the people working for him act the way they do. While there are very

---

[61] Note once more, "Satan" (and those among his chosen elite), contrary to Dante's depiction, is not himself "below," but should rather, and at least as a practical measure and aid to our understanding him, be seen as inhabiting a sort of heaven in the sky. Herein I use "down below" more so and with respect to the rank and file. These separations by locations it is not necessary to insist on as true and accurate in every and all instances, but are intended as general guidelines to help set up a preliminary broad picture of how these people might be conceived.

concrete similarities between our world and theirs, they frequently go by a logic and way of understanding that seems to us quite alien and bizarre. This is one reason Hell is hard to describe because in many respects their ways are like that of a strange and foreign country we never heard of, and yet they are in other ways very like ourselves. They, whether we are speaking of a spirit or else a flesh-and-blood Hell person, are often very confident of themselves, and sometimes a fanatical conviction will sometimes accompany their wrong doing. Otherwise, much of their beliefs and carryings on are simply the result of fear, egotism, superstition, greed, stupidity, and brainwashing, and which are not all that hard to recognize and identify.

Yet for all the conformity they are still individuals like anyone. While there seems an effort by Satan to get them to think the same, the accumulation of souls from various times and places around the globe has apparently to some extent made it diverse, though bland uniformity is more often the rule. Some are smarter than others, some dumber. Some are morally better than average (for their sort), some are worse. Some are gifted in certain ways; others seem to be fairly useless. Some try to avoid doing the wrong thing. To others the thought seems to hardly occur to them (except as the wrong doing might be an inconvenience.) Some are brainwashed into thinking doing the wrong thing is doing the right thing; while others knowingly do the wrong thing and, in their ignorance, simply just don't care.

There are some real "characters" among these different groups of people, some of them are actually quite amusing. But, generally speaking, and as I understand it, the lower down you go the more repugnant they are, and any attractive personality trait those might have otherwise have possessed diminishes with their guilt, inasmuch as their vileness drowns out their positive attributes. Now this is not always true, because some of Hell's very worst are able to make their way up here, but these should be seen as exceptions.

Given the way they act, many of the people of Hell might well be viewed as mental patients. They really are people who are insane; however, the fact is they came to this pass by doing evil, and it was their ongoing doing of or involvement with evil that has driven them to this state.[62] This said, they are not entirely without thought or free choice; it's just that they are inclined to deny that they have; typically claiming that they are *forced* to do something wrong. But really, in such instances, ordinarily a better term would be to say they are pressured rather than forced. To further make denial easier for them, it is common practice of Hell also to orchestrate wrong doing, so that the guilt (at least in their own minds) is disseminated, and in that way watered down, among various different persons as a means of avoiding individual blame.

---

[62] St. Augustine, in his "Letter against Julian, Defender of Pelagianism," makes the interesting remark: "perversity of heart comes from a hidden judgment of God, with the result that the refusal to hear the truth leads to commission of sin, and that sin itself is also punishment for the preceding sin."

They will tell others they are somehow excused, if not justified in their wrong-doing. But really they know what they are doing is wrong; only it always someone else who is more to blame than themselves. They often say they don't know what they are doing as a way of denying their guilt. But again the question can then be fairly asked, why then don't they just as soon do the right thing rather than the wrong thing? I have heard them say things like "He doesn't understand why he has to do these things," and "He's done so many things he doesn't think it matters." They frequently think that the more people they can get guilty, the less guilty they themselves are. Put differently, the more guilty others are the less their own guilt stands out. Guilt then does have meaning for them; only they have a variety of ways of absolving themselves from it in their own minds, and often in the minds of others as well. If devils see themselves as enlightened by testing the limits of morals, and no less be extravagant about it, let them do so. Only I think sensible people will agree that they have no obvious *good* reason why *they* should be able to force their peculiar ways on someone else.

These people will often claim what is wrong is that they don't have nice things and circumstances like others. Yet if they had it good, say were granted a paradise or demi-paradise to live in, their dispositions are such that it would be just a matter of time that they would turn it into Hell. If the devil won't seriously try to be good, no amount of riches and miracles will ever be able to make him truly happy.

Some more bitter sorts of spirit people, whether from down below or above, seem to think because their own life is "over," that those who are (flesh and blood) living don't matter by comparison. There is a certain strange resentment toward sex, not arising out of love of chastity and, at the same time, no doubt originating with an Orkonist's hatred toward love in any form; which makes them believe that those who are born after them do not really matter. The hypocrisy of such belief, of course, is not anything that apparently bothers them. As it typical, these same kind of people habitually believe they are not guilty if they don't know what they are doing. On this basis, they will then deliberately do things in such a way as to minimize their knowledge of what they are doing, by being purposely irrational, for example. This same behavior, certainly, is not unknown among some regular people as well, but with these kinds of spirit people it is even more pronounced, habitual, and ingrained.

Hell spirit people may be subject to rules which have some-kind of sensible reason behind them, or else which may be only superstition or blind custom. Some spirit people may have been Hell people before, but have since opted out of Satan's organization and try to stay out of trouble as best they can, though still prisoners in one way or other. Being usually under heavy pressure for doing so, it must be understood that staying out of trouble is not always such an easy thing to do; so that these "retired" sorts of devils it could be said are deserving of our sympathy and prayers.

All active spirit people (as opposed to those *at rest*) evidently retain a certain amount and kind of appetite. The worst of the Hell people are cannibals and vampires of a kind. I don't mean that they eat flesh or blood as such, but that they will leech on a living person's (or animal's) energy or well being like a parasite. This behavior will perhaps in some measure explain why Christ offered himself in the Eucharist, i.e., as food, so that people would consume him in his self-sacrifice rather than eat others. Observe also in this regard that the ancient Egyptians, for example, believed that in order to keep the departed from coming back and causing trouble, they needed to have food left in or near their tombs in order to appease their cravings.

Probably the most powerful force driving the behavior of Hell people is fear. It is fear that, one way or another, probably most underlies their excessive arrogance and cruelty. Jealousy (or envy) as well is sometimes a very strong motivator working in them, and very typically – almost religiously – made a pretext for wrong-doing. This said, jealousy is also used as a phony excuse by them, and the claimed jealousy or envy merely a masks some other motive or purpose.

Although themselves usually guilty of whatever it is that is wrong, they will blame or viciously attack an innocent person or party for what they themselves are culpable for. It seems to be the case that the more a person is engulfed in sin and wrong doing the more they are given to being irrationally aggressive. Otherwise, it is hard to say where the attitude of being belligerent toward another ultimately originates and finds its greatest source of power. According to the Bible, the first known act of aggression came from the serpent, in his desire to attack Adam and Eve. If we accept this account, we might reasonably conclude that belligerent aggression is something we acquired from our contact with the serpent. Be that as it may, it is helpful in understanding Hell behavior to conceptually isolate aggression from the person themselves, and to then determine the origin and source of the aggression in that person. In nature we do see competition and fighting, but these even if allowed as behavior that is genuinely "natural," is quite different from the psychotic and insatiable belligerency we sometimes find in Hell people, and Hell as an all powerful force in the world generally. The logical conclusion would seem to be that Satan's and or the serpent's enmity is, by one means or other, instilled in a person and then finds realization and expression in our own and others bad behavior or attitude.

These Hell people though they may in certain ways be advanced technologically and adroit at conniving, are (usually) extremely backward (and primitive) intellectually and culturally speaking; though there will be ways in which they will pretend otherwise, and sometimes successfully, including, on more rare occasions, creating enthralling, though at bottom meretricious, visions or mirages of heaven or paradise. But in most instances, the sort of cultural sophistication they display is predictably shallow and superficial. It is not always easy for some to see this however; for the tricks or psychological manipulation they can display or impose (say by a proficient sorcerer) can be quite dazzling and blind the judgment of the un-scrutinizing, timid, and unwary.

Immaturity, the reverse of innocence, is very pronounced among these people, and is often used to deny their accountability. They regularly live in various kinds of fantasy worlds as a means of escaping their miserable circumstances, the "sorrows of Satan" ostensibly being the most painful thing one can suffer. The resorting to fantasy, of course, is neither unusual nor in itself necessarily a bad thing. Yet with Hell people it takes the form of a regular lifestyle and mania; which becomes outrageously and tragically expensive to feed and maintain, and this at the doleful and cruel expense of others.

Reality as they understand it is, for many of them at least, reflected in the Peter Pan story, about the boy from "never, never land" who never grows up; whose world is alien to our own, who tries to impress people with his magic and by that means obtain our acceptance and admiration. He is friends with Captain Hook, yet at the same time, he fights him. Yet as much as he might fight Capt. Hook, he can never truly escape from him. Games will be played in which someone will *act* as (someone like) Peter Pan "fighting off" Capt. Hook (Satan), and in this way fool others into thinking "Peter Pan" is protecting them. In reality, what Peter here (in reality some personably agreeable devil) is doing is fooling them, so that they will serve him or Satan. "Don't worry about Capt. Hook!," says Peter Pan, "I'll take care of Captain Hook. You just attack [so and so as I told you earlier]."

The mad school children in William Golding's *Lord of the Flies* give another illustration of the kind of childishness one might find in some Hell people. They are fond of slogans, and will far more often form their points of view based on these and trite catch-phrases (or if you will "catch-thoughts") rather than rational arguments.

There is a sense and a degree to which some of these Hell people are what they are a result of being deliberately isolated; with the corresponding reaction of seeking some kind of community to which they may belong. Some of these people are so alienated that they will do anything to get attention. Meanwhile, they are fed a kind of "dope" (perhaps taking the form of false comradery or humor) which they are led to think is love, while at the same time incited to wage war on real love. They then are conditioned to seek the "dope" (in whatever form it might take such as servants, magical powers, drugs, money) thinking its love, and this conditioning then becomes an addiction which then enslaves them to those whose possess greater powers of deception and manipulation, in addition to these other things.

The phenomena of multiple personalities is something out of Hell which one might find in some of these people. It is a result of a person both having multiple demons (or sprites) in them, combined with an effort on their part to compartmentalize reality. Note in this regard, that spirit people themselves can most certainly have demons in them; "possessing" them, just like regular people.

If heavily involved in doing witchcraft, they sometimes have fits of what I have called "damnation fever;" where they go into a kind of frenzy in which they are more violent than usual. In addition, a strong atmosphere of sleaziness (like a kind of radioactive fallout) might be found among them on these occasions.

As mentioned earlier, they want to get people to participate in their guilt. In some ways it is almost a traditional thing, handed down from by gone ages. If they got themselves guilty, it is only fair that others be made guilty too, because after all, they were only innocent themselves (really.) It sounds silly, and it is, but if there are enough people thinking that way, people who have little or no scruple in getting what they want, they can become a powerful force to reckon with. To get others to be like them therefore can both gain them power, while making their guilt seem more common, and therefore more accepted, than it otherwise would be. Very commonly, Hell people will threaten others that if they don't do the wrong thing (or if they do the right thing) they will cause trouble; such that over time people can become conditioned into thinking that doing something right is what causes trouble – not the Hell people who are the real origin and source of it.

They have often a lot of time on their hands will, in a given instance, take great pains and efforts to do some entirely pointless and useless thing, to hurt people or animals, typically under orders from a sorcerer or warlock. At times there is no accounting for their bizarre behavior. At other times there may be something like an understandable reason for their carrying on which is lost on us. It should be noted once more, notwithstanding, that one of the main reason's for their constant, gratuitous misbehavior is that people from Hell are deceived into believing a certain amount of wrong doing must be done, like meeting a quota. If such a quota is not met, it is argued, things will be worse. This doing wrong frivolously and without seeming purpose, and getting people not to care about it, therefore, is just a kind of subtle brainwashing designed to get people to be more guilty, and thus place them all the more under Satan's control.

The outrageous and gratuitous maliciousness of these people often has to be seen to be believed, and even then you will still have a hard time believing it. In the cruel prank played on Arthur Grimsdyke (Peter Cushing) in the film "Tales from the Crypt," and the various sadistic machinations of "Baby Jane" Hudson in "What Ever Happened to Baby Jane?" are found very correct illustrations of both their methodical conniving and incomprehensible cruelty.

Having said all this, it is very sad to think how it must have looked, way back when, to have seen one of these people crying as a child, before some evil person warped their innocence into immaturity, and to now reflect on how these wretched people are today.

While it is right then to pity these people, it is altogether wrong to tolerate too much of their wrong doing or let down one's guard when it comes to their sometimes relentless tricks and schemes. Some of their troublemaking can and should be put up with; much the way we would excuse a mental patient for behaving badly, but only up to a point. It is generally understood that they are normally not even supposed to be up here, and if they cause serious problems one should not put up with them if one can help it. With respect to powerful spirits like so-called "fallen angels," for whatever power they show or possess we have every right to insist that they behave morally and legally as much or like anyone else if they are going to come into or be in our midst.

As children of Adam, sin is a collective illness, and in that sense we all share some blame, howsoever small, of any given person's guilt. It certainly behooves us then to be forgiving and tolerant to some degree. On the other hand, in individuals where the wrong-doing is of an incessantly ongoing and unrepentant nature, a given person is responsible and answerable for their guilt. While of value as mitigating factors, claims of being deceived, or acting in ignorance are not, by themselves, legitimate excuses for repeated criminal behavior. And even if out of some compassion we should separate the person from the act, this is no reason to casually make light of the act; all the less so as the act is of a heinous or malicious character.

~ * ~

Who Comes from and Works for Hell

"You know from the repeated declarations in the Law that the principal purpose of the whole Law was the removal and utter destruction of idolatry, and all that is connected therewith, even its name, and everything that might lead to any such practices, e.g., acting as a consulter with familiar spirits, or as a wizard, passing children though the fire, divining, observing the clouds, enchanting, charming or inquiring of the dead."
~~~~~ Maimonides, *The Guide for the Perplexed.*

People of "geographical" Hell do not always stay down there. Some make their way into our world, and what might surprise some people to learn is that the vast majority of the world's serious problem arise or have some connection with the presence of these (usually invisible) Hell people in our midst, and the still living (or again "regular") people who work with them. Together they infiltrate family, businesses, culture, media, churches, government, whole nations and their influence is very much like a disease; often taking the guise of good fellowship, patriotism, benevolence, advanced learning, religion, or enlightened progress. Progressive, conservative, religious, atheist, black, white, Christian, Jew, rich, poor, criminal, police, etc., etc. – you name it – you can rest assured that the Evil

One usually has some way of making himself accepted or otherwise inveigling himself into their company.

The principle known means by which Hell is carried up here is witchcraft and sorcery (a more sophisticated, and to that extent all the more sordid, kind of witchcraft). Witchcraft is at least as old as civilization itself. It has since existed in every era to some degree or other, sometimes being treated as an official religion, and often times in antiquity took the form of paganism; though it would not be fair to assume that all forms of paganism involved witchcraft. An atmosphere of lies, deception and corruption is that in which witchcraft most thrives, and secrecy and denial of its existence has been perhaps its greatest protection for ages. Those who find out about the real truth underlying these things are regularly told that if they tell someone what they know they "will have something done to them." And indeed, it is not so surprising that many people have given into this because people from Hell are, after all, often extremely frightening, terrifying, and extremely violent. As well, those who do try to tell the truth about witchcraft, and its dangers are made (sometimes by means of witchcraft itself) to seem reprehensible (for one reason or other), delusional or fanatical, and thus discredited.

It is my understanding, and so I have been informed, that the ability to summon up spirits from Hell involves money. As well, the ability to perform feats of more impressive forms of witchcraft and sorcery invariably is going to require sacrificial victims. Exactly how all this is supposed to work I frankly do not know. M. R. James gives some suggestion in one of his fictional ghost stories, "Lost Hearts." The warlock in that tale is given to write down:

"It is recorded of Simon Magus that he was able to fly in the air, to become invisible, or to assume any form he pleased, by the agency of the soul of a boy whom, to use the libelous phrase employed by the author of the *Clementine Recognitions*, he had 'murdered.' I find it set down, moreover, with considerable detail in the writings of Hermes Trismegistus,[63] that similar happy results may be produced by the absorption of the hearts of not less than three human beings below the age of twenty-one years."

It cannot be emphasized enough, I don't think, that the intensity of witchcraft and sorcery going on at a given time and place could vary considerably, same for its character which might be anything from intensely evil, to oppressive but distant, to friendly and comical, with other shades and combinations of these. So that people and the uses of witchcraft and sorcery can take on a wide divergence of appearance and are not at all simply *this* or *that* as

---

[63] It is unclear why the warlock speaks of "Hermes Trismegistus," 3st century A.D., who was rather a neo-Platonist mystic, with teachings believed to have originated from the Egyptian God *Thoth*. The title Hermes Trismegistus (i.e. "Hermes thrice great") itself means Thoth, who was seen as having an affinity to the Greek god Hermes. Possibly the suggestion in this passage is that the warlock is drawing from spurious writings.

they are sometimes treated as being. Hell people, *depending on one's powers of discernment*, can appear as anyone and everything.

There are spirit persons who are warlocks, witches and sorcerers, as well as flesh and blood people. Indeed, the spirit person sort are in the vast majority of cases, if not always, going to be more expert and proficient in witchcraft and sorcery than their flesh and blood counterparts (due to their greater experience.)

This raises an interesting point. Though we perhaps are normally given to think of spirit persons as being other-worldly; such that are sorcerers are truly among the most worldly people in the world. Part of this disposition is seen well in their typically obsessive desire to keep tabs on all that seems to be of importance going on, say in a given community, town or nation. Witches, of course, traditionally have been seen as busy bodies.

Professional witchcraft and sorcery, as I understand it, are a perversion of the laws of God and of nature, and are at their most powerful when fueled by murders. They bring the filthy and criminal dead from Hell among the living, as a means of gaining power to certain individuals. Traditionally, there has been some very obvious aspects of slavery and cannibalism to it. 17th century Protestant theologian, Jeremy Taylor, was correct when he spoke of witchcraft as being a "damnable and impious" practice. Yet we can find admonitions and denunciations of it going much further. The ancient Egyptian story of Setna (from the 19th Dynasty, Ptolemaic period), as translated and retold by Flinders Petrie in *Egyptian Tales* (1913), paints a both vivid and troubling picture of someone's life wrecked by sorcery and magic in a plot somewhat similar in its basic premise to the Faustus legend.

Certain of the beliefs of the Wicca school of witchcraft are praiseworthy taken by themselves. But the practitioners of white magic, such as the Wiccans, as a whole are very deceived and irrational people, and their attempt to prettify witchcraft does far more harm than whatever else good they do. While they would like to think that their white magic can ward off black magic and evil, the high powered sorcerers, warlocks and witches in black magic can in most (if not all) instances trounce a Wiccan or practitioner of "White" magic if they need to. The ability of Wiccans then to combat the power and cunning of the truly baleful spirits then is really after all very feeble, and even something of a joke. They are like children given a toy gun. They say "Bang" and the adult pretends that he is shot by them, both to encourage their false belief and to amuse himself. At the same time some of their magic will be assisted by these other spirits to all the more foster the Wiccan's credulity, and thereby use them to promote witchcraft.

In terms of what is presented versus what is actually going on, witchcraft and sorcery by theirs nature invariably are fake, and or are used to trick and cheat people. Depending on the "captive" audience, they can be used to produce all kinds of illusions. Not all witchcraft is overtly diabolical. There are things like falling in love spells, which can (under the circumstances) be or seem quite pleasant. Yet it must be borne in mind that while something might *seem* harmless, in reality it might be very harmful. Like noxious drugs or drugs used inappropriately, certain witchcraft can give a person a certain "high;" however, it only ends up doing far more harm then good, all the more so as it is used in excess. At the same time, a sorcerer can take very good, indeed most valuable persons and things, and use them as lures and props in his deception.[64]

The reasons spirits are summoned up by Hell, by a given witch, warlock or sorcerer, are various. The witch, warlock, or sorcerer may in a given instance be pressured by someone else to do it. The end result, however, is invariably the same, causing trouble, which could be anything from senseless mischief to something completely evil and malevolent. Where the sorcerer or witch is an experienced veteran it is a fixed and constant rule that the demon they call up, no matter how disgusting and horrible, is invariably a better person than the sorcerer or warlock summoning them.

A "spell" can refer to a number of very powerful different tricks which a sorcerer or warlock might do. One involves changing the feel and appearance of the atmosphere around you. Another might be a full-blown illusion. Yet another might be a combination of the two. In all instances will spirit people of one sort or another be used to achieve the desired effect, combined perhaps with a power of some sort to change atmospheric conditions. Sunlight, cleverly manipulated, seems to be used by them to create very dramatic effects or atmosphere.

There is something like a calendar used in doing certain witchcraft; so that some spells or devices will work better under particular, say for example, lunar conditions. The details of such a calendar I would also not know to tell you. Ordinary changes in weather, such as wind storms, heavy rains, can interfere with witchcraft operations. Saturday night, being before the Christian Sabbath of Sunday, similar to the way Halloween is the night before All Saints Day, tends to be a night when witchcraft people "whoop it up" and are more active. This is true to some extent also of Friday's, the day before the Jewish Sabbath.

"Fallen angels" aside, it certainly would be mistaken to say all spirit people come out of (geographical) Hell necessarily, but the more regularly violent ones assuredly do. On the other hand, sorcerers and warlocks will on occasion

---

[64] Which brings up the point that good or innocuous things can be used for bad purposes, and one must be careful not to mistake the thing (or say person) used for the person using them. References to pagan gods for example in the space program, or Proctor and Gamble corporation's use of a crescent moon with stars as a logo, are instances where the invoking of such names or symbols might seem promoting of paganism or mystical symbolism. While granted in the given instance this may be someone's intention, ordinarily one should not assume a bad or reckless motive unless there is specific, clear, and substantial evidence for doing so. Part of the reason for this is that you can use anything (or anyone), distort it in the minds of others, and thereby use it to serve some bad purpose. But the apparent fault (if there is any) is invariably in the use of the thing (or person), not in the thing itself necessarily.

have a spirit person working for them who apparently is not from Hell as such. One of the "little folk" (to be considered later here) might be an instance of such a person. How and why this might be possible is not clear, but bribery, deception, or violent coercion seem to be the most likely reasons.

A spirit person generally appears like the ghosts in say the films "Topper" or Charles Laughton's "The Canterville Ghost." You can see them, and at the same time (to varying degrees) see through them. To do witchcraft costs a certain amount and kind of energy (usually measured by the number of spirit people employed in the given scheme), and depending on how much the warlock is willing to spend, a spirit can person can be presented or present themselves invisibly or fully materialized. I had one experience of a full manifestation presented to me; that caused me to stutter (so frightened I was) when I saw it. One should normally avoid looking at or conversing with spirit people. One is, for practical purposes, being handed a kind of forbidden knowledge that while perhaps not bad of itself cannot be contemplated without some amount of violence accompanying doing so, and which is not so easy to avoid. Also, they are after all the departed, and or possibly of dubious intention (you don't know yourself after all what they really want) and it is like exposing oneself to radioactivity I've found to welcome or be too casual about their presence. Circumstances over time might well change all this state of things. But for most people and in most instances direct contact with visible or audible spirits should be avoided and rejected.

Sometimes spirits are manifested in such a way that anyone could see them. In most cases they seem to walk, though I have seen some fly (such as an angel and a spirit girl.)[65] At other times, whether or not one can see them depends on the person's (what for lack of a better term) physico-spiritual sight. Spirit in this instance refers to "carnal" spirit, as opposed to pure or true spirit which is invisible. On one occasion in the presence of some sprites, while I couldn't see them looking directly at them, I could see a quick glimpse of them out of the side or corner of my eye. One can also detect spirit persons presence visually in the way infrared will pickup heat in darkness. You cannot really see them as such, but your eyes can sense their pronounced presence in the form of energy taking up a given space. Other times you might be able to see them (i.e., by means of the naked eye) in a kind of indefinite, hazy form.

When speaking with spirit persons, they do not move their mouth – at least not in your sight. They will stand before you and you will hear theirs, or someone else's voice in your head. Yet there are spirits one can hear quite audibly, specifically sprites, the smaller spirits, on a regular basis will have talked to me from under my pillow, and I can feel the vibration of their speech in the pillow itself (or, alternatively and admittedly, this may only be an illusion created to

---

[65] Though I have seen "angels" with wings I have never actually seen the wings used for the actual purpose of flying. In other words the angel moved in the air but without the wings flapping as we might think they would.

make it seem as if you are hearing them audibly.) Demons or sprites can also be got to make (or be made to seem to make) noises, such as gasping, hissing, heavy breathing, screams; which needless to say can be quite disconcerting, if not outright terrifying, if one is not already familiar with these phenomena. This all said, overt noises from spirit people are evidently very unusual.

Ordinarily a spirit is not all that strong physically. Even so, there are ways that certain "energy" can be spent on them to make them so. They can even be made so strong, as to hold down a person temporarily, but this kind of thing is also relatively rare, as it is expensive. Less rare, but still not all that usual, I have had them do things like make scratching noises on the bottom of my bed, rattle my bed, knock things off of shelves, and turn off lights.[66]

Not all spirits one might see are actual spirits necessarily. A sorcerer can conjure up apparitions, which, as well as I understand the matter, are a kind of visual image projected in or on the air. It is possible then that there is some kind of ether, co-present with the air, by which medium spirits travel and apparitions are projected. How does one tell spirit person from an apparition? Usually it is only a spirit person that can be conversed with, and will react to your own speech and actions. Spirit persons, however, can be completely disguised and I have seen (with the naked eye) spirit persons of living people whom I otherwise knew to be alive and well.

~ * ~

Demons and Sprites

"And forth he cald out of deepe darknesse dred
Legions of Sprights, the which like little flyes
Fluttering about his ever damned hed,
A-waite whereto their service he applyes,
To aide his friends, or fray his enimies:
Of those he chose out two, the falsest two,
And fittest for to forge true-seeming lyes;
The one of them he gave a message too,
The other by him selfe staide other worke to doo."
~~~~~ Edmund Spenser, *The Fairie Queene*, Book I.

The people of Hell, as opposed to spirit people generally,[67] take on various forms. The common term for a typical Hell (spirit) person is demon. Our normal use of the term "demon" does not always fairly apply to an ordinary demon.

---

[66] On not a few occasions, I have heard references by these spirit people to bows and arrows sometimes being used, at least against and among themselves, and that is when there is some obvious conflict going on of course.
[67] And about whom admittedly I could little speak from experience otherwise and in any case.

When we use the word demon, what we usually mean is a fiend. But really not all demons are fiends. Most demons are better viewed simply as criminals or convicts. A fiend on the other hand connotes a fanatical, violent and completely amoral maniac; which not all demons are. As well, there are other "kinds" of demons, for example those which are unusually lethargic, lascivious, or themselves in some way distressed or other.

As general rule, demons are insufferably and detestably filthy. The Bible refers to "unclean spirits" which is what they literally and frequently are. Depending on how bad a person they are they will exude a certain unmistakable filth and heat, is very pronounced in some individuals. Some are so repulsive, that those who've had the unfortunate chance to meet such, will know how little wonder it is that the Gadarene swine drowned themselves rather than suffer their presence. By filth I mean the kind of sensation or feeling that might arise out of contact with or proximity to an open garbage can, though without the smell.[68] If demons hang around for too long a time, they can cause vegetation they are near to wilt or dissipate. For a time, the demons sent to bother me would regularly take a certain path to my back door, and over time, they burned a distinct trail in the lawn where the grass could not grow. After I became aware of this phenomenon, they were directed (by the sorcerer who commanded them) to take a different route.

There are two kinds of demons: regular size people and sprites. Sprites or homunculi, are people who, for some reason not entirely clear, are spirit people who were shrunk in size. These people can range in size from a few feet, to quarters of inches. Not all sprites are of the demon sort. Some are nothing more than little mischief makers, or not even this.

After these are souls living in various degrees of degraded states, who are in reality or can be made to take the shape of the most frightening kinds of beings imaginable; including what we commonly know as ghosts and wraiths. These people are customarily kept so down low in Hell that is relatively rare for them to make an appearance up here, and to have them up here is very expensive. Goya's nightmarish murals will give one some idea of what some of these far, lower down persons are like. Not all ghosts are from so far down necessarily, but certainly these more frightful ones are. A ghost as such would seem to be a kind of spirit person who, for one reason or another, is chained to a past (or perhaps relationship with another) that haunts them emotionally and psychologically. That, at least, in addition to perhaps an extra paleness to them, is how I would distinguish them from spirit people generally.

---

[68] Though I have had perfumes (of some kind or other) carried or sent my way by spirit people, I have never known spirit people to actually have a smell, nor to my knowledge do they themselves need to breathe. They are, on the other hand affected by temperature to some degree that is not entirely clear. Spirits seem normally to walk; though some apparently can fly or hover in air in some way or other, though I haven't seen one actually fly by means of literal wings.

Many might think that a demon's being inside a person is rare. In point of fact, probably every person has had a demon (or sprite) in their body at some time in their life, and some will go through much, if not most, of their entire life with one or more sprites or demons in them without even knowing it. In this way, such spirits like to make themselves at home in a person or animal, and will sometimes be able to stay or go as it suits them. There are some lethargic demons who do nothing more than sit in a person or else in a location, say on a bed or sofa, without actually doing anything as such. Yet one will feel their unhealthy presence all the same.

Augustine's claim that "demons" cause illness will be taken as superstition by modern science. But in point of fact what he was saying is to some extent actually true. Yet this does not mean that all illnesses are caused by demons (or sprites.) Keep in mind also that Satan and Satan people can, by mutating nature, very possibly if not certainly create conventional diseases and parasites. A person can be made ill through natural causes, a demon, or both, and it is not always easy to discern which is the true cause of a given illness. In medical diagnoses, as in situations where we are looking for an unknown cause, it is best to prefer a more "natural" explanation. Nevertheless, one should not be so closed minded as to dismiss entirely the possible intervention or participation of spirit people in a given illness. This said, such positing of the "supernatural," of course, should be done with level headed caution and circumspection.

Some sprites do bother animals, and can on occasion be the cause of animal attacks, excessive barking and howling. When Dracula says, "children of the night, what music they make!" albeit indirectly, he is referring to such spirits, not so much to the wolves themselves. On a couple occasions I have had a sorcerer get a large flock of crows or ravens to circle my house repeatedly and for a prolonged period of time while cawing very loudly. This kind of thing are, as far as I can tell, is done with sprites who actually ride on or go into the birds. Also there may "spirit" radio waves of some kind and used in some instances to command people or animals. The main thing to keep in mind about these and other instances of strange animal behavior is that even where aggression is involved one should not blame the poor animal; as they are just being used much like a puppet. This will remind us also, in passing, that not infrequently sprites can be the cause of domestic turmoil or unruly behavior (especially in situations where the trouble in question seems without apparent justification.)

Goon (as in "hoodlum") sprites, a kind of little folk, on the other hand, are, at least in their appearance, actual Hell people. These usually look like bald old men with pointed noses and sunken eyes, and are usually of a gray or greenish color. Some though, not all of these, can look like little fiends. One of my cats was choked and beaten to death by these same people; despite my determined efforts to fight them off. As much as I would pull them off, they kept coming back to attack and there was no keeping up with them. I have, on not a few occasions, grabbed one of the smaller goon sprites and I could feel, *in my hand* holding them, a pronounced rage with which they were filled. These kinds of sprites can go into a person, and cause these same effects, except that the person entered thinks it is themselves that feels this way when really it is the presence of one of these kinds of sprites that is causing the feeling. Negative thoughts can also be triggered in the same way, and this is sometimes done to a person in a subtle and sophisticated way, in order for the given sorcerer to manipulate a persons beliefs.[69]

A number of times in capturing some of these goon sprites, and others like them, I put them in closed jars, but was afterward not able to keep them. The reason I was not able or did not want to keep them is that because (to make a long story short) I was I had been systematically isolated over time, and persons were kept away from me who might have helped me, who I might have given these sprites to for study. At the same time, some of these sprites are so dirty that to keep them around would be like keeping around a disease. As it turned out they either ended up being released by regular people accomplices who would break into my home when I was out or asleep, or else I threw them (sealed in containers) out with the trash. While the latter might sound cruel, do understand that this was a regular war going on, and there would be times that my house might be flooded with these sprites or demons sent over by a warlock – a more horrible experience, rest assured, one can hardly imagine, and such was as much as anyway I knew of getting rid of them.

~ * ~

Little Folk and Other Spirit People

"And did they move upon the stage a thousand years ago,
In some play in Paris or Madrid?"
~~~~~ Al Stewart, "One Stage Before"

Hell puts on a show, and those who come up here invariably are brought up here on the basis of certain abilities and characteristics they have. What we see up here of these people then is not always necessarily representative of the many and various people there are down below. Even so, it seems not unreasonable that one could encounter enough of a variety of them to get a reasonably good idea of what they seem to be like generally.

The following is list of some of the spirit people and "types" who, uninvited, "visited" me in my home

A bearded and robed "prophet" figure.

---

[69] There was an occasion where I was visited by a goon sprite in my bedroom one night – and, rather interestingly, saw his back reflected in a mirror as he passed it.

A young "nun" with a presence I can best liken to fire and brimstone from the lowest nether regions (the full physical manifestation mentioned previously.)

A "bed-sheet" ghost. These types of apparitions are particularly weird and frightening because their form is more like an unraveling sheet then the body of a person, yet they possess a presence and emotions of an actual person all the same.

A horned devil with goat-like beard. He looked like a common Latino, and I inferred from circumstances that he was "dressed up" to make him look like a devil. Like most demons he exuded a kind of dirty heat. With respect to being dressed up, spirits can be made to look more ugly than they actually are. In order to save on "expenses," a more or less regular looking spirit will be dressed as monster to frighten people. Some of the alleged meetings with purported space aliens would seem to have been instances of dressed up demons.

A woman of decidedly dark complexion, and very dark eyes, dressed in 19th century outfit who while forbidding had a certain unusual attractiveness about her.

An approximately fifty-foot tall "angel," with a rather aristocratic visage, standing outside of and next to a church. He did not say or do anything, but seem to just stand there awing the vicinity with his presence.

An "angel" (seen in the middle of a church) swooping down from above on a child, as if to touch it, at the moment it was being baptized. His body was rather fluid, and his movement was so quick that much of his extension was lost in his motion, such that he appeared more slender then he actually was.

A girl who I called "Amelia" or at times "Undine." This girl was both quite pretty and very funny, and somewhat reminded one of "Jeannie" in the "I Dream of Jeannie" television series. She told me things like she lived in the sun, and with her sister (who I called Sylvia, also very pretty, but different in personality), owned a candy company in Colorado (which made fruit candies). I took the latter to mean that some owner of a candy company in Colorado took her and her sister on as partners of some kind. She was ostensibly nude, but you could not see any details of her body. Instead all one saw was the outlines of her body, while its details were rather vague and indistinguishable. Her face and blonde hair on the other hand, were quite vivid. There was a decidedly child like quality about her, and at times her face reminded me of an intelligent infant.

The first time I saw Amelia was both funny and interesting. I was watching the noon news here in Seattle, when I saw this anchor person, who looked directly at me (the viewer) and nodded, as if in recognition of me. There was otherwise nothing at all unusual about the news program, and someone watching it would not have thought there was anything strange going on. The anchor was Amelia, however, and only later after she showed up as a spirit person, that I understood the joke. On this and other occasions, some of these spirit people were able to appear on my television set; as if it was a regular television program. Whether this was simply a manipulation of my perception, or something done with the television, I'm sure it all amount to nothing more than a kind of magic trick. I mention Amelia was somewhat like "Jeannie," but in actual looks, she more resembled Pat Priest on "The Munsters;" except that she was much more rambunctious (though usually graceful) and given to playing jokes. In fact some of the jokes she played on me were somewhat hostile, but like with the little folk, I tended not to be too offended by them because she was otherwise so appealing, and usually I would joke back at her. When I objected to her being associated with Magus (or the Magus, of whom more later), she said she was forced to do these things, but didn't want to do them. Exactly why this was so, and exactly who she was I have never been able to figure out. They told me that she was Norwegian, and had died at a young age in a car crash without having believed in God.[70] Whether this was actually true or not I am in no position to say.

Dora Lee and Escoban were two little folk sprites who for a while sat in my living room, along with two or three other sprites. They usually just sat, Indian style, under a table doing nothing, and when instructed were (by Magus or a warlock working for him) told to do things like put something in my drink (say to make it taste bad), throw little stink bombs, or in some way cause trouble. They did these things in a quite routine way, and without any personal hostility toward me. I would admonish them, but I didn't usually hold these things against them personally. On the contrary I rather liked them for the most part, and sometimes had little chats with them. On one occasion I was playing a Chieftains' CD, and Dora Lee and Escoban got up and did a little reel or jig which was quite funny and gracefully done, going back and forth, then arm in arm in a circle. The names Escoban and Dora Lee were names they themselves gave, but with the understanding they were not their real names. For some reason, these spirit people are not normally supposed to give you their real names, apparently because if you or someone else has it, it will make it possible for you to have power over them. Dora Lee by the way was Escoban's companion, but did not herself cause any trouble. At times, she expressed a sudden and frank affection toward me.

These little folk like this are often quite amusing, and have peculiar ways of seeing things and expressing themselves. They sometimes speak in riddles, often with a high-pitched voice, and make the most odd kinds of remarks, observations and jokes. Sometimes they would take things, and put them where I couldn't find them. I accused Escoban of this one time, and demanded where it was. He replied "You will find it where I put it." I was told these little folk have their own little communities, where in addition to homes, they have little ships which they can sail in, a certain kinds of "sprit" liquids they can drink, and musical instruments which they play. As well (or so I was informed), there is a

---

[70] I was later told after writing this that Amelia did not die in a car crash but it was someone else, so again I was told afterward by someone.

little folk army or militia that is sometimes called up when a need arises, and they feel they can act effectively. Some of these people in their regular lives were soldiers who served in wars. I was told time and again these little folk do not like these witchcraft people, and detest people like "Simon Magus" (of whom more later), but are sometimes forced to work for him and his like under "brainwashing," duress or other intimidation. These little folks were dressed, but I can't (with the exception of Dora Lee) seem to recall exactly what they wore for some reason, only their faces.[71] They do not normally wear the medieval type costumes, such as pointed or coned hats, we usually associate with "little people," but as the situation requires they will sometimes dress up like this for deliberate effect. Apparently, as with us, they are not restricted by age old clothing styles, and prefer to dress more casually. In size, Dora Lee and Escoban were about two feet tall. Escoban seemed to have a somewhat dirty face, and had black hair. Dora Lee had on a red plaid dress, was a blonde, with a cute face, but a face with features somewhat out of proportion to her body.

In Christopher Marlowe's play "Doctor Faustus," Faustus has "Helen of Troy" summoned. I myself have seen some extraordinarily beautiful spirit girls or nymphs, brought to me by these people. At one time in his career, if not currently, Simon Magus reportedly had an attractive companion named Helena of Tyre. The obvious question then comes to mind how did such hideous devil people get such pretty girls to work with them? The answer is what one sees is an illusion, and that the girls, while actual spirits, are "done up" to look better than they do, or else and alternatively they might simply be seen as well treated slaves. A given spirit girl may then be genuinely attractive or appealing in her own right, but is forced to work for these people under some kind of duress or perhaps is deceived herself that by cooperating with the sorcerer they are serving some good purpose. Remember, spirit people also can be fooled by Hell's operators, not just regular people. Are then these girls from Hell? Not necessarily. There is apparently a spirit realm outside "down below" from certain spirit people will participate with Hell, similar to the way regular people willingly or are duped into assisting Hell (which is to say not always out of bad or malicious intention).

A number of these pretty girls I have seen are real as one could think. They have very feminine and attractive presences, and it would be untrue to say I didn't like them or enjoy their presence (usually.) Indeed at one point and for a season early on, I had even fallen in love with one of them. I've had them lie next to me while resting or sleeping, and the experience is a very pleasant and enjoyable one. And even though I have been put through many horrible things I rarely if ever blamed the girls for anything. Like the little folk, they also seemed like slaves acting for others using them. This is not to say they aren't guilty as accomplices, but I never knew of they themselves, doing anything particularly bad. They are simply employees of a kind for the organization. Yet though a sorcerer might be able to get them to work for him this doesn't necessarily mean they like him. Indeed although the sorcerer could in certain circumstances get

---

[71] Dora Lee was the exception. She at least on one occasion had a red plaid dress.

them to work for him, he cannot have them for himself. So while they might at times be slaves, they usually have a higher status or are better treated than other Hell accomplices.

In the course of a sometimes truly agonizing and sometimes ludicrous ordeal, been visited by a number of famous personages with whom I spoke or had conversations with. Although these spirit people bore what, as far as I could tell, were exact resemblances to famous people, they were not actually these famous people. At least I do not believe them to be so, though at the time of meeting them I could not help thinking that they were who they presented themselves to be, so forceful was the impression they made. Some impersonators, on the other hand, were much more believable than others. The following (in no particular order) are some of the famous historical personages I was visited by and or who I spoke with, with the understanding that they most likely were not who they appeared or claimed to be.

- Washington Irving
- William Wordsworth
- Henry Wadsworth Longfellow
- Mabel Normand
- Lew Cody
- William Desmond Taylor
- Graham Chapman
- St. Mother Cabrini
- St. Gemma Galgani
- William Blake
- Edie Sedgwick
- Natalie Wood
- Dean Martin
- Sir Phillip Sydney
- Napoleon
- John Paul Jones
- Lucile Ball
- Harpo Marx
- Peter Sellers
- Krishna
- Bast, the Egyptian cat goddess
- Chaim Weizmann
- J. P. Morgan
- John F. Kennedy
- Robert Louis Stevenson
- Fulton Sheehan
- Von Richtofen
- St. Jean Marie Vianney
- Mary Magdalene
- Charles Nungesser

Marshal Boufflers
Frederick the Great
Robert E. Lee
Kant
Jascha Heifitz
Tiepolo
Rachmaninoff

And others...

Typically they would act either in cooperation with the sorcerer-ghost or else as if in opposition to him. They might make comments on something I was doing, or else tell me to look up something in a book and read from it. The indicated passage then would have some sort of message they wanted to impart. Just in passing, I would remark that they entered the room where I was in as these pleased. This is typically how these Hell employed spirit people will act; which is to say rudely, walking right into your house without invitation or announcement. Yet unlike the more usual violent demons and fiends, they would leave when politely told to – most of the time at least, since some of them were, after all, are demons of a sort.

On more than one occasion, I had a spirit person come to me who presented himself as "Goethe." I was given to understand that he had gone to Hell after he died, but that, these days, things were not so bad for him, at least by comparison with how things were before. I was discussing a theory of colors with some of these people, arriving at the conclusion that inasmuch as color is a frequency of waves or vibrations, that color was motion. "Goethe" told me that there was no point of telling these things to people, because truth is not allowed in this world.[72]

For a couple weeks I received somewhat regular visits from "Ammonius Saccus." "Saccus," the teacher of Plotinus (the pagan thinker) and Origen (the Christian thinker.)[73] I was led to understand he had managed to survive the many trials and travails of the spirit world by his knowledge of sorcery. He appeared like a withered old man, with very darkened eyes. When he came to me, he told me he would help me fight Magus, the sorcerer who was making it his business to bother me. I don't recall much of what he told me. But they were things like I should regularly change my soap because Magus put a certain dirt on it; so that when I used the soap it would dirty rather than clean me. Of course, later I realized that these help "tips" were mostly a lot of nonsense intended to make a

---

[72] This same thing. "the truth is not allowed," was also said to me by the ghost-magician "Simon the magician" on another occasion.
[73] Some scholars question whether the Origen who Ammonious Saccus taught was the Christian theologian or another person with the same name. Eusebius quotes Porphyry, Plotinus' pupil, who seems to be speaking of the same Ammonious in what he writes about Origen. In the same parts of Eusebius' text it is asserted, with accompanying anecdote, that Ammonious became a Christian.

fool of me. "Saccus" was living a wandering kind of life. Having traveled though many galaxies, he had seen much of the universe. But it had got to be a somewhat hard life, and he had grown rather tired. I told him he should be a real Christian and do the right thing. He said there was something to what I said, and would think about it. Meanwhile, he invited me to be his companion one day and fly around the stars with him. This way, he (in effect) said, I would be free and that he could be a great assistance to me with the phenomenal powers and knowledge at his disposal. There were bandits in the universe and in traveling among the planets, stars, and galaxies one had to be shrewd and resourceful in order to protect one's self. Strange to describe, "Saccus" would enter my head (from behind) from where he spoke and from where I could see him. A that juncture I was so interested in combating Magus, and "Saccus" acted in such a friendly way toward me and in assisting me in my fight, that I didn't at that time mind him doing this.

While some of these things might sound quite dramatic, and certainly a given sorcerer like "Simon Magus" loves playing the showman, if you have experienced them long enough they can also be alternatively very painful and annoying, or very stupid and pointless. At the same time, a ghost actually from way, way down below is rarely going to lose all its scaring power; no matter how many times you might see them. Even people of Hell themselves are sometimes frightened of seeing them.

And though some of these people sound interesting, and some of them are, one should keep in mind that I never saw one of these more likable people without the offensive and obnoxious characters around, like a sorcerer or some demon. The girls and little folk were often pleasant company of a kind, and even seeing a regular ghost intriguing in their way. Yet such experiences were not worth the inexpressible amount of pain and suffering others and myself were put through by the incorrigible sorcerer and his more violent henchmen. If one could see the girls or little folk, by themselves, without the demons and warlocks around I suppose one might enjoy their company without trouble. However, that is not, in my experience how these things were done. The sorcerer or warlock in one sense or other paying them to be there. The purpose of a spirit person sorcerer and those working for him doing these things was usually to either tempt, torment or annoy me. The more amiable kinds of spirit people were present in order to gain my confidence, and encourage the belief that these other witchcraft people were somehow benign. This is one way that many "regular" people are fooled into thinking people like Simon Magus and other sorcerers are somehow friendly. Nothing could be further from the truth. Sorcerers like Simon Magus are as vicious and sadistic as you could possibly conceive anyone to be, and if they don't want to hurt you as such they want to get you guilty. That's what it is all about. What typically occurs is that Magus, et al. remain "friendly" with and helpful to a person if they are willing to hurt someone else. A deluded person is then gradually led to think it is not wrong to hurt others. While "Magus" acts kindly and provides him with the necessary sophistry to believe serious wrong-doing is no big deal, a more credulous than usual dupe, might actually murder a dozen people, and think little of it.

~ * ~

Devils, False Gods, and the Damned

"Beloved, do not trust every spirit but test the spirits to see whether they belong to God."
~~~~~ I John, 4: 1-6

"Moreover, if sorcerers call forth ghosts, and even make what seem the souls of the dead to appear; if they put boys to death, in order to get a response from the oracle; if, with their juggling illusions, they make a pretence of doing various miracles; if they put dreams into people's minds by the power of the angels and demons whose aid they have invited, by whose influence, too, goats and tables are made to divine, – how much more likely is this power of evil to be zealous in doing with all its might, of its own inclination, and for its own objects, what it does to serve the ends of others! Or if both angels and demons do just what your gods do, where in that case is the pre-eminence of deity, which we must surely think to be above all in might? Will it not then be more reasonable to hold that these spirits make themselves gods, giving as they do the very proofs which raise your gods to godhead, than that the gods are the equals of angels and demons?"
~~~~~ Tertullian, *Apology*, ch. XXIII

Did spirit people in ancient times interact with "regular" people more commonly than they do today? Did political and social conditions for spirit people change over time as did that of their regular person counterparts? Was, for lack of a better description, the political realm or distribution of power among spirit people different than it is today? Were or are there benevolent spirit people (separate from "conventional" Heaven), who interacted with regular people? Are their angels, distinct from spirit people, whether for good or bad, who did or do take a role in what goes on? For example, the "god" who is reported as bestowing the Babylonian legal code on Hammurabi it might be reasonably was a spirit person or angel. Are those who have been identified as "gods" subject to change, and if so what kind and to what extent? Similarly, if the gods of old were real persons, what happened to them? Augustine makes reference to one pagan view that saw demons as intermediaries between gods and men, (the gods, from Augustine's perspective being fallen angels, see *City of God*, viii. 23.)

Now, of course, some have ready clear answers for these questions based on revelation and their religious faiths, which must be respected. Even so, for someone who wants to examine these matters scientifically, one will have to assume ignorance until something more like conventional empirical proof is forthcoming on which to make a more objective determination. Yet based on what I have discovered about the spirit people through personal communication and interaction with them, and then examining and reassessing historical facts, it is easy to see how certain spirit people from Hell may possibly have influenced certain cultural practices and beliefs in ancient and primitive societies.

It is more than probable that the gods and beliefs of ancient religion were a result of: inspiration (both divine and otherwise); imagination; superstition; and, at the same time, direct contact with spirit people. Human sacrificing Moloch of the Carthaginians, and Huitzilopochtli of the Aztecs were more than likely persons from Hell. Priests and priestess of temples, such as those at Delphi, for example, were probably (at least much of the time) in regular communication with spirit people. A number of Church Fathers took the view that gods of the Greek and Roman pantheons were "demons."[74] It may be, however, that the Greek and Roman gods were actually former regular people empowered by and made dupes of Satan, but were not necessarily persons of malicious intention themselves (or else not so malicious than the one ultimately controlling them.) Alternatively, the pagan gods of antiquity (and elsewhere) were or at least could be seen as "fallen angels" who made it their business to use and deceive people, and or who put on the costume of being a certain god. The savagery of the god and his worshippers in Euripides' *The Baccahe* is very Satanic in character and reflective of both the trickery and claims to piety often so characteristic of Hell methodology and thinking. Even if we take the position that there are no gods as such, what does and has it meant to have spoken of them?

Satan and his people delight in mocking and making fools of others, and it may well be that some of the stranger ideas of some primitive religions were originally false notions inculcated by them. Many of the bizarre and cruel rituals and religious rites referred to in Frazier's *Golden Bough*, for instance, take on a different light when considered with respect to the possibility of Hell's influence. In ancient Egyptian medicine drinking urine and eating excrement were sometimes prescribed as remedies for illness. Such obvious foolishness is just the kind of deception or sick practical joke that a Hell spirit might suggest as a way of degrading or having fun at regular people's expense.[75]

---

[74] Yet much earlier in the Old Testament we find:
"Do you indeed pronounce justice, O gods;
do you judge mortals fairly?
No you freely engage in crime;
Your hands dispense violence to the earth." *Psalm* 58: 2-3.
"All that takes place around these gods is a fraud: how then can it be thought or claimed they are gods?" *Baruch*, 6:44.

[75] It is worth observing when they engage in practical jokes against a regular person of this sort, some spirit people will, (rather ludicrously) feel a certain pride and amusement in their "superiority" at being able to carry out such pranks and deceptions, not aware that they themselves are invariably being made a fool of by someone else.

In Babylonian, Egyptian and Greek mythology, the reigning God is often portrayed as someone who overthrows someone else who was "head" God before them. In Babylonian myth, Marduk overthrows, Tiamat, who attempted to usurp the power of Anu. In Egyptian myth, Horus overthrows Set, who attempts to steal the rightful authority from Osiris. In early Hindu myth, Indra, kills his father, Tvastr (the latter closely related to Vrta, a dragon and or demon), as part of the steps necessary to his obtaining power. With the Greeks, Zeus, defeats Kronos, who had taken Ouranos' place. While it might be argued that in these kinds of stories "Satan" is represented as the first rebel, i.e., Tiamat, Set, Vrta, Kronos, he may nonetheless have been satisfied with these religions in the sense that in some way they legitimize the idea of rebellion against God – though, in fairness, unwitting believers of Marduk, Horus, Indra, and Zeus understandably saw theirs belief as a virtuous point of view – which, at least in part, it certainly was.

In more developed and traditional Hinduism, Shiva is the power of destruction opposite to Vishnu the power of life. Yet destruction as we noted is obviously to some extent natural, and Shiva is sometimes seen as life affirmative in his destruction (as when trees lose their leaves in autumn.) If we posit say a Satan, this would not then necessarily imply either that Shiva was Satan or Satan was Shiva. It is perhaps in this gray area between natural destruction and unnatural destruction that we might say Satan or a Satan is able to conceal and legitimize himself.

By the time of the events of the New Testament, the religion of Zarthustra had been much corrupted with superstition and magic. The original Magi in fact following upon Zarathustra espoused a more pure teaching. It seems very possible that the first creed and belief were a reaction to the heavy involvement of Babylon in sorcery and witchcraft, which dominated the region before the rise of the Medes and the Persians. So much did the Zarathustrians detest the Hell people that failure to leave a corpse on a rooftop to be eaten by birds was punishable with death. The idea being they didn't want their bodies to be buried and therefore nearer to the earth (or "down below"), and further from God who, as fire, was high above. Although Zarathuistra is known as one of the earliest exponents of cosmological dualism – that is the view that the world is divided between good and evil – it was probably adopted originally more as way of seeing good as distinct from evil, as opposed to later dualisms, such as found in gnosticism and manicheanism where it might be made to sound as if good and evil were merely complementary opposites; thus perhaps suggesting they were of equal worth, something Zarathustra certainly did not have in mind.

It may be that amulets and talismans, used in primitive and ancient societies, worked to ward off evil spirits since the user was giving themselves to a kind of idolatry which Satan or a Satan approved of, and hence protected them. This is to say that is by overly respecting a mere object or idol, Orkonist or Goomerist spirit people would not bother them, though of course such a method hardly could be said to guarantee protection.

Of all Satan's powers, he perhaps packs his biggest punch in his use of false "gods." These are spirit people, possibly including "fallen angels," who are presented as persons of great authority, knowledge, and historical background. It is made to seem that they have been everywhere, seen everyone, known everything. They are not without their faults, but otherwise it is implied that if they are not themselves "God," they are gods of some kind. Coming into a person's life, and often displaying to that person phenomenal powers of some kind the person is lead to believe they mean well and are someone of "divine" consequence and importance. What the given individual doesn't realize is that this spirit person they are lead to think is God or someone connected to God, is really nothing more than a Satan or Mephistopheles – even though they present themselves as being, say, "Jesus."

When these characters first came into my own life, I will confess, I was very stupidly taken in, not having ever had any contact with spirit people before. There was a period of about two months, when I was bombed out of my mind listening to them. In some ways they ran rings around me, using all kinds of sophistical arguments, joking, witchcraft, spirits, advice, "inside" information, to believe some of the most absurd and idiotic things. Again, I do not have the space or time here to get into the personal experiences with these people and how they, uninvited, came into my life; as it would go on for quite some length. Suffice to say, these people did a number on me, which though at first somewhat successful, I was ultimately able to overcome through *my love of and insisting on the truth*. These false gods, or agents of Satan, aside from Satan himself, are some of the world's ultimate liars and con-artists, and typically they are armed to the teeth with means of deceiving, enticing, frightening, cajoling, intimidating, and coercing people. How many regular people had the misfortune to meet these characters, I do not know. Yet it is apparently not all that uncommon for them to be in welcome (or unwelcome) contact with some of society's most influential persons, as well as more ordinary people who might somehow be politically or strategically situated. The frequency of such contact, however, may vary according to the moral state of a era and or locality.

Meeting these "gods," if one hadn't met a spirit person before, is in itself, needless to say, quite astonishing. At the same time as they come in to make one's acquaintance, odds are one – unaware – will have assorted kinds of witchcraft or sorcery used on them to make them believe these spirit people are credible (more of how this works later). On top of this they will use all manner of subtle, specious reasonings, humor, and sometimes as well, displays of fantastic power to make a complete fool out of you. They are usually very extremely well-informed people, and, it is not hard to believe that they have been around for centuries as they claim. Even so, they certainly don't know everything. Indeed, paradoxically, one comes to learn that as well as being some of the most knowledgeable and clever people you ever met they are also, in certain ways, among the dumbest and most irrational: wisdom certainly not being one of their strong points. And even if they have been around for ages, and despite their possibly pretending the contrary, there are people, things and experiences which can be "new," and un-thought of, to them also.

There are at least three of these agents (or three types of these agents) that I myself am aware of. It might be argued that they really are one in the same person (or warlock from Hell), taking on different guises, or that one is simply a kind of "prop" or actor working for another much more powerful. To make things more confusing, there might well be multiple versions of the same person. However, in the interest of simplicity and furnishing you with a general and sample idea I will refer to them as though they were three distinct and separate individuals.

The three are:

1. "Jehovah Jira" - or as I prefer to call him "Gyro."
2. "Simon Magus." Though called Simon Magus – i.e., Simon the Magician – it is very much open to question whether he is the same person as the one mentioned in Acts of the Apostles. But for convenience sake we will denote this devil, warlock and sorcerer by this name.
3. "Gomez," also "Satan Jr." or "Dr. Insane." This is a person who is made to look like Jesus Christ and or is said to "represent" him. These names I refer to him by were ones I simply made up for my own convenience, as it isn't quite clear exactly who he is.[76]

Although they will typically appear affable and sympathetic, one should know that they have been involved in some of the heart wrenching of crimes, and behind that mask of geniality, clowning and seeming good nature lies an abyss of such sadness and sorrow as such as you could never fathom. They are very indifferent to others suffering, particularly Magus and Gomez., and will casually exhibit the most monstrous and cruel kind of insensitivity, rationalizing some terrible thing that is going on as if to say "that's just how it is," "it is better if things are that way," or "it doesn't matter."

The lies they tell and the roles they will play will overlap. For example, in my own experience, Gyro (who looked somewhat heavy set and had a trimmed white beard) for a period pretended *as though* he were "God the Father;" while Magus pretended he was "Jesus." The lie or method of deception they use depends a lot on who they are trying to take in. Though often acting together, they are not always in complete agreement, it being generally understood for example that, relatively speaking, that Gyro is a moderate devil compared to Magus'

---

[76] In other writings of mine, I will sometimes refer to this person as the ghoulish magician, but again it is not necessarily always possible to know what specific individual one is dealing with; so that after a while I came to use "the magician," pretty much generically, for just about any of these more high powered, sorcerer type spirit persons.

(relatively) more extreme devil, and there may yet be others who are worse than both.

Typical propaganda they will use is to disparage both Christianity and/or Judaism, and or set one off against the other. "Jesus" is not what he used to be, I was told. Since the Reformation things have been hard for him and he's become confused (of course it is assumed, contrary to orthodox Christian belief, and in the vein of Arianism, that Jesus is nothing more than a powerful spirit person). As well, Jehovah Jira himself, as "God the Father," is also not what he used to be, and spends half his time ruling Hell as well as Heaven (this phony Heaven, by the way, to those who already know what it is, is sometimes referred to as "Gyro Heaven," – an illusion created by sorcery and spirit people). The "House of Israel," these spirit people will say controls Heaven, and that the Jews are selfish, dirty, greedy people intolerant of others. "Jesus," it is said, has a hard time dealing with them, but he does his best to try to work things out for non-Jews. Some religious figures, such as King David will be presented in a positive light, however, only for the purpose of making Jews look bad in general. For example, they will represent him as saying the Jews won't accept him as leader any more, and have in general made a mess of Heaven, despite his best efforts. It may be the case that there are people living in Hell who are led to think that they ended up in Heaven; trusting as they did in one of these false spirit people, and to this day do not know the difference. This is not so strange a possibility as it sounds because a powerful Satan person can create illusions that seem (to more unthinking and unfeeling people) quite heavenly.

The universe is presented by Gyro and Magus, as essentially a dreary place. "God Almighty" (somehow different from "God the Father") is a very distant, tired old fogey, mad and senile; who doesn't know what he is doing. He bought us a bargain basement kind of world, which is very finite, and which will end up wearing out and collapsing, like an old house. "Jesus" is sorry things are this way, and does his best to distribute good, but there is only so much to go around for everyone.

All these kinds of arguments and explanations, are complete rubbish, of course, and originate entirely from Hell. Nonetheless, people, both below and above including devout and well-meaning religious, are still taken in by these kinds of ridiculous stories and explanations. Yet in fairness, one should understand that a person who they try to brainwash, will have various kinds of witchcraft, demonism and drugging done to them, that they might be more susceptible to Gyro and Magus' influence and manipulation. What's more they can be very heavy handed in doing this.

Characteristically, they might tell you something that may well be true, yet not the whole truth. They will, for instance, relate historical anecdotes and information, which to say the least, one has to take with a large grain of salt, as they are notorious liars, even though what they might tell you sounds very plausible or have some amount of truth to it. Likewise, for example, they might say so-and-so should not be listened to because he is an immoral person. But what they neglect to tell you is that by immoral they mean that is that so-and-so cussed, jay-walked the other day, or did something wrong long ago and which they've outgrown. This using actual facts to misrepresent someone or something is a regular practice with them.

They always try to justify themselves by the outcome of whatever happens. If things turnout badly, they were right all along, and that's just the way things are. Or if things turn out well, they will say, "you see, I was trying to help everyone all along" – always trying to have it both ways in this manner.

Outside of the less subtle hellions who work for them, they cannot openly take pride in playing the villain. What then they might do instead is attempt to justify themselves by playing the "tempter;" that is to say, they are (so to speak) merely "finding out" those who are not truly worthy of God's blessing, and in this way make to seem as if they are serving a good purpose. While personally I do not accept this version of their carrying on, I am at least willing to concede that there may, in theory at least, be something like a plausible argument to back it up. Howsoever bad as Hell is or gets, one must remember that there is probably some truth, albeit a very small truth, behind what they are doing or believe. It is their taking things to an extreme or distorting them, nonetheless, that actual evil comes about.

While the same is probably true of most other Hell people, certainly with these "gods" taking advantage of and or robbing you in some way is not only routine and predictable, but mandatory. Indeed, inmost instances, it would be remiss and irreligious of them not to do so.

~ * ~

Gyro

"In the German stories we read how men sell themselves to – a certain Personage, and that Personage cheats them. He gives them wealth; yes, but the golden pieces turn into worthless leaves. He sets them before splendid banquets; yes, but what an awful grin that black footman who lifts up the dish cover; and don't you smell a peculiar sulphurous odor to the dish? Faugh! Take it away; I can't eat. He promises them splendors and triumphs. The conqueror's car rolls glittering through the city, the multitudes shout and huzza. Drive on coachman. Yes, but who is that hanging on behind the carriage? Is this the reward of eloquence, talents, and industry? Is this the end of a life's labor? Don't you remember, how when the Dragon was infesting the neighborhood of Babylon, the citizens used to walk dismally out of evenings, and look at the valley's round about strewed with the bones of the victims whom the monster devoured? O

insatiate brute, and most disgusting and brazen, scaly reptile! Let us be thankful children that it has not gobbled us up too. Quick. Let us turn away and pray that we may be kept out of the reach of his horrible maw, jaw, claw!"
~~~~~~~~Wm. Makepeace Thackery, *Roundabout Papers*

The name he actually goes by is "Jehovah Jira." Although the name may be legitimately his, I resented and took exception to the fact that it carried with it a possible religious significance. Hence, I called and do call him "Gyro" instead; which he did not like me doing. It is claimed, by the way, that at some point in the history of the Jewish people, that he was accepted as "Yaheweh" by many Jews, and that he will still manifest himself to some individual Jews as such. [77]

It should noted here that the name "Jehovah" itself is a misnomer. In Hebrew, which doesn't have written vowels, the name of the Lord is Yhvh or Jhvh is usually translated Yaweh or Ya-he-vey. In Renaissance and Reformation times, theologians translated Yhwh, by arbitrarily using the vowels of the word "Adonai" ("Adonai" being a word the Jews would use as "Lord" in prayer) for the missing vowels in "Yhwh." Consequently these theologians somewhat mistakenly translated "Yhwh" or ("Yhvh" or "Jhvh)") into "Jehovah." Now if "Jehovah Jyra" was really who he claimed, why would he himself use this artificial, modern translation of his name?

He is the more traditional devil, and is often the consulting devil of business executives. The exact extent of his own influence is difficult to gauge. No doubt he has had some notable clients in his time. On the other hand, he and others working with him will claim to have had contractual agreements with famous persons, which claims one should view with skepticism. A very well known English rock band was reported to have made a deal with him; when as it turns out what really happened was, they didn't really know who he was and did a small favor for him, by playing certain songs at a concert (or something trivial like this). More than this, he, or else some Hell people, will tell completely fabricated stories about how such and such a famous person made a "Gyro deal," i.e., a pact with the devil. Again, when and if one ever should happen to hear such claims, one should be view them with an extreme caution and skepticism.

There are and have been some persons who thought they could accomplish good by doing a "Gyro deal." It should be obvious that this is not a very smart idea. At worst, such well meaning pacts with the devil, will end up in tragedy, and at best the person will only end up making a total fool of themselves.

One of his favorite phrases is that if things aren't done a certain way then "Nothing will get done." Exactly what particular significance it has, given that it could be interpreted in different ways, is not entirely obvious. It can be

---

[77] In one talk that I had with him he told me he was younger than the Peloponnesian War (that was before his time, he said), though he may actually be a good deal younger.

mentioned, however, that when I have heard this phrased a certain humorous sarcasm is seemed to have been implied.

Both in the past and present, a number of powerful and influential people will have listened to him to some extent or other. He will tell such people that if things are not done with a certain amount of wrongness to them, then they will not turn out all right. For instance, if people do things ethically and legally, "nothing will get done," and the US will deteriorate into a backward economy, such as one might find in Central or South America. Truth is, those countries in Central or South America are in as bad shape as they might be in not because they didn't listen to Gyro, but because they *did* listen to him.

In some ways, and if he doesn't abuse you too much with his company, he is a regular fellow, and just a rogue. Contrasted with, say Simon Magus, he comes across as Long John Silver, or someone like that; and which you might say thus makes him something of a moderate. He's been in on any number of killings and plunderings, but he hat least has *some* actual (as opposed to false) sympathetic human qualities. This said, he is not someone one should talk to or deal with, and certainly he is not someone one should trust.

Compared to other Hell people he and Magus apparently live materially well and have their own homes, if not mansions or estates.

I was told by spirit people that at some time during World War One, Gyro was captured by the Germans, but later managed to escape. Whether this is actually true or not as such I would not know. Yet it brings up the point that these people such as Gyro and Magus are not invulnerable, and it would, I think, be possible, in the right settings and circumstances, to stun or otherwise incapacitate them through loud or high pitched sounds, or certain energy waves or impulses (such as electricity). I know this from my own personal dealing with them, particularly Magus, and have on occasion found these or similar measures effective in temporarily getting rid of him. The main problem of course would be keeping them pinned down without their knowing that an attempt was going to be made on them – a not very easy thing to do naturally.[78]

~ * ~

---

[78] In late August 2003, a story was reported in the news of a man from Erie, PA., apprehended by police for allegedly being engaged in a robbery, who had collar strapped around his neck which contained an explosive device (of some sort.) The man said he had been told that if he didn't carry out the crime the collar would (by unknown accomplice) be detonated, which it subsequently was, and, as a result of which (while in immediate police custody) he died. I mention this incident because at one time Gyro spoke of being burdened with a very similar sort of device, but one implanted in his shoulder (presumably by "the House of Israel" or else "Satan," with whom he claimed to have sparred unsuccessfully with on a number of occasions.)

Simon Magus

"Simon Magus so deceived the City of Rome that Claudius erected a statue of him, and wrote beneath it in the language of the Romans Simoni Deo Sancto, which is translated 'To the Holy God Simon.' While the error was extending itself Peter and Paul arrived, a noble pair; and the rulers of the Church; and they set the error aright. When the reputed god Simon was about to show himself off, they showed him for a corpse. Simon promised to rise aloft to the Heavens, and came riding in the air in a chariot of demons. The servants of God fell on their knees...(and) launched their like-mindedness in prayer against the Magus, and struck him down to earth. It was marvelous enough, and yet, no marvel at all; for Peter was there...for Paul was there...and they brought the reputed god down from sky to earth, to be taken away to the regions below the earth."

~~~~~~~St. Cyril of Jerusalem, *Catechetical Lectures*

"This Simon, who perverted many in Samaria by magical arts, was convicted by the apostles and denounced, as is recorded in Acts; but afterwards in desperation he resumed the same practices, and on coming to Rome he (again) came into conflict with the apostles; and he perverted many by his magical arts Peter continually opposed him. And as his end in Gitta drew near, he sat beneath a plane-tree and taught. And now, being almost discredited, in order to gain time he said that if he were buried alive he would rise again on the third day. And ordering a grave to be dug by his disciples. He made them bury him. So they did as he instructed him, but he has remained (buried) to this day, for he was not the Christ."

~~~~~~~St. Hippolytus of Rome[79]

The spirit person "Simon Magus," (or "Simon the Magician"), is, without exaggeration, simply one of the most hateful monsters and criminals there ever was. Even so, people who have known him, will think he is not so bad. Fact is, Gyro is a saint by comparison. Many people, and animals as well, have gone through horrible pain and suffering so that someone like Simon the Magician and Faustus, can live a life of play, luxury and power.[80] At times he has reminded me of an ancient convict brought back to life; while being deceived into thinking he can somehow live his life over again. As well, he has reminded me of a middle aged ne'er do well who come back to hang out with the kids in the high school parking lot, gain the their confidence with his "street smarts" and sell them drugs. Many people of his own time have told him on not a few occasions to stop playing with children, and act his own age.

What takes some people in about Magus is that in addition to his talent at magic, he possesses or pretends to possess a great amount of knowledge on

---

[79] As quoted in *New Testament Apocrypha*, vol. II, edited by Wilhelm Schneemelcher.
[80] He has told me that if he didn't do things "this" way he would otherwise be reduced to the drudgery of digging ditches down below.

various subjects. He can carry on interesting conversations, all the more so the more ignorant the person talking to him is. Except when it comes to hatching plots and schemes, he is impossibly irrational. This would not in itself be such a terrible problem; except when it is combined with his shameless arrogance and his acting like he knows everything.

The actual Simon Magus of history was one of the founders of Gnosticism. It seems to me likely that the historical Magus became what he is because he believed Jesus was just a powerful spirit person, or a person in regular contact with spirit people, so that Magus then had a mind to achieving something similar.[81] Whether the Simon Magus of whom I otherwise speak here is the same, is not certain. Like the historical personage, this Simon Magus is a lover of attention, a showman (indeed a megalomaniac), and a person who will claim secret profound knowledge of things. He, as a sorcerer and magician, it can be conceded he is quite proficient, and can do some spectacular magic. He has powers of reading one's thoughts as one is thinking, and can make one see images in one's head (I call this phenomena "witchcraft or sorcery TV") which he chooses to project. Like with any magic trick, the magic trick of witchcraft have rational, physical explanations which if explained to you, you would see how it was done. I have, in fact, met different Simon the Magicians, and the one who most actually seemed like the real one, and yet whose visits to me were rather limited compared to the other "Simons," had a manner of speaking that reminded me of the voice of Wile E. Coyote.

Magus' sort of character was probably isolated, gotten to reject (true) good, and then gloried in self partly because he felt rejected by others; which the evil one very much encourages. But the evil one is not much of a friend so it ends up leaving such a person very lonely. Yet the Magus sort of character has power, and he will use that power to force himself on people (in one way or another) and thereby find for himself a social circle he finds acceptable to a person of his "importance."

The following passage from William Gilmore Simms *The Partisan* in many ways I found to be a good description of Magus: "The insane man usually exhibits the possession of no little vanity. A diseased self-esteem is apt to be an active condition in the mind of most lunatics, and has contributed not a little to

---

[81] St. Irenaeus (as quoted in Eusebius): "Simon [the Magician], we are given to understand, was the prime author of every heresy. From his time to our own those who followed his lead, while pretending to accept that sober Christian philosophy which through purity of life won universal fame, are devoted as ever to the idolatrous superstition from which they seem to have escaped…Their more secret rites, which they claim will so amaze a man when he first hears about them that, in their official jargon, they will be wonderstruck, are indeed something to wonder at, brimful of frenzy and lunacy, and of such a kind that not only can they not be put down in writing; they involve such appalling degradation, such unspeakable conduct, that no decent man would let a mention of them pass his lips." (*History of the Church*, II.13.) For more on the historical Simon Magus see ,for example, the entry contained in *A Dictionary of Christian Biography*, by Henry Wace and William C. Piercy.

their mental overthrow. The madman's vanity is delighted when he can show you that he schemes and contrives. He loves to startle you. He anxiously seeks to extort from you acknowledgments of this character, and would seem to be pleased with complicating his own purposes, if only to compel your admiration. The lingering reason still strives to maintain some of the shows of its authority – of its presence, at all events – in the brain of the unhappy man, in which it harbours, like the fiery volume in the core of the volcano only for explosion. Feeble, willful, and deprived of all its best auxiliars of steadfastness and judgment, it still seeks, if not to establish, to assert its supremacy. How it plans, with what effort; how contrives; how chuckles over its contrivances; and with what grotesque ingenuity it will combine and create! This cunning of the madman is, perhaps, the true key – if there be nay – to his disorder. Properly studied, and you may find in it the true key – if there be any to his disorder."

He has sorcerer associates, and sometimes Magus will be identified with something they have done. While this isn't always fair, it is, given his character and for practicality's sake, understandable and excusable. If it is not quite clear to me who it is, I will sometimes make up a, usually derogatory, name for whoever the sorcerer bothering me happens to be.

"Satan" (or someone like this) has a strong hold over Magus, and in fairness, it seems that if Magus could be got away from the former's influence it might be possible for him to stay out of too serious trouble – at least to some extent. As it is "The Evil One" works him like a marionette, sometimes using him to perpetrate some of the most inhuman and sickest kinds of crimes worse than any nightmare or movie. Part of Magus' wrong doing stems from the fact that he wants to be "a great man in a great show." He loves power and can't get enough attention, and will, as necessary, stoop to the lowest evil to get these. He and his witchcraft associates are so habitually vicious and violent that they regularly have to take their aggressions out on people or animals, really it is almost a religious thing with them.

There are times he expresses something like remorse, or at least regret about the rapes, murders, and atrocities he has participated in, and occasionally one gets the sense sometimes that he is like an alcoholic who can't help himself. This said, one should not feel too sorry for him, for he is also one of the most hypocritical and egocentric persons who ever lived. Were it possible to get rid of him a phenomenal amount of the world's problems would be solved in one fell swoop. Magus and his associates will sometimes say they have done so many things they don't think doing more wrong matters. While in a way this is true, they can, even so, make their situation more difficult by being more arrogant. It is such incorrigible arrogance, rather than guilt as such, which acts as the greatest pressure on them.

He will try to be funny and genial, and on rare occasion (and if you ignore everything else going on), it can be admitted that he actually can be. But taken all in all, he is the most insufferable and detestable monster one could ever encounter. Many of the tragedies of the holocaust are ascribed to him, and after having been forced to get know him, this is not at all hard to believe.

Not uncharacteristically, he will act as though to befriend you, and seem to do one favors. At other times he will play the clown, and seem like just a silly character. One is reminded of the 19th century fiction creation, Varney the Vampire. Yet one should not be deceived. He received his great powers, by betraying the entire human race, and, truly, a more loathsome and vile individual you could not possibly imagine. He says he has suffered terribly. But the truth is he apparently hasn't even been properly punished! If he really had been he would not still be able to be about doing the wretched things he does, while still, as well, being able to play miser and despot.

As well, as being a Satanic fundamentalist in his incessant wrong-doing, Magus is an insufferable gossip and celebrity hound. One gets the sense that because "he has nothing to live for," he is trying to live his life up here all over again; except that the only way he can be up here is if, under Satan's aegis, he is willing to cause trouble. He lives on a kind of dope that makes him think the hideous things he does don't really matter, and he can just kick back and not worry about. He lives his life like someone who steals money all the time, while never thinking they will have to pay it back.

Magus will routinely go around impersonating or at least looking like "Jesus." This is part of his way of doing things wrong, while naturally it makes Jesus look bad (to some.) When Magus or one of his cronies impersonates Jesus, he will look like the "Jesus" person in the Edie Sedgwick film "Ciao! Manhattan." Incidentally, it is my strong belief she was bothered by him (or his cronies) as well, and may have in some way been responsible for her early death. What he actually looks like without his "make-up" is something out of a horror movie, and is as dreadful and filthy as it gets. With very darkened, and somewhat sorrow filled eyes, truly, there is damned look about him, as if he carried around the void wherever he went. He has told me himself that he contains with him a large number of smaller demons that it is impossible to rid himself of.

In impersonating Jesus, he will sometimes seem to speak out of the sky and say things like "I am the Lord your God who brought you out of the land of Egypt," and "I am the Resurrection and the Life," "I will be with you always even until the end of the world." In this way he and Gomez (another Jesus impersonator) will get people to think he is Jesus.[82] On one occasion, while sitting alone outside my home, I saw Magus (pretending to be Jesus) standing in the clouds looking down at me. At the same time speaking to me in my head (strange

---

[82] Many Greeks thought of Zoroaster as a magician (see for example Plutarch and Pliny the Elder), possibly because a similar sort of spirit person went around posing as Zoroaster, and it was in this way and for this reason (i.e., a spirit person imposter) perhaps that Zoroastrianism became most corrupted and associated with magic.

as that sounds) he directed my attention towards the east and displayed a huge vast whitish orb in the sky that looked like the sun. He claimed this was "God the Father." Even so, by that time, however, I had figured out who he was and told him, in not so polite terms, to beat it and get lost.

Commanding spirit persons like Gyro and Magus, can very easily enters one's head with sounds, thoughts, and images, in a manner which brings to mind the refrain from the Moody Blues song, "Melancholy Man":

> "The beam of light will fill your head,
> and you'll remember what's been said,
> by the all the good men
> this world's ever known."

Sometimes, though more unusually, they can create this or a similar experience from within the very center of your head! Of course, this can be quite overwhelming or simply horrible depending on how you take it.

The impact of these tricks and devices on an individual, and they can be both quite frightening and intimidating, and give a person to think they are dealing with someone from God. Magus and Gomez (of whom I will be speaking presently) also have what I call "Wheels in the Sky keep on turning," or the "religious machine" which, upon "turning it on" will make you think you are at the center of some great religious experience, usually coming from the sky. These kinds of charades, typically involving spirit people (say for instance an "angel") accompanied by unusual lighting and atmospheric effects, have proved very effective in fooling and scaring people into going along with them.

As spoken of earlier, among his and a Satan's many tricks is bring spirit people into your midst, and these might not only be devils, but beautiful nymphs, angels, saints and historical figures. Some are so engaging it is very difficult to know if they are who they are presented as being, or else another in disguise. As a general rule, nevertheless, if such are ever manifested to you, it is safer to assume they are false, as Satan's powers of deceiving are quite extraordinary. One sure way of protecting yourself, is to make sure they do not encourage or incite you to any kind of wrong thinking or doing. I have had seeming angels in my presence, only to realize that they were just Satan (or Satan-like) people trying to get me to put my trust in Magus or Gomez. Truly, such spirits can seem as if they came right out of Heaven, when in truth they were, or at least in the employ of, the opposite camp.[83] The question might be asked, if Satan is so awful how can he know about what a beautiful or Heavenly person might be like? I believe the

---

[83] It will ever (presumably) be among Hell and criminal spirit people's tricks to present and pass themselves off as all Heaven and all goodness. And invariably many will be taken in by this. But the wise, however, will always and courageously see through the fraud by scrutinizing and subjecting such divine displays and pretensions to and according to the standards of honesty, right reason, and right, just and common sense morals.

answer to this possibly is that having been at his business so many ages, he managed to somehow "photograph" (for lack of a better term) such divine persons and divine like events and thus is able to produce false copies. To expect to see actual Heavenly angels for most people, in this fallen world, is rather presumptuous and perverse, and it makes much more sense to first look to see God in more ordinary goodness and circumstances

As an experiment to get a sense of what *one* of these kinds of experiences can be like, have someone sit in a chair, while instructing them under no circumstances to move their head. Then from behind them, have another individual, of whom the first is entirely unaware, speak them from behind and say something like "I am the King of Kings, and Lord of Lords,"[84] "I am the resurrection and the life," or "I am the God that brought you out of the land of Egypt." The second individual, preferably someone gifted with a little dramatic ability, could do this in person, or by means of a loud speaker. Check then to see how the seated person reacts, and then further see how the "god's" voice sounds yourself if you are placed in the seated position. Keep in mind that when Magus or a spirit person actually does this kind of thing, they will use additional effects (and spirit people) to both frighten and deceive a person.

To use or properly do hypnotic suggestion, a sorcerer must keep subject from awakening, yet at the same time also keep them from going into too deep a sleep. In this twilight of conscious can he best achieve his behavioral effects. To illustrate, one unusual kind of torture/trick Magus does is that he will have you dream you are trying to solve a puzzle, or else reading something to comprehend it. Then he will interrupt the flow of your thoughts (most likely with some kind of sprite), and in this way will repeatedly frustrate your solving the puzzle or comprehend the reading. I have had this done to me numerous times.

Magus will, if you allow him, do this strange thing in which he (or at least seems to) place his hands on your head and does something that feels like he is splitting your brain. There is no physical pain involved in this, but it does feel almost as if your brain (less so mind) were being split in two. This will have more affect on some people than others but, in any case, it is not something one should permit him to do, its obvious purpose being to promote double-mindedness.

Brain washing, which is his stock and trade, involves changing one's world view, and in this way the rules of the world are seem to change. How we see something can be changed by our apprehension of it when it is perceived. Brain washing works on the same principle, except that what is changed is not merely one's emotional "a priori" apprehension of a particular object of perception, but such apprehension of the world as a whole.

Below are some mind control techniques someone like Magus might use:

---

[84] This was also a favorite title among the Pharaohs of Egypt and the Kings of Persia, among others we might name.

1. Change a person's world view, and by this means disorient them with respect to someone or something else.
2. Give a person arguments against doing good that will inflate their pride, excuse their selfishness and fear.
3. Fill them with great fear of doing good.
4. Make them think that doing good is really foolishness, while at the same time permitting them an excuse to give into their deep set fear.
5. Give them to make a virtue of a weakness, and a vice of virtue.

In my earliest dealings with him, Magus did this thing where he told me you to look out your window or to go outside, directing me, in either case, to see something. He then would point out, say, a cloud or a street light, or just about any object, and tell me there was some secret meaning behind it. At first, though puzzled, I was taken in by him in this, only to later realize that one can read secret meaning in any possible thing one might look at if one is disposed to do so. So there was nothing particularly significant in what he was directing me to look at, as far as a secret or prophetic meaning; only that he wanted to use such a device to get me to believe his propaganda, and be dazzled by his seeming profound knowledge. Poets see meaning in otherwise ignored objects, but this was different inasmuch as Magus had you look at objects not to see poetic meanings in something but rather literal or prophetic ones.

To relate one story, it was late at night and there was a full moon. At this point in my ordeal, I was getting quite sick of Magus and those with him. He ("dressed" as he always was as "Jesus") told me to come with him outside, he wanted to show me something. I followed him outside and he pointed up at the sky at the full moon. I looked up at the moon and saw (the head of) Abraham Lincoln, drinking liquor out of a bottle (i.e., this animated image appearing on the full moon itself.) "You see," Magus, in effect, said, "that's how things would have ended up for him, if he had done things his own way." Of course, "doing things his own way," meant not listening to spirit people.

He might tell someone to pray for another and if enough people pray for that person he will lay off hurting them. Again the purpose of this trick is to make it seem like he possesses divine authority and is somehow really helping people for all his otherwise obvious wrong-doing.

Magus can be fooled by "false cause" tricks, just as he can fool others using the same tricks. By "false cause" trick, I mean a deception by which someone is made to believe that something or someone causes something else to happen when in fact the seeming cause is no such cause at all. Talented sorcerer as he is, Magus' own tricks don't always work, and when they don't he will sometimes adopt the expediency of pretending that they did.

Because Magus has come up so frequently in my discussions on Hell, he may well be viewed as a type and not necessarily a single specific person; for which reason I refer to this type also as "Archimago;" taking that name from the sorcerer in Spencer's "Fairie Queene" who in certain respects is very like Magus, and vice versa.

~ * ~

Gomez

"The hour is coming when everyone who kills you will think he is offering worship to God. They will do this because they have not known either the Father or me."
~~~~~~~John, 16:2-3

"Then there's no reason for a god to speak falsely...a god then, is simple and true in word and deed. He doesn't change himself or deceive others by images, words, or signs, whether in visions or in dreams."
~~~~~~~Plato, *Republic*, Book II

What makes "Gomez," (or as I have also called him "Satan Jr.," and "Dr. Insane") so dangerous is that he appears to be someone very benign and kindly. With Simon Magus, one can ultimately detect the air of the pit about him. With Gomez on the other hand, the disguise as "Jesus" is more complete. Since he seems to be the most powerful among the three "gods" mentioned here, Gomez may well be Satan himself, or perhaps Simon Magus acting with Satan's own greater powers at his disposal. It should be mentioned that neither Magus nor Gomez overtly say they are Jesus. What they actually or might say is that they "represent him," or otherwise in some way loudly imply they are him (without directly stating so), all the while going around looking in appearance as if they were him. In the case of Gomez, he will act as though he cares about you, and that he is against what Simon the Magician and Gyro are doing, but is somehow powerless to stop them.[85]

It must be understood that there was as short time in which this Gomez person fooled me, even after I realized Magus was a fraud. Like Magus, he can know the most obscure private facts about you and "predict" things. Such powers of prediction, it is conceivable, may arise from the use of some kind of

---

[85] In his *Ethic*, Spinoza wrote: "If we imagine a certain thing to possess something which resembles an object which usually affects the mind with joy or sorrow, although the quality in which the thing resembles the object is not the efficient cause of these effects, we shall nevertheless, by virtue of the resemblance alone, love or hate the thing."

"astrology" (or, as best we know, something very like),[86] knowing certain patterns of events and behaviors in the grander scheme of things, or orchestrating events so that a certain prediction turns out to be true or inevitable.

Regarding predictions, one could say a number of things. For one, It must be remembered that many of these spirit people have been around a very long time. As a result, they can see patterns in regular peoples lives, behaviors, relationships, outlooks and beliefs, but which to us are entirely new by comparison. What seems new or inexplicable to us (such as a fated meeting) is familiar enough to them. Knowing these thing they can manipulate regular people who do not even have the faintest idea of such otherwise regular and natural patterns. These predictions work very effectively in fooling people, and entire nations, as well as individuals and families, have been ruined by such deceptions.

Or similarly take, for example, names or identities come joined up together under seeming extraordinary circumstances. Now let's say a "god" (this might be a spirit or regular person of great power) whom we'll call "A," might (for whatever reason unknown to us) be interested in promoting the fame of say a family name, say "Smith." Now when a Smith achieves notoriety, other Smiths will think themselves perhaps in someway honored. But though the promotion was intentional by god A to only one particular Smith or Smiths generally, certain *other* Smiths took the intention and ascribed god A's interest (as being directed) to they themselves personally somehow – that is, if they attempt to read something mystical into the coincidence of god A's promoting Smith's name with their own name. Now a devil might well know something like this, and encourage the other "Smith" to think that god A intended his effect for him personally, not, say the Smith name in general or a separate individual Smith, and thus perhaps start goading him to conceit on that basis. But this again is one isolated example for which many could be raised or given.

Since Gomez probably goes around calling himself "King of Kings, and Lord or Lords;" while others pray to him under the same title (mistaking him for Jesus), this might explain why he has such great power. In other words, his megalomania combined with the enthusiasm of deluded fanatics perhaps generates a very powerful and intimidating persona, even if, at bottom, he is only a fraud masquerading. Alternatively, perhaps he is someone who has been deceived into believing he is who he says he is, such that when he makes divine claims he believes them true – but only because he himself has been deceived.

Gomez's powers are far greater than that of Simon Magus. He can create experiences and feelings in a person which seem like the Holy Spirit. Similarly he can cause a feeling in one that makes one "feel" Catholic (something I am at a

---

[86] Augustine: "We must confess that when the truth is foretold by astrologers, this is due to some most hidden inspiration, to which the human mind is subject without knowing it. And since this is done in order to deceive man, it must be the work of the lying spirits." *De Genesi ad Litt.*, II, 17

loss to describe but which many other Catholics will be familiar with). He can ostensibly cause astounding changes in nature, such as the weather.

If this person can do these things, one well might ask, how does one know he is not actually Jesus? My response to this question would be:

1. He himself doesn't claim to be Jesus, but says he represents him. This after all is quite silly, inasmuch one would think if Jesus Christ sent someone to represent him it would be a saint or an angel, not a celebrity look-alike.

2. The very nature of the look-a-like approach is a mockery of Jesus.

3. Orthodox Christianity views Jesus as the Son of God. If this person did represent Jesus, such a view would turn out to be false. Now I of course understand that not everyone is an orthodox Christian, and will not find such a rebuttal compelling. Nevertheless for those who are Christian I offer it.

4. He is mostly indifferent to and trivializes suffering cruelty. His attitude is like, "don't worry, things will work out all right in the end." While in a way this does make sense, on the other hand it reflects a very shallow attitude about life and directly conflicts with the Jesus of the Gospels who wept at the death of his friend of Lazarus, and mourned the people of Jerusalem.

5. The actual Jesus was a noble person of great courage and compassion. There is no sign of anything like this in this person.

6. He is a Pharisee in that he nit picks at relatively small things, emphasizes things like diet and religious formalities, or things which given the circumstances are of relatively minor importance. Simultaneously, he has a rather cavalier attitude about great moral outrages, such as murder and tortures.

7. While on occasion he can offer some helpful advice, or make you "feel good," he really is usually useless when it comes to things that really matter. Simon Magus, incidentally, is also like this. There have been many times when I have wondered if he represents who he says he represents, what good is it for him to be showing up in the first place, because the actual benefit of his presence, aside from some occasional small advice, is nil. If he does any great good, it will only be so he can reap an even greater amount of evil out of it, say for example, by getting a church person to put their trust and confidence in him.

8. In his *Critique of Practical Reason*, Immanuel Kant makes the point that Jesus was recognized as who he was because he embodied the moral law people were already familiar with. In this person, Gomez., you would recognize little of any moral law in him. He is just a person of great power, and it is this alone really, along with perhaps some kindliness, which makes people think he is or is connected to the real Jesus.

9. I have spoken about and denounced him to others, and he reportedly tells those people something like, "that's all right, he (referring to myself) means well, and I forgive him, but he just doesn't understand how things really are." All the while he will tell these same people to lie and keep secrets. One of the reasons he gives for such attitudes and behaviors is "If things aren't done a certain way, it won't work for anyone."

10. The regular people I have known who have been taken in by him are secretive, in important matters dishonest, and general are, to some extent or other, hypocritical and unjust people.

Neither he, nor Magus (as "Jesus"), in my own experience at any rate, will ever say "praise" or "thanks be to God" if you ask them to. Quite simply, he is *not* honest, forthright and to these extents *of the truth*.

Essentially, the "Jesus" which Gomez. acts as, is a Jesus who (mostly) tells people merely what they want to hear, and not the honest, albeit compassionate and just, truth. This is one aspect of his presentation of Jesus, especially to morally weak or corrupt people, that makes his claim to divine importance so persuasive.

In *The Interior Castle*, St. Teresa of Avila states: "A very learned man said, the devil is a great painter and if the devil were to show him a living image of the Lord, he wouldn't be grieved but allow the image to awaken his devotion, and that he would thereby wage war on the devil with that evil-one's own wickedness. Even though a painter may be a very poor one, a person shouldn't on that account fail to reverence the image he makes if it is a painting of our very God." [87]

While there certainly is wisdom in what she relates, the images of Jesus created by Magus, an Archimago, or Gomez can be a real bane to one's prayers and religious meditations. The effect is like thinking of Charlie Chaplin only to be reminded of Hitler. This interfering with people's thoughts by Magus and Gomez about true Jesus may well help explain the zealous iconoclasm of Islam and the Reformation, and makes it all the more imperative for Christians to think of Christ not in imagistic terms, but in the Spirit of honest Love and Truth. This does not mean there is anything wrong with religious pictures and art, only one naturally must be careful not let them to overly dominate their subject in a person's thoughts.

In another part of *The Interior Castle*, St. Teresa argues that if the vision brought uplifting spiritual feelings of devotion it is from God. For myself, I would insist that such feels must also be moral, as Satan can produce all kinds of false raptures and ecstasies, and the moral sense will protect a person from being led astray, even if the vision or high feeing is from the devil. As it is, some religious people can get caught up in visions and raptures, yet ignore basic honesty, decency and justice, thinking that devotion by itself is all God wants. If devotion without due respect of morals is all God wants, then why would he have given us morals and reason, the protection and shields of love, in the first place?

While I have great respect for the writings of St. Teresa, I have much less for that of St. Catherine of Siena. I am thoroughly persuaded that her purported dialogues with "Jesus," were in fact based on conversations with Gomez, or

---

[87] VI.8

Simon Magus. The manner and message, and kinds of reasoning in which "Jesus" is given to speak in these writings, is almost identical to how Gomez and even Magus will talk. For example, in The Dialogue, "Jesus" speaks of the glories of suffering, but there is no distinction made between rational and irrational suffering. There is the distinction made between suffering under persecution and suffering for one's misdeeds, but how exactly one (of one's self) is able to distinguish the one from the other is hardly clear. The idea that suffering is good for its own sake, makes for an easy way to excuse the violent misdeeds of an Orkonist and his torturers. It is one thing to speak well about suffering nobly and with devotion to God. It is quite another to say suffering of itself is a good thing. The latter can easily be made a cloak for being indifferent to or encouraging the suffering of others, whether or not the violence is persecution or "just" punishment.

Many of the more bizarre and peculiar reported mystical and miraculous experiences of saints and others would clearly seem to have been impostures of Hell. This is not to deny that all such reports of extraordinary mystical experiences, spirit person/angel encounters or miracles are not genuine. Yet I believe a sane and sober person will have to admit that those which are genuine are (arguably by far) the exception, not the rule. Further, this is not to necessarily impugn or malign the sincerity or integrity of those who report such experiences, but to merely observe that even the most devoted and well-meaning are not always above being imposed upon by Hell's tricks.

If God is not understood as expecting us to act on a moral basis, how is he to be known from Satan (or some powerful "satan"), since Satan can produce wonders also? God has established the moral law for all to follow, and it is wrong to say that whatever God says is the moral law outside this pre-established understanding; since how does one know what the person or spirit tells us is from God, if what they say does not fulfill or is not in the true spirit of the moral law to begin with? Again, Satan himself is capable of producing wondrous feelings and visions, so these of themselves cannot necessarily credit an injunction or message from someone as being from God. No, common sense would seem to suggest that any messages, visions or feelings of divine origin be consistent with the moral law; which we know from right reason, our hearts and consciences. This is not to suggest that mere legalism reflects the moral law, but it is a simple insistence that we always respect and have due appreciation for decency, honesty, justice, innocence, fairness, mercy, rationality, sincerity, etc. based in love. Not even the breaking or wrenching apart of the earth, or the planets being hurled out of their orbits should ever deter us from seeing God, and our duty to him and our fellows, as anything but moral in character. God should never be merely seen as awesome power, since Satan and any number of others are easily capable of such – all the more so as people are irrational and credulous.

Aside from those already given, two other reasons why people are fooled by Gomez are these: persons are commonly led to argue, if he (Gomez) is not God or Jesus, then where is God anyway? This point, of itself, I will respond to later. But in addition to this argument, the other reason a given person is fooled by Gomez' (or Magus') Jesus impersonation is that by the time a person is even willing to consider of this spirit person's actual identity, they are already so appallingly guilty, and as well, now so frightened, that they find it easier to believe that it's Jesus they have been listening to all this long while, rather than some Satan, and that "Jesus" needs to do these things the wrong way. In other words, they continue to rationalize their guilt, even if it requires making themselves more guilty. It's the case of someone doing something wrong, who all along thought it was no big deal, only to find out it could land them life in prison. Rather than face up to the fact, they will continue doing wrong, in the way of arguing that they didn't know what they were doing was so wrong in the first place. While a defense of this kind might have some just grounds to it with respect to some minor offences, one would be amazed how people who have involved themselves, or make themselves complicit, in the most sickening and horrible murders, rapes and tortures will use this childish and phony excuse. The point I make to pawns and dupes is this. "Yes, I understand how these spirit people were able to fool you as they did, given their extraordinary powers of deception. On the other hand, how could you not have known that witchcraft, judicial railroading, torture and murder, etc. were wrong?"

These things said, it will then be argued, that perhaps Jesus Christ is just a big fraud anyway. Some, of course, do and will contend this. If so, it is very remarkable then how he should be so closely aligned with Satan given the extent to which Satan vehemently tries to undermine and destroy true Christianity.[88] A proof of this, I think, is found in the extent to which some people's rejection of Christianity or Judaism takes on the form of vitriolic hatred rather than a mere reluctance to believe.

~ * ~

How Hell Characterizes Judaism and Christianity

"Let no one disqualify you, delighting in self-abasement and worship of angels, taking his stand on visions, inflated without reason by his fleshly mind, and not holding closely to the head, from whom the whole body, supported and held together by its ligaments and bonds, achieves the growth that comes from God."
~~~~~~Colossians, 2:18-19

---

[88] There is another spirit person called "The Angel of the Lord" who sometimes might make an appearance who is nothing more than a demon or fallen angel who will act in a supporting role to Gyro, Magus or Gomez. I should take this opportunity to mention that a character of this sort does not need to appear visually to interact with you, as you might also merely "hear" them, and or feel their presence.

It is sometimes claimed that but for the Judaism and or Christianity, Hell would not cause as much trouble as he does. The historical truth is, Hell was long at work rabidly ruining and destroying people ages before Judaism or Christianity became faiths as we know them. One needs only to read ancient writers like Herodotus to realize the truth of this. And even if there were no Moses or Christ, you can rest assured that Hell's agents would (and do) impersonate other well known persons and authority figures in order to mislead people. St. Justin Martyr, one of the early Church fathers, makes the interesting remark: "Before the Lord's advent Satan never dared blaspheme God, since he did not know his condemnation.'"[89] This may in part perhaps be said account for Satan's pronounced anti-Christian belligerence.

Among the spirit people I had conversations with were "Percy Shelley," and "William Cowper." While welcomed in Christian Heaven if they would adhere to Jesus' rule, Shelley and Cowper did not like Christian Heaven so they chose instead to fly around in the void among the stars. While this got to be rather dull over time, nonetheless, they preferred it to Christian Heaven; which was presented as dry and stifling. Though Shelley at least had his wife with him, Cowper was alone and apparently not very happy. He, in effect, said being a Christian had turned out to be a great disappointment. Oftentimes, the complaint from these people was that Jesus was either a tyrant (e.g., Jesus wanted people to be crucified) or powerless to have any real say in what was going on. But even worse, were the Jews, who, it was said, really owned Heaven, and it was regular practice among these ruling Jews with all the wealth they possessed to hold people, even very famous people, as slaves. So awful was their mismanagement of things, that Heaven, unlike in times previous, was in many places strewn with litter.

After I got to the point when I realized these people had nothing to do with the real Jesus I was told that the "real" Jesus was living in the "underworld" with the Egyptians. The "underworld," as it was described was meant to refer to a place somehow separate from the rest of the universe, like an island far off from everything else. Jesus, had been there for some time and was preparing along with some others to have a go someday of assaulting and taking over Heaven from the Jews. Stephen Boyd, the actor, was one such follower; who when he realized who the false Jesus was, left Christian Heaven to be with the "real" Jesus in the underworld.

One can easily infer that the effect of people like Simon Magus and Gomez on religion has been disastrous and tragic. Indeed, I would go so far as to say people should seriously consider the possibility that the Bible may have been tampered with by Orkonists (and persons under the spell of such) during the course of centuries. After all, we don't have anything like original manuscripts available to us. I realize that to some to suggest such is to seriously challenge theirs basic religious beliefs. So do understand, that as a believer, I do not say

---

[89] *Against Heresies*, IV.12.

such a thing lightly. Yet there are things in scripture which really don't make sense, and seem inconsistent with God's character. For example, the portrayal of God as the ravaging destroyer, and the use of magic by Moses and Aaron in Exodus. These gratuitous anomalies *might* better be explained as false texts inserted by people under the influence of spirit persons posing as persons from Heaven. This is by no means to impugn the Bible as a whole, but merely small portions. This said, the issue is one capable of other interpretations; so that this is merely one explanation. However, it is one well worth being at least open-minded to.

Note also how the "Our Father" is recited in church liturgies. It says, "lead us not into temptation," when the actual scriptural text we have is more accurately translated, "keep us from the final test." The church translation implies far more than the truer one, and better describes Satan than God. [90]

It is strange to think that what some people have taken as holy day-glow from on high, is really nothing more than some ancient pervert molesting them, interested not in instilling God's love, truth, faith, or the moral law, but rather awe and intimidation for Satan's ravenous and manipulative purpose.

False miracles and magical powers are what Satan frequently uses to fool people. Intelligent persons will then all the better realize that true miracles come from love, truth, faith, and wisdom – not the other way around.

~ * ~

"Regular" People who work with Hell and Why

"In the last times some will turn away from the faith by paying attention to deceitful spirits and demonic instructions through the hypocrisy of liars with branded consciences."
~~~~~~1 Timothy, 4:1

"The man who strays from the way of good sense
will abide in the assembly of the shades."
~~~~~~Proverbs, 21:16

Hell cannot do its more serious work of destruction unless acting in conjunction with "regular," i.e., flesh and blood, people. Indeed, Satan's whole empire against humanity can only function if there are cowards and traitors from among mankind who will betray us to him. At the same time for many people and as they see it, lying and deception can prove very profitable. In fact, there are people who make their living (and more) from these practices. And when they do

---

[90] Epistle of James 1: 13-16.

they inevitably risk being connected to and or themselves becoming Hell employed people.[91]

There are, and just about always have been, persons who know about these spirit people, and accept them as a normal part of life. Such people are lead to think God doesn't exist or isn't pertinent, and that it is these spirit people who really have the power. Some of these will attempt discredit or ridicule those who try to tell other about the dangers of Hell and witchcraft, even though they know full well such warnings are based in true fact, and are no delusion. Alternatively, some persons, well meaning but deceived, will be lead to think these spirit people are from or somehow represent God, and will go on to practice some bastardized form of some otherwise legitimate religion, while denying or keeping silent their direct knowledge of these things. Both types will want the benefits of justice, mercy, truth, humane treatment, freedom, yet they will be prepared attack these very principles and those who strive to uphold them when Satan (or one of his representatives) informs them that their selfish or religious interests are at risk if they don't.

There are other people who have listened to these spirit people for so long that the idea that these people are from Hell is unthinkable to them. By the same token, the extremism of Hell people, is exactly that so extreme that many are incapable (usually out of fear) to comprehend it, and cannot believe others would be so relentlessly cruel and vicious for years and years on end. Yet that is how many Hell people are, and if not always like this, sometimes; as like anyone their temperament and disposition can or might change with circumstances.

Persons under Hell's influence are not necessarily people who have contact with spirits or witchcraft, and they are not at all necessarily people with bad intention. Often times, people with very good intentions, at least in their own mind, will be led to think that if they do something wrong it will help people; even if the nature of the wrong is felonious or of a vicious character. Even some of the worst devils of Hell can think of themselves as acting out of some good or higher purpose.

---

[91] Elsewhere I have written, and worth adding here: "If the world is in turmoil and this bothers you understand that for certain people it is desirable that it be so. Though you might find it hard to believe, mankind has a very real and implacable enemy. This enemy is basically made up of spirit people who, at least for practical purposes, you can think of as angels or else aristocratic spirit people. They enlist regular people to betray their fellow men, and these latter, and others more or less like them (and to a greater or lesser degrees) cooperate with the spirit people. Thus the enemy is let in through a back gate and permitted to make his home in our midst.
Now the story behind how this animosity toward humanity got started is not entirely obvious and open to some interpretation. But suffice to say these very hostile spirit people prefer to have us live as second class citizens, similar to how minorities of color have been treated in some human societies. These kinds of spirit people I am referring to in effect say this. 'These humans are not 'righteous,' therefore they should be kept down and got to know their place.' That at least is the kind of excuse they give, though they can give such a reason without it having to be their true motive (again just as certain racists might refer to certain ethnic groups as hopelessly lazy and therefore not worth assisting or educating, when he really makes his claim for other reasons.)"

For some, acquiescing to Hell is a matter of that's just how life is supposed to be. In the strict sense this is true of most if not all people, due to the fact that all "fall short of the glory of God," and "the spirit is willing, but the flesh is weak." Even so, what is remarkable is the extent to which some people will compromise themselves in the most shameless, irrational, cruel, disloyal, and pusillanimous manner, so in awe they are of Satanic or Orkonic power.

For many, in lies and murder lie the door to the success, power and riches, and it is lies and murder (in various forms) which these spirit people seek (on various and diverse levels) to encourage. So those who will murder, or will be accomplices to murder, these spirit persons will assist, while those who are against murder and for truth, these same spirits, and their regular flesh and blood person will crucify (in various possible ways.) These are, in other words, simply people who make a living out of witchcraft, torture and murder. They carry on a double life, and rationalize what they do based on specious arguments Magus might give them. Satan can make cannibalism and human sacrifice profitable, while at the same time allowing people to keep the practice disguised and hidden. Many in society are party to this, especially the more greedy, and will also be party to an attack against those who would expose what is going on. This they will do; while perhaps reasoning that what they are doing or assisting is actually in everyone's greater interest.

There is a psychological tendency, most certainly encouraged by Hell, to blame others when we ourselves are guilty. Hell will get a person to be guilty, and when the person realizes their guilt, Hell will then guide them into blaming others. Often the more atrocious the guilt, the greater the tendency to blame and fault find with others. It gets so bad that often in society those who do not participate in some kind of wrong-doing going are actually punished or penalized.

Some will make the case that they can't stand these witchcraft people so much that they would rather behave immorally than have to deal with them. Through pressure of this kind, Orkonists often finds a way of getting people to do the wrong thing. It should be asked of such who bow down or give in, if you can't in some serious way defy Orkon people in this life, what assurance do you have that you will you be able to defy them in the next?

These spirit persons can be so adroit at manipulation and deception, and display great seeming control of a given environment, that often a given person is prepared to abandon all morals and reason listening to them in order to further their own career or interest. Usually they don't see their behavior as wrong as such; rather they see it as enlightened and clever. Others are so scared to death by these spirit people, and or so intimidated by wealthy and powerful Orkon people, that they comply and give in out of sheer terror; indeed, will consider themselves righteously *justified* in doing so.

As mentioned, Orkonist strategy is to infiltrate every walk of life he can. The more people who reject wisdom, truth, justice, proper religion, and doing the right thing, the more people he will have under his rule. Thus, people under Hell's influence can be found in practically every area of society, among the rich, among the poor; among government people, among educators; among overt criminals; among religious, non-religious the educated, the uneducated. Some of Hell's people who stand out most are those in cults. Many of the cults that became well-known over the years are people who have regular or significant contact with these Hell spirits, such as the "Moonies," the Jim Jones' "People's Temple," the San Diego Flying Saucer cult ("Heaven's Gate"), and Scientology. These religious people mean well, only they believe that intimidating spirit people they have encountered, directly or indirectly, have ultimate authority, rather than truth, logic, morals, reason, or true religion (which is based on faith in what is *unseen*). As a result, given agents from Hell are able to run rings around these people, deceiving them into believe the most absurd kinds of folly and madness.

This said, it must be emphasized that Hell's influence is by no means restricted to the fringe. In point of fact, Hell's influence is, especially in the last few decades been very much made part of the "mainstream." There are some very wealthy and influential people in a given society who listen to these spirit people as authority and who set the policy to others what the attitude should be towards such spirit persons. It is not the desire of spirit people to have themselves be discussed openly since, like with organized crime, it is in their interest to be seen as they prefer to be presented, and not as truth would show them to be. Stephen Spielberg and people from the Microsoft corporation are some of those with very close ties to Hell people,[92] and much, if not most or all of their phenomenal wealth and power, stems (in my opinion) from that connection. Once more, these kinds of regular people don't necessarily mean bad, it is just that they are given to believe ridiculous arguments from Hell's agents that if things are done the "wrong" way, it will work better for them and for other people. The methods Hell agents use to fool people are very numerous and varied, and often times reflect the most amazing cunning. Moreover, the more morally "dirty," dishonest and selfish a person is, the easier it is for a satan (or someone acting for him) to exert his influence on them, similar to the way a disease might more likely infect someone with poor hygiene.

Now the sprit persons I have described, in particular the false "gods," not everyone is able to thoughtfully reject or disbelieve. In fact, many will believe what these spirit people say without any mind to being skeptical about the assertions being made by them. Instead, they take what these spirit people say at face value, and then on some level or another assist or in some way become

---

[92] Hell people have spoken of them to me *many* times; which is to say they have been frequent topics of conversation; with the accompanying suggestion that they are being (unwittingly) used by and or cooperate with Hell people. In addition, and even if I were not directly and thusly informed by others (of such relations and involvement), there are ample circumstantial grounds, one could list, for strongly suspecting such ties and involvement to be the case.

involved with these spirit persons in murdering others. These "murders" then becomes the key to their success and riches (involving both ousting the competition and de-sensitizing people to murder generally). These activities happen on various levels of cooperation, and needless to say not everyone has conscious direct contact with spirit persons. Some will act in cooperation with these people on the basis of bribery, "what everybody thinks," fear, psychological manipulation, etc. without necessarily having any awareness of acting in cooperation with a group which is being led or dictated to by a spirit person. Meanwhile, those who are contact with the spirit persons might become involved in the most lurid and gruesome crimes (such as serial killings). These then become the ones who set policy for the other people, who are criminally implicated in gradually less severe degrees of culpability. Yet taken all in all, the group becomes a powerful force to capture power and command policy in a given society. A crucial aspect of making such aspirations to power work is in attacking and undermining reasons and morals; since it is by means of reason and morals that their criminality can be unmasked. Keep in mind also, that these spirit persons are both ultimate con-artists while, at the same time, the most unbelievably callous and cold blooded people imaginable, despite efforts they make to put on a humorous and or benevolent mien.

Such who go along with these people, do not necessarily intend to be criminal, yet are given to think that is the way life is: reason doesn't matter, rather what "everyone" thinks matters, or that's the way real life is. Really it can be quite ridiculous the kinds of things they can got to believe – and believe for a lifetime. Should they come to realize the gravity of the criminality they have involved themselves with, they perhaps understandably, fear scandal and criminal prosecution, and therefore become all the more desperate to hide or justify their complicity. That Hell should not infrequently invoke "what everybody thinks" as the standard of truth is very understandable seeing their often tremendous capacity to manipulate, bribe and deceive whole communities and large groups of people if they feel it is worth their while.

One thing I have noticed over the course of time about those regular people who cooperate with Hell is that the more they give into Hell's enticements, commands and demands, the more immature, dishonest, self-defeating, irrational, and self-indulgent they are. This does vary according to the individual, nevertheless, the tendency to be selfish, dishonest, cowardly and immature is very pronounced. It is usual of such persons, especially those who have in one way or other actually met or spoken with spirit people, that no matter how actually ignorant and irrational they are, tight-lipped, they will behave as if they know everything, though perfectly incapable of adequately explaining or justifying their beliefs.

For some they allow these Hell people to rape and molest them, mentally or otherwise, because they don't have character, intelligence or principle to resist, and rather than admit to their shame, will then try to justify their giving in saying, "well, that's just how things are done;" obviously dreading to admit to the horrible disgrace and humiliation which they – to some extent (depending on the individual) – have agreed or given in to. And if you then challenge them with the truth of what is going on, they might see you as a mortal enemy, so devastating is the possible shame and disgrace they might have to face should you be right.

Is a person, who does something wrong, not guilty if they are "possessed" (of which more later)? It depends on the person, the nature of the wrong-doing, the degree of pressure the demon(s) is exerting, and other circumstances under which the individual in question is acting. It should go without saying, one ought to be extremely careful in judging. If someone is under Hell's influence long enough it does seem possible, at least in some instances, that they could be led into wrong bad behavior without the immediate instigation of a devil. Then again, I would be inclined to think that such cases are the exception, all the more so as the bad behavior is of a severe nature.

Is a person who was fooled by these people not a Christian or believer in God? The answer to this is "not necessarily;" though a Christian who lies about their contact with spirit people should be thought lamentable. Yet while being ignorant and fooled can mitigate an offense, very rarely will it actually exonerate someone, since guilt and innocence are measured more in the heart (that is in deepest motive and intention) than the head. Although Adam and Eve were fooled, and the serpent blamed, this did not spare them culpability. Being fooled or deceived then is not necessarily an excuse for doing evil – all the less so as the nature of the wrong-doing is both very serious and pre-meditated in character. This said, there is repentance and forgiveness for making up for one has done, but the repentance and forgiveness assume sincerity on the part of the person repenting. Someone like Simon the Magician or Gomez, for instance, when he disguises himself as Jesus will tell someone that they can do something wrong now, and he (or Jesus) will forgive them later for it. Such an approach, of course, is intended only to make a mockery of forgiveness and get the person guilty.

Why would God have blessed us with morals and reason only to have these things turned upside down by some not-accountable spirit person; who, after putting on a show, merely tells us that "they know about these things in a certain way?" Even if we say that God allows or even encourages spirit people to play tricks on regular people or use them for divine purposes, that does not absolve flesh and blood people from their obligation to be rational, moral, honest. And even if we allow that these spirit people know what they are doing, it doesn't at all necessarily follow that those who listen to them do.

~ * ~

Faustus

"'That puppet yonder,' thought Mother Rigby, still with her eyes fixed on the scarecrow, 'is too good a piece of work to stand all summer in the corn-patch. Why, I've danced with a worse one, when partners happened to be scarce, at our witch meetings in the forest! What if I should let him take his chance among the other men of straw and empty fellows who go bustling about the world?'"
~~~~~ Nathaniel Hawthorne, "Feathertop: A Moralized Legend"

"Crush a fool in a mortar with a pestle along with crushed grain, yet his folly will not depart from him."
~~~~~ Proverbs, 27:22

Of the different kinds of regular people who might act in cooperation with Hell, certainly the most dangerous is the "Faustus" type. It is this kind of individual who poses the greatest threat to a community because he is someone who, consciously or not, sells his soul for worldly riches and control. Many of histories most notorious tyrants, criminal despots, and cult leaders have been "Faustuses" of one kind or other, and some in our own day; or else have as an important minister or functionary such a person. They themselves are not necessarily warlocks (though to some degree they might be.) More likely someone else takes care of the witchcraft, while they play the executive. Now there are different kinds of Faustuses. Some are people who are Hell bent for power, and don't really care what happens to them in the final resolution of their lives. Another would be someone who believes he will ultimately earn for himself a place of prestige in Hell (though under the misguided impression it is Heaven.) Yet another is someone who believes it somehow helps people if someone like himself does the wrong thing. Not because he really cares about others – he doesn't – but because it at least serves as a good rationalization for his crimes.

Not everyone can be a Faustus. They typically must make themselves eligible for the massive power they might possess by committing, or be willing to commit, so many murders. These need not be committed by himself necessarily if he can get someone else to do it for him.

There may be multiple lesser Faustus, yet whose presence in society together effectively amount to one Faustus proper.

These sort of people will typically act like they know "how things really are," but, for all the inside information they possess or potentially have available to them, in truth they hardly know what is really going on, and, in this way, end up making fools of themselves.[93]

---

[93] Similarly, someone like Simon Magus might know many extraordinary things, but at the same time there are inevitably many more extraordinary things he *doesn't* know, yet which are commonly known among many people (for example, the value of equal human rights, or in what real happiness lies, despite his own beliefs to the contrary.)

Faustuses are not likely to make policy or act in isolation. They will invariably work with some kind of Mephistopheles, i.e., one or other of the false "gods" described previously, and possibly others like himself. Not unusually these Mephistopheles will have more than one client, though not all are necessarily all-out Faustuses. Some might be less power-pretentious warlocks or witches, or else religious persons who can't tell the difference between God and the Devil. With Faustuses invariably it is the Mephistopheles who is mostly in charge of what goes on.

In obtaining his power, which really amounts to nothing more than having demons and other spirit people to work for him, Faustus gets himself guilty in any number of ways, ranging from lesser to greater guilt. In the course of time, people from Hell will come and tell him he must be punished for what he did (otherwise "he will be having it too good.") Faustus is given a choice for suffering for what he did, or, alternatively he can find sacrificial victims, human or animal, to appease Satan's or an Orkonist's claims on him. Not unusually, Faustus will not want to suffer for what he's guilty or pay for what he owes, so instead he agrees to victimize others, and thus (as he sees it) make it easy on himself. What purpose does it serve Hell having Faustus find more victims? For one, it makes him more guilty, and thus puts him that much more in Satan's debt. Faust is caught between having to suffer for his wrongs or else go deeper into debt, and it is characteristic of a Faustus (though not necessarily always the case), that he will prefer to defer payment, and engage in more victimizing – similar to the way a chronic alcoholic will rather go buy himself a bottle, rather than give up the drink which is killing him. The additional wrongs Faustus might do, also include endeavoring to get others to do wrong (thus spreading Satan's influence and control.)

Worth noting is that a Faustus and his Mephistopheles might approach and bring problems to a corporation which has been acting in a morally dubious way, and blackmail it. By this means, Faustus is able to extort money and power from them, and deceive, threaten, or incite them to further cooperation.

~ * ~

Hell at Large

"Why would the spirit people try and make a fool of me if I did things the wrong way just as they told me to?"
~~~~~ A modern Faustus reasoning.

Many of the world's most absurd and horrendous problems stem from persons of wealth and status who believe these spirit people, or who believe other (regular) persons taken in by these spirit people, whether through fear, false benevolence, or allurement to easy gain. For instance, someone in, say a food, corporation might be under the gun of some of these people. In order to avoid trouble with these people he will have some pointless ingredient added to the food the company processes. Likewise, a company might downgrade its product due to similar extortion. Though as much as the spirit person's decision will have affected the whole proceedings, any number involved in the extortion, whether as criminals or victims, will not even know of his existence.

In this manner, whole societies themselves may in some way compromise themselves or even have out right pacts the devil. Orkonism is, on some level or other, embedded in most major societies and cultures like a disease, a disease which rejects practically all good, except that good which the society can use for its own selfish, mercenary and prideful ends. Money in a society will attract high-powered Orkonist the way sugar attracts flies – the more sugar the more flies. I was told by a spirit person that Industrial Revolution had caused Hell to me more active, presumably because of the dramatic increase in available wealth.

It will help to get at the Hell problem on the society level by roughly classifying people in a given society according to how they might or might not be affected by Hell's influence. Such classifying we will want use to create a tentative and plausible picture of what's going on, rather than to judge anyone as such.

With this understanding, people as affected by the Hell problem might be placed into the following categories or classes.

* Faustuses, Magusites, Goonatics, i.e., Hell's regular direct contacts with living (flesh and blood) people.
* Pseudo-religious, who believe that whoever seems at the moment to have the most power is or represents god.
* Pseudo-religious but who mean well, but whose irrationality and unconcern for morals is so great as to make them depraved.
* Unreligious who are easily manipulated by Hell forces, irrational and or uncaring.
* Unreligious who are more sophisticated, and perhaps more caring people, who while they might be easily manipulated by Hell people have a better chance of realizing error than those who are simply irrational and or uncaring.
* Those who would be opposed to Hell people, but who because of fear or deception imposed on them deny there is any such problem.
* Indifferent people, who will generally go along with what everyone else wants.
* People who are aware of and against Hell, but are nevertheless, because they are insufficiently rational and or insufficiently concerned about morals, are credulous and more or less easily deceived by Hell people.
* Intelligent and or moral people who are against Hell but don't know or understand it as a real and literal threat.

* People who are aware of Hell, and make earnest efforts to protect themselves by keeping rational and moral.
* People who are aware of Hell, and make earnest efforts to protect themselves by keeping rational, moral, and religiously devoted.

Now lets say we were to rate each of these groups on the basis of the social power and influence they possessed, such ratings in turn might be gauge by a) the number of people in the given group, b) their level of rationality and intelligence, c) their level of moral integrity, (including courage), d) their financial resources, e) their presence in institutions and various important societal professions and organizations.

Taking such ratings for a given time period we could compare the different groups with each other and then arrive at a good idea of the extent of power Hell has over that society for the given time in question. This kind of approach could not pretend to exactness or anything like perfect accuracy - there are just to many variables, not least of which the role of Providence. But it provides a reasonably accurate tracking of the Hell problem, and a convenient overall grasp and understanding of society's state with relation to Hell's influence.

Another but more simple way people in a society might be sorted as the Hell problem affects them might be:

>those who fight Hell
>scapegoats, sacrificial victims
>do nothings
>accomplices with Hell
>active agents of Hell, e.g., felony criminals, killers

When a society is under the domination of Satan people, there is invariably some degree of scum or residue that attaches itself to those prominent in society and their work no matter how well-meaning or talented they might otherwise be. In other words, one can't act in cooperation with Hell or with Hell's approval and not have some of the curse show on you and your work. There may be exceptions to this, such as in the case of providers of essentials, such as food, clothing and housing, or else in the case of those who openly protest the Satan "regime" (or status quo.) But otherwise as a general principle this holds valid. For this reason, I am inclined to believe original sin is more a social malady (social in the worldly sense) then a biological one.

The following are *some* of the factors which work to create or influence a given one of the above listed groups:

Pro-Hell factors:

Secrecy, lies, wholesale denial of the problem
Disruption of communications
Fear
Infiltration of institutions by Hell operators be they regular or spirit people.
Getting people guilty then putting in a position where they can be blackmailed, scandalized, terrorized.
Deceiving people that they somehow represent a divine purpose
Selfishness
Irrationality, the rejection of reason
Presence of professional sorcerers, witchcraft people
Guilt (of varying kinds and degrees)
Materialistic greed

Anti-Hell factors:

Free Speech and communications
Sincere religious faith
A strong respect for reason and rationality
Morality founded on a sense of justice, fairness and mercy for all
Charity and a desire for altruism
The need and desire for things to work right
Courageous and dedicated people willing to take up and face great risks and dangers

As an Orkonist empire grows stronger the strength of its members *as individuals* weakens. Meanwhile, those who oppose them grow stronger as individuals. The question then becomes will there be enough of the latter to topple the empire or ruling hegemony of the former. It may be that under certain circumstances that the regular people a Satan has working for him are so incompetent and inept that his own power among us becomes jeopardized. Likewise as the power of a Orkonist empire increases, the value of the happiness it offers to its citizens goes down in value, and people then are given the choice of being reduced to a literal Hell person or fighting the enslaving regime.

The following describes how a given group of people might be viewed as being under Hell's influence (rated from bad to the worst):

a. Hottentots who are deceived, and do mean well, but are simply in error.
b. Hottentots who are deceived, do mean well, but are in reckless, gross error.
c. Hottentots who are deceived, but who care only about (what they see as) their more immediate self interest.
d. Hell people disguised as Hottentots.

Since deceiving is a essential means of promoting Hell's policies, lying will be encouraged among people and built into the social fabric. Those who do not lie are not allowed to participate, share wealth. An honest person to the liars is seen as an offense to their self-esteem. The liars conscience may be offended by the honest persons candor, yet in feeling the hurt they will not blame their own lying, but the honest person for reminding them, perhaps unintentionally, that they are doing wrong. The social structure, which is nurtured by devils, will then reinforce the liar's resentment. "Everyone lies," so the liars will reason, "and if I don't I will be ostracized too. This honest person is asking to much of me, and therefore they are to blame, not my lying." Rather than own up to their guilt then, they will be disposed, again; especially if Hell people are active in their midst) to believe that, it is not they, but the honest person that is actually wrong. Indeed, they may take things so far as to treat the honest person as an unpardonable, arrogant enemy. Add to the honest person's burden, the disdain some will feel toward some who is "down on their luck" (thanks to the efforts of Hell to bring them to this pass), and you will see how difficult the honest person is going to have it. Have the Hell people, as well, attack the person with violence, character attacks, demons and other crimes, and then see how insufferably extreme that honest person's situation will have become.

Why doesn't the government do something to shut high powered Orkonists? Because Orkonism is actively working at various levels of society, and often has taken a grip on society like a cancer. Further, they might do this in a guise that presents them as one of humanity's great benefactors. Often times a person is involved with these Hell people, only to find out too late what they are really about. And when they do, they do anything to cooperate with these them lest they become victims of brutal retaliation. At the same time, "Satan, inc," has a religious branch which pretends to promote religious faith, but is really a holding tank for frauds, hypocrite and the insincere, but who nevertheless sometimes are very influential in religious affairs. Such people are told not to combat witchcraft, because, for example, that is how God wants things done. And of course, there are any number of devices for hoodwinking and controlling people. For instance, someone might be threatened that if they don't cooperate they will be dragged down to Hell after they die. I had this done to me in the way of a dream. It was very frightening, but after I woke up I realized this was not God's way, and for this and other reasons, I remained defiant. Many others, unfortunately, are not so careful to question what they see or are being told, and see no alternative but to give into these people; though in doing so are making the biggest mistake of their lives. What happens to such people sometimes is that the very thing they feared ends up happening to them; not because they *didn't* go along with these spirit people, but because *they did.*

It is my belief that anywhere from 75% to 90% of the most serious pains and problems we suffer in life come from the direct or indirect presence of these Hell people. Why should this surprise us? Someone like Simon the Magician or some Archimago, when he is up here and has a Fautsus or Faustuses working in cooperation with him, can be a person of extraordinary power and influence. With fantastic amounts of money and magic powers at his disposal he or they can get away with the worst crimes conceivable. At the same time, know this is someone who is an unrepentant mass murderer; who sees it as his primary business to cause trouble. If someone like this is, along with his Faustus(es), at large exerting a major influence on society, what wonder that this is or becomes a formula for individual and communal disaster? Combine the extent of the sorcerer's power with his inordinately depraved character, and he and his salves and henchmen can infiltrate and work like a malignant tumor on society, directly and indirectly, creating no end and manner of problems.

Quite simply, until Hell spirit people are better understood, and rationally and scientifically based civil laws and measures are enacted to protect people from their criminals deeds and machinations, we can never seriously expect efforts to eliminate crime, disease, and other social problems to ever really succeed.

The criminals are told that for anyone to really get ahead in life they need the big money. To get and keep the big money, they must feed the ghost. To feed the ghost requires victims to be murdered, tortured, tormented, or some way or other ruined. These most usually are the poor, the young, the isolated, outcasts, dissenters, sick and elderly. Now since these last often never have money or say in society, they will be rarely heard, and even more rarely helped. Probably the most numerous of Hell's murder victims are the disenfranchised, the outcasts, the homeless, those involved in witchcraft; because such as these have less say in society than everyone else. This reminds us how all the more imperative it is for these things to be discussed, for without such discussion, the cries of such victims will continue to go unheard, or else will continue to be tragically misunderstood.

Further the deviousness and relentless cruelty of the ghost are something few people can deal with. The horror of it is just too much. Victims are swallowed up, the ghost is glutted, the criminal maintains his power, and secrecy, lies, and darkness cover all (or seem to.) In fact the more lies, guarantee more brutal excess from Hell; for these people are typically unappeasable in their cruelty, and usually the only thing that hinders them from doing worse than they do is the risk of getting caught, exposed, or openly attacked – three things they are very careful to avoid. People react to the above dilemma in at least two ways, either they hide out with the money, while looking the other way to what's going on, or else (as best they can) they fight the monsters. Of course, the latter are less numerous than the former, and the extent to which fear, dishonesty, and irrationality prevail in a given community is the extent to which Hell prevails. Yet if these things were discussed openly, and their were people of bravery, determination, and intelligence to fight Hell, Hell could actually be fought. But all this of course will have assumed that we have evolved beyond superstitious and irrational ways of thinking. Yes, the problem is the most difficult one imaginable, but how can we afford to ignore it and then otherwise pretend we know real progress in anything?

Ideally and preferably, and when at all possible, only well-informed, responsible and intelligent, scientists, philosophers, higher educators, jurists, and also more cultivated clergy (i.e., rabbis, ministers priests), should be authorized by civil authority to deal with spirit people of a more powerful than usual order. Otherwise, what will happen is that more manipulative and powerful spirit people will intermingle with some of the populace, confer secretly with them, and in this way end up wreaking havoc on the community – that is by coming through the back door so to speak. Once spirit people are better understood, civil regulations should be put in place to determine the civilly appointed "priest" representative's responsibilities in their real or potential dealing with spirit people on behalf of the community. In short, if spirit people really need to deal with regular human society it should be done by way of a (more or less open and) public forum; where only the community's most wise and best educated deal directly with influential or imposing spirit people; while at the same time empowering such "experts" in a way that duly conforms with public accountability, human rights, democratic principles, and those of a free society in general.

~ * ~

Specific Methods and Devices Hell Attacks With

"In the misfortunes of our best friends we always find something that does not altogether displease us."
~~~~~~~La Rochefoucauld

Hell wants very much to get people to gradually rearrange their priorities so that their ideas about what is good are "backwards." This might be accomplished in diverse ways. To give a quick example, they might try to frighten you with something bad, while tempting you with a false or inappropriate good (such as over eating.) It is not that eating is bad, it is just you are led to misunderstand its role, and, under the circumstances, place an undue importance on it. The same can be done with money, material wealth, other's approval, etc. The more you become addicted to the false or short term remedy the more it is possible for a clever devil to exert control over you, at least if he really wants to.

They can cause a seeming defect in someone or something in a way that can be either subtle or obvious, and, in our fallen state, there is almost no way a given individual cannot be made to look bad. Think twice then about thinking badly about someone based on how they look, mere looks as always can be potentially very deceiving. Sorcerers and warlocks have "spells" and other psychological devices which can cause the person being cast at to look like a devil, even though the person (or animal) may be not nearly so bad at all, or even be a more or less good person. This "demonizing" someone is a very powerful tool to deceive and get unthinking people to give into them, while eliminating opponents. Consequently, when a person "looks bad" to you (for some reason or other), before thinking badly of them, you should ask yourself if the other is possibly being made to look bad by someone else.[94]

Hell has beauty "spells" to benefit those cooperating with them. This is sometimes why those who love money look better than poor people. For instance, because the idol of money is more dear to a person than it should be, this idolatry might be rewarded with a beauty spell of some sort. Meantime, demons that otherwise normally leech on to that person and make them look bad are kept off – at least in public.

When Hell's atmosphere or influence dominate, "good" people, animals, things can be made to look and seem not only "not good" but positively bad or even virulent. One of the worst dangers these Hell people pose is that a person may fall victim to seeing life, and seeing the world the way these Hell people see it. For example, the callousness and indifference to the suffering of others is something Hell people take for granted, so that its effect on ourselves and those around us is extremely harmful, and at the same time can be extremely subtle.

Even though there may be a truth, even a salient and valuable truth, to what these Hell people sometimes to say, as a practical matter, one should never (unless in a group with other responsible and intelligent people) listen to them; as it is more than reasonable to assume and as a general rule that they are "up to something." In addition, sometimes they will impart falsehoods which they themselves, genuinely believe, such as matters pertaining to life, religion, happiness. Yet if you believed the same you might ruin your entire life. One should be very careful therefore that what one thinks and believes was not something unconsciously inculcated by Hell, while seeking the ultimate source and support for their thinking and beliefs in right and moral thinking.

On the Hell level of seeing things something truly wonderful can seem trite, or even detrimental. For example, if we lived in a grimy, raucous ghetto, Mozart's music might seem quite out of place, and being out of place seem to be something we should dislike. Something out of place suggests alienation - say our neighbors don't like Mozart – so that to avoid alienation from others we decide we don't like Mozart either. We see Mozart at fault, and (perhaps subconsciously) see liking his music as something which will cause us to be alienated from our neighbors. But the greater truth is, is that it is not Mozart that is to blame but the degraded character of the environment in which we live, or else in how we are dealing with and uncreatively allowing ourselves to be manipulated by that environment. This, incidentally, is usually why God and traditional religion are

---

[94] Indeed, keep in mind that that person, whoever they are, might look a hundred different others ways than they do -- either for good or for bad, but that present circumstances aren't different than they are. Further consider the possibility of whether they look bad because of something wrong inside you, or whether it is something wrong inside them, or possibly something somehow in between the both of you (acting as a negative filter) that is causing the problem. More often than not what we see good or bad in another is a result of something going on in ourselves, and usually relates to some value related assumption we have.

often seen in a bad light. Having said this, it is obviously true that something may simply be inappropriate for an particular occasion; without implying that there is something seriously wrong going on otherwise, for example, playing the piano while someone is running a vacuum cleaner, or vice versa. If someone or something looks or otherwise seems bad, the reason may be not that they are bad, but that they are out of place given the present time and circumstances. This is important to remember, because one way the devil will mock and denigrate religious notions is by placing them out of context, perhaps ridiculously so, a thing, after all, very easy to do.

Orkonist or Goomerist spirit people will sometimes try to confuse you as to their intention by doing you gratuitous favors, or else make you feel their presence when you are enjoying a pleasure derived from someone or something other than themselves. In this they try to condition people into thinking good comes from them, and thereby gain others' trust. Alternately, they might confer a good as means of humiliating or making a fool of that person. To illustrate, they might romantically set you up with someone, only to then defile and make you look and feel bad when you are with that person. They will then do things in a way such that you may well be receiving a benefit, yet nevertheless you are still being robbed. Of course, they may also rob you without benefiting you at all as well. Either way, they would never think of benefiting you without robbing you also. Devils are not interested in fair deals. Their business is to rip you off; your (real) value in exchange for their junk. The duress they cause may, by your resisting it, admittedly strengthen you. But there's no reason to assume that they are needed for such a purpose. One could strengthen one's self by creating voluntary stresses of their own, as in physical exercising for example, or in activities requiring stamina and courage.

It is puzzling why they will bother you with some things and not with others. The reason for this seems to be that by causing you trouble in some little thing, and refraining from something either worse or else similar, it encourages you to think that they need to get so much wrong done. It is this influencing your thinking that they aim to effect; the immediate annoyance or pain itself is otherwise without purpose.

One of the most pernicious and disgusting use of spirits is having one or more of them enter someone's body, or what is commonly spoken of as "possession." Not all possessions are of the convulsive, agitated kind we are most acquainted with. There is practically *no* part of one's body a given spirit *can't* be got into, and the effects they might produce on a person are almost as numerous as there are parts of the body. They can cause bodily organs not to function; can suppress normal mental and bodily flows; can invade and triggers reactions in the brain.

The manner in which demons or sprites are used in possession and in influencing actions and behavior are so many that no doubt a large book on the subject could be written. There is one use of them, however, which I will attempt to illustrate. Let's say "A," a good decent person, is going to testify against criminal person "B," before "C," a judge. What will happen is that a demon will be sent into "A," perhaps to make him stutter, or feel sweaty and uncomfortable. Another demon will be standing in front of him, so that when people look at "A," they will only be able to see him by looking through the invisible demon, whose presence they nevertheless pick up. A third demon will be sent into the judge to make him suspicious or impatient of "A." A beauty spell perhaps of some kind will be put on "C" to make him look good. The end result? "A," the good person is made to look bad, perhaps even scandalized (given other machinations added on to his embarrassment); "B," gets off the hook, and a miscarriage of justice has taken place. Using demons to set regular people against each other, in this and other ways, is a very common practice, and can be the source of much altercation and disagreement. This is brought out very accurately, by the way, in Hieronymous Bosch's painting "Christ Carrying the Cross;" in which, among other personages, a sorcerer is depicted as setting three people on each other in a heated quarrel.

A sorcerer (using very small sprites) might a) trigger the thought of someone, then b) trigger another thought of something bad or offensive; with the result that you associate the two. Such transference of thought may (as best as I can hypothesize on the matter) be something created by means of the sorcerer somehow transferring his thoughts (perhaps through an intermediary demon or sprite) to someone in the way similar to how sound is carried on waves to a radio. This, however, is admittedly only rough speculation. In any case, the sorcerer might make you think someone is supportive of you, while simultaneously have you think badly of that person by, for example, highlighting their possible faults. This both encourages false pride and ingratitude in you. Likewise, they can make you interpret a genuine kindness as something cheap; as well as make something trite and affectatious look valuable and sincere.

Demons (again, not necessarily fiends) can be sent into both good and bad persons; the difference is that they can generally remain in bad persons longer (i.e., it costs less.)

Possible symptoms of demonic presence in a person are the following:

1. Sweaty face and or glazed eyes
2. A look of carnal or hedonistic vulgarity, perhaps unusually protruding nostrils
3. An aged look beyond the person's actual years
4. Something about the person makes them look some how dirty and or guilty
5. Unusual distress or agitation

A demon can artificially instill bad tendencies, or else he can amplify bad tendencies already there. In the case of the former he can do this, if for no other reason, because of our ties with fallen Adam. A person then can be attacked by a demon and not be aware of it. Oftentimes a person will blame themselves or someone or something else for feeling bad or thinking bad thoughts when in point of fact it is a demon actually causing the particular problem. In the same way, demons can hit you in a most vicious manner without you realizing that is what they are doing: hurting you in a very bizarre, abnormal and yet subtle way while getting you to think that things are else, more or less, "normal."

Many times then great pain, such as fear, embarrassment, or anxiety, one might feel is just a literal demon infesting them; especially if the demon is very old, and exceedingly guilty. Some demons will act in a brazen way, while others might be used with great finesse and cunning (by a sorcerer) to bring about a particular effect. They can attack you on various levels of your well-being; so the individual pains are not so noticeable. But added together those individual pains can put you in one great agony. This having been accomplished they will then attempt to get you to deal with your pain or discomfort in a way that will make you guilty.

A demon can literally "beat up" a person from the inside, quite physically; though the person might not know that is what it is that ails them. The feeling is like someone violently pushing you around except that it is coming from within rather than from without. They can be very vicious when it comes to beating someone up, on top of one thing they will add another, and another, and another, and another…

Demons or sprites will sometimes be used to poison or taint food. Another thing they do is draw essentially invisible (to us) five pointed stars on household objects, such as pictures. The basic purpose of this is to defile one's surrounding in some sort of Satan related way. There is also something which I myself refer to as "witchcraft dirt." It is a kind of filthy dirt or mildew, brought out of Hell, that can be spread over furniture and clothing. Like the aforementioned stars, to the normal eye it is invisible. Yet one nevertheless can very much feel and actually see its polluting effect. This dirt can usually cause skin problems, such as rashes and other breakages, if put on a person's flesh.

There is no end to Hell's stratagems, and methods of attacking someone. After all, usually all that is required is to take something and turn it upside down or inside out from what it is supposed to be, or put it where it shouldn't be. This said, the following are more which might be mentioned:

They have spells where they can get you to like them (no matter how horrible they are or what they did to another, perhaps even to a loved one.) I suspect this is simply done by manipulating certain emotions and the cognitive associations (or triggers) we have that might accompany them.

A sorcerer, like Magus, can induce and create personalized dreams or nightmares while you are sleeping. In addition, one can be hypnotized while asleep, and possibly be made to say or do things in that state. When I first experienced this kind of thing, I thought it was regular people using some sort of technology on me. What the actual experience apparently (and as best as I could at present surmise) involves, however, is a kind of hypnotism brought about through a more sophisticated than usual use of demons or sprites. [95]

By means of witchcraft, demons, and put-ons paid for with money, a sorcerer working with regular people, can seemingly turn the universe upside down on you – over night, like some of the ludicrous happenings in the Peter Sellers' film "The Magic Christian." Again keep in mind that Satan has people working for him round the clock, and they will sometimes spend great time, planning, effort and energy to execute the most elaborate and preposterous charades. These kinds of things range from somewhat funny, though typically in some way mean and cruel, to something extremely vicious and sadistic. Some of these people Satan has working for him, are like Nazi scientists who are skilled in the art of deception, torture, psychological manipulation, and earn their pay doing such things to Satan's opponents. What we ordinarily think of as witchcraft is often times just a kind of psychological warfare technique, combined with (what one might denote as) spirit "technology." Such devils, who do these things see themselves as sophisticated specialists, and can find it both entertaining and interesting to engage in such practices regardless of who is hurt.

Hell people will not only do harm to others, but prevent others from doing good.

They will spend a great amount of time and go to great lengths to get someone guilty. And when and if they have gotten that person guilty, they might then hound that person for years; based on the excuse that the person is guilty (of whatever they previously prompted or enticed him to do.)

If you are religious the odds are that they will more likely be extra hard on you.

They will accuse or blame say a good or decent person for what's wrong (or get others to do so) call out a posse and a lynch mob against that person, when all along it is they themselves who are guilty.

---

[95] As well as sleeping dreams, there are possible sorts of waking visions. But without going into the subject at any length here I would mention that it seems very possible that what was or has been written down as history – say perhaps in the Bible or other historically related work – originated in whole or in part from such a vision, show, or a waking dream imparted by a spirit person. I say this because I have had stories myself related to me in this fashion, though naturally, being of a skeptical disposition about such sources, I am apt to question the veracity and accuracy of their content, and the credibility of those imparting them. As a general rule, I take the view that often dreams and what is commonly referred to as the subconscious is simply ordinary conscious, yet operating on very low energy levels; as, for instance, when we speak of a mechanism running on low batteries, it runs imperfectly if at all.

They will start a war, and afterwards found a charity for disabled veterans; rob the charity of most of its funds, then take credit for helping the veterans.

They will take over a hospital. Then make you sick, so that if you go to the hospital it will make it possible for them to make you suffer more, all the while pretending that they want to "help" you. They need not replace the whole hospital staff but only one or two individuals (with one of their own.)

They might learn in advance a person's good fortune, then torment the person about one things or other, get them, say, very angry, perhaps cursing; so that when the good news comes they will find the experience soured by their undeniable guilt.

They might throw you a party in your honor, at which you given freely to enjoy yourself, while unbeknownst to you they themselves are off elsewhere destroying your family or murdering your loved ones. This sort of combination they find droll.

In one Three Stooges comedy, Moe, Larry and Curley, secretly plant bugs and mice in a well-to-do woman's house by means of a back entrance. They then go to the front door and offer the woman their services as exterminators. This kind of thing is something a Hell person like Simon Magus (as "Jesus") or Gyro might do, pretending to rid or rescue people from witchcraft problems which they themselves have instigated.

They will encourage a person to be religious in an otherwise genuine, though superficial way; only so that when a crisis or the moment of truth comes they can use them for a hypocrite. For instance, an Orkon or Goomer person might give a huge donation to some church or charity, and have their picture taken with the church or charity as its benefactor. The Orkonist (or Goomerist) will then get themselves involved in the most heinous crimes; which will then not be publicized or even prosecuted for fear of embarrassing or discrediting the church and charity which so warmly embraced and welcomed them.

They will misrepresent an opponent by masquerading as him. For example, purposely ludicrous, offensive, and preposterous articles written against witchcraft might be published by these Orkon in order to discredit those who do sincerely combat the problem of witchcraft and summoning up demons.

They will trivialize and attempt to even make cute the practices of sorcery and witchcraft by means of entertainment, thus desensitizing people to the real dangers of such practices. This is not, however, to say that all such frivolous depictions of witchcraft and sorcery arise out of malignant intent or are without value. Even so, if people knew how many truly heart-breaking and tragic things have been brought about by witchcraft and sorcery, they would light hearted portrayals of them a good deal more distasteful than they do.

The use of demons and other sprites to cheat in gambling is an ages old practice, hence the traditional antipathy of many religions to gambling. It is difficult to say how common it is, but certainly if they need or want to clever sorcerers and spirit people can easily fix gambling.

Lastly, in this overview of some of Hell's methods of attack, one can mention horror shows. These are displays of one kind or another of unbelievable evil, and will sometimes involve regular people cooperating with Hell persons. For example, let's say a young man is deeply in love with a girl. These people will actually do things like have the young man see her gang raped by a mob of men, and witchcraft might be used to make it seem as though she enjoys the experience. This sort of outrage has been reported to have happened literally, or else can be made to take place by projecting it quite vividly, in a visual and "audible" way into a person's mind. The purpose of such displays is total degradation and the glorification of evil, and I could mention to you other examples, some of which I have been put through myself. For the time being, l hope this admittedly token illustration will be sufficient for making the point.

The situation I personally have been placed in by these Hell people is very strange for a varied number of reasons. For instance, at the same time that I am attacking "Hell's" influence as a communal problem (as I do in this writing), I am and have had these people attacking me personally for over a decade. This means that while I am put in a position to try and get people to take a look at something which is, for many, very difficult to both accept and understand, I also have to defend myself, my reputation, and my work against the dirty tricks, physical violence, smear tactics, deception, and other problems, both small and great; which these Hell people have and do put in my way. Really, unless you yourself have been put through something like it, you cannot imagine the very extreme and numerous pressures these people, that is spirit people, their ("flesh and blood") henchmen, and other dupes, can place upon a person.

There have been so many obstacles and hardships that have been thrown my way in the course of my war with "Hell," that it is impossible, I feel, to give you even a general and properly adequate idea of what they are in scope and in character. Some of these things are so crazy, as they certainly are intended to be by these Hell people, that I would hardly know how to begin to describe them to you. Yet it will be not undesirable to touch on a few troubles which I have had to deal with which are not too difficult to explain or understand.

Among the most difficult things I have had to contend with is others being able to contact me, or myself others. This is not due merely to these Hell people interfering with my mail, phone, etc., but to their quite elaborate methods of making me look bad, frightening others from the subject, or deceiving people into believing that I somehow don't know what I am talking about. These methods have been used successfully not only against people for whom my acquaintance was only casual, but also against people who I have known for years. It is a well-known truism that when one is having troubles that friends are hard to find. Indeed, and especially when these Hell people are involved, someone being victimized will be made to seem the cause or source of whatever is wrong, rather than (as in this case) the Hell people who are actually causing the trouble.

I have had people I have otherwise known for years suddenly betray or turn their back on me for no apparent reason whatsoever. Not so long ago, a person who I had on a prior occasion spoke to in a friendly way and who I had known for years, acted as though I had committed some horrible crime, and told me not to call him ever again. "Have a good life," he said rather sarcastically, in a hurry to finish our conversation. His attitude came as a complete surprise, and when I asked that he allow me to speak to him about why he was being this way, he refused any further discussion. Even family members have treated me similarly, and in addition have accused me of being crazy, when I knew in undoubted point of fact they were under the influence and brazenly lying to me. While it grieves me to have to admit these things publicly, I feel the benefit others can gain from my candor will have made the bother worth it.

~ * ~

Ways to Fight and Cope with Hell

"St. Michael the Archangel, defend us in battle..."
~~~~~~ prayer to Saint Michael

"Be ye therefore shrewd as serpents and meek as doves."
~~~~~~~Matthew, 10:16

I am a Catholic, and in this section will tend to give a mostly Christian view on how to handle these kinds of people and events. In doing this, however, it is by no means to be taken as somehow a slight of other religious faiths; which no doubt have good remedies of their own for addressing the problems of Hell. What I can do though, is try to share with others what – after over ten years of brutal conflict with these monsters – has worked for me, and worked well.

In dealing with attacks by Hell, the first thing one should most guard against is doing something wrong yourself. It is the main thrust of Hell's strategy to make people guilty. The more guilty a person is, the more it is perceived that Hell is justified in hurting them or otherwise claiming them as one of their own. In addition, a guilty person is that much more susceptible to being brainwashed then a non-guilty person. In sum, those who explicitly reject truth and right, ever risk enslavement by those who know how to do the wrong thing better than they do, and there is *always* someone to fit that description.

This is not at all to say that because one is doing the right thing Hell people will not hurt them. To be sure, on the contrary, people that aspire to faith and virtue are very much their targets. Nevertheless, a person that hugs close their faith in God and moral virtue is infinitely better protected that someone without these. Take therefore any of their possible challenges or threats as opportunities of proving your faith. The more one thinks of others rather than self, the more they are strengthened to combat Hell's assaults. As well, the converse is true. The more one is wrapped up in their own self-interest, the more painful Hell can make things for them. For all Hell's frightfulness and horrors then, the one thing one should fear most above all is doing evil one's self. "It is not what goes into a person that defiles them, but what comes out of them." The very worst place to be then is not to be oppressed by them, as unspeakably awful as that can be, but to either be assisting or otherwise be one of them in their wrong doing. They will cause you a problem, then get you to deal with it in the wrong way; thus creating for you (if you will but cooperate) *two* problems! The challenge then often becomes fighting a war without making oneself guilty. Bear in mind as well that the more you do the wrong thing, the less able you are to do anything truly positive, most especially those things that require heart, talent and imagination.

No one else can do more for you than you can do for yourself by doing the right thing. If you can't do the right thing, there will be disharmony and strife within you and you will not therefore be able to make another truly happy. And if you can't make others truly happy, of what value then is your love? Even if you had all the riches in the world, what good would they be to you if you could not make another truly happy?

It is well to remember that fears, apprehensions, worry, confusion, concerns like "why does God allow evil?" are in most instances not natural to life, but a result of our fallen state. If we were not fallen, these doubts and negative feelings would seem irrelevant and meaningless. Conditions where you really and inexplicably feel "down and low" very much might suggest the presence of a demon. Yet you might all the while mistakenly think there was something wrong with yourself (instead of knowing the possible true facts.)

One's own circumstances depend considerably on how much Hell people are hitting you with, combined with how strong you are morally, intellectually, spiritually, physically, and socially – in that order. Having others to help you is very important, and if they are after you bad enough, Hell will go to great lengths to try and isolate you. In my own case, they very well succeeded; which is partly why it was allowed for me to be come acquainted with so much of these phenomena. The idea being that isolated as I was and am, people would all the less believe such incredible stories from me.

When one is first openly attacked with witchcraft, demons, etc., it is normal and understandable for one to throw one's self into religion. This is not always such a good idea, because Hell is expert at messing up and confusing a person's religious ideas. Reading aloud the Bible as an immediate defense I would not readily recommend, unless one has a good idea already what passages they are going to read. A Simon Magus, for example, has ways of imposing perverted interpretations on scripture in one's thoughts that might only make your situation worse. He can do the same with regular prayer. Consequently, one should approach Bible reading and prayer with a certain amount of caution; generally reserving such for occasions when one is least under assault. On the other hand, someone like Magus might encourage you to be religious or devout, and you should be, yet you obviously resent the suggestion coming from him, and then make the possible mistake of seeing religion as something bad.

The best approach at first is to keep level headed, moral and rational. If proper "religion" then is not already a regular staple, they should bring it into their life in a consistently scheduled way, without feeling pressured to overdo things. In this way it is like physical exercise. One doesn't build muscles by lifting weights for a week. Such things take time to grow, and must be looked to regularly. Remember, religious faith was not created to fight off witchcraft, but to love and serve God and others. One ought make sure they understand its true function, before invoking it as a means of dealing with Hell. Religion in its proper sense simply refers to our love of what we see as the highest form of our highest good (Love, Truth, Innocence, Wisdom, for example.) A church or temple then is properly a place where we celebrate and make sacred the importance of Love or whatever else we consider to be among the most pure and highest goods, "God" being the supreme unity, embodiment, and purpose of those highest "goods," in an experience that we share with fellows of our faith.

Thoughtful and intelligent religious books written by saints like St. John of the Cross, St. Francis de Sales, Jeremy Taylor are always helpful.

Saying the Rosary on a 6-7 times a week basis, has proved of tremendous help, and I cannot recommend it enough – Hell attacks or no – as a means of keeping one spiritually strong and focused. I've tried to make it a point, once a week, of saying one of the mysteries of the rosary on behalf of people of down below and Hell who want to stay out of trouble. Among other benefits, the rosary fosters a healthy humility. For a quick extemporaneous prayer try: "God make me good, keep me from foolishness." Prayer, in general, takes us out of ourselves by making us think of others and our obligations to others; which can only be to one's advantage when dealing with Hell.

As mentioned, use Hell's attacks as an opportunity to prove and demonstrate your faith. Yet do so without in the least suggesting that their attacks are somehow needed (or justified) for this purpose. If they, for instance, cause you jealousy, use that jealousy to praise the envied person's special virtues, and by doing so thus find yourself that much more commended to those virtues. In this way, jealousy can be used to make us grow and mature; instead of being something that might otherwise degrade us.

Worship need not always be extemporaneous. As long as we schedule it in our life's efforts can we still fulfill both what is our due and sing His praise; just so long as our expression is not unnecessarily formal, phony or insincere – this based on (at least) better-than-common standards. I have been in situations where some demons were literally trying to force me to sing praise to God. Of course, my objection was not praising God, but their needlessly forcing me to do so. To have just went along with them otherwise, in those peculiar circumstances, would (to my mind) simply been condoning and giving an excuse to gratuitous and violent bullying.

If one's religious faith is not working for them, it is probably because they are not being as dutiful a person of faith as they should or might be. But if you can't be a better Christian or a person of faith than you are, for goodness sake, don't blame Christianity or religion with not making your present life easier; for to do so would make as much sense as blaming a proper health food diet for not guarding you against an accident or assault by a criminal that caused you to break a limb. Be, as well, willing to suffer for different ways you adopt for pursuing the good and fighting for the right. He who lives by the sword will die by the sword. Though there are, after all, good soldiers who are Christians, it yet must be admitted that as good soldiers they might not theoretically be as good Christians as they ought. Still one can be a good Christian and a good soldier, if they are willing to suffer the soldier's burden, discipline, and sacrifice, all the while prepare to meet death in the face. Philosophers and scientists can do the same, though, in their case, the burden and sacrifice might be public denouncement and a cup of hemlock.

After regularly and sincerely attended to "religion," the study of philosophy and mathematics is an enormous help in keeping one's thoughts clear and forming better understanding to offsets Hell's efforts at mind control and deception. As a matter of fact, after my Christian faith, my reading of Plotinus and Bertrand Russell may have saved my life in dealing with these people. Geometry is good also because there is something about witchcraft that involves circles, and knowing that a circle is really nothing more than a series of tiny connected straight lines does something to break certain spells. This is also one of the reason the cross is so effective in dealing with witchcraft. Also, for some odd and inexplicable (to me) reason, in my own particular experience, I have found saying the number "046" would break certain spells.

A little poetry and music, now and then, is good, though be aware what might sound nice and pleasant in one kind of circumstance might be made to sound foolish or silly while Hell is attacking. This said, poets I have found particularly helpful are William Wordsworth (particularly his longer poems), and William Cullen Bryant, (to name of course a few.) Singing religious hymns is very good if they try to create an atmosphere of fear, and there have been occasions when I have had multiple ghosts coming into my bedroom and I enjoined them to sing along with me Isaac Watts or certain passages in Handel's *Messiah*. Martial music (such as fife and drum music), Native American tribal music, on the right occasion, heavy rock music played on a music player, can be effective in dealing with prowling spirits, and inspiring in their own way. Indeed, one can say this is true of good music in general where circumstantially appropriate, and the selection is a thoughtful and well-considered one.

Abstaining from meat, I also feel has helped to make me much stronger than I otherwise would be.

One of the key ways a satan can harm a person is by getting them to look at things too much from the perspective of time rather than eternity. The more one is caught up seeing life and their experience from a "time," rather than an "eternity,"[96] frame of reference, the better Satan is able to instill fear and worry into them. The same with anything that is finite versus that which is infinite. Life was always better before Hell showed up, so think as if you just started having these problems today or yesterday, and in this way start afresh. This approach does not always work, but it nevertheless can still be used regularly over time. For all his extraordinary persistence, time is working against the devil – so get through that bad experience today, looking ahead to tomorrow. "Those that persevere to the end will be saved." Do not confuse what is going on now, at the moment, with the long-term situation, the two are not at all necessarily the same thing. The moment for example may indeed be insuperably painful or horrible, but, it doesn't necessarily follow that your long-term situation is. Some of the greatest pain of Hell comes not from individual assaults but assaults which are ongoing, perhaps over the course of years or even a lifetime. Remember not a few of these people are insanely obstinate in their trouble making. If Hell's attacks do continue on an ongoing basis, do not spend too much time worrying and concerning yourself with tomorrow. "Let tomorrow take care of itself," and just aim at getting through the given day.

There is a big problem of seeing God as chastiser of the good; because if the devil afflicts a good person, the conclusion seems to be that the devil represents God, or worse that God is the devil. On a general level it is all right to say God permits and cooperates with evil, but in individual and particular instances we should at least suspend judgment, and ascribe blame to where it ostensibly and is most clearly due; which is to say in the more immediate people present and circumstances.

Whenever things are excessively bad then, and as a very practical and useful rule, blame the devil. To blame God is to presume a greater knowledge and understanding of Him than we possess. At the same time, we know in point of fact the devil is a murderer a torturer thief and liar who will want us to hate God,

---

[96] Or also "finite" and "infinite" if you prefer.

while at the same time doing all he can to manipulate and deceive us. Avoid being more angry toward Satan (or a Satan) than you are grateful, happy and obedient toward God. Though this, said outrage and indignation toward Satan are quite justified. However bad things get, do know that even if all the world disappeared, God, true goodness, the infinite, would still be there. And that even if all the world became evil, you will still, in some measure, have the ability to do the right thing.

Often the standard and criteria of judgment of our judgments is that of who has the most power. This is common to both good and bad minded people. But with good mind people power is with God or the truly divine, whose edicts for us are always consistent with what is rational, moral, just beautiful. For a Satanist or Orkonist all that ultimate power requires is brute force and perhaps some kind of beauty.

"I saw the snares that the enemy spreads out over the world, and I said groaning, 'What can get through from such snares?' Then I heard a voice saying to me, 'Humility.'"
~~~~~~~St. Anthony the Great

"If he had been Antichrist, Creeping Jesus,
He'd done anything to please us:
Gone sneaking into Synagogues
And not used the Elders and Priests like Dogs,
But humble as a Lamb or Ass,
Obey himself to Caiaphas.
God wants not Man to humble himself:
This is the Trick of the Ancient Elf.
Humble toward God, Haughty toward Man,
This is the Race that Jesus ran."
~~~~~~William Blake, "The Everlasting Gospel"

In dealing with a possessed person, remember that they can only go so far in their thinking from where they are psychologically tethered. They may, with a bit of good reasoning and kind persuasion, be got to go a little distance, that is to say in a direction that will help them to objectively understand the given situation at hand. But beyond that point they can go no further, and will presently return to their bondage (i.e., their brainwashed state); under pressure from the particular demon (or demons) who "has" them. In the Gospels, when Christ expelled a demon he spoke to the demon, or directed his command to Satan himself. Sometimes the demon would speak back. But interestingly, it was if the person who was actually possessed had no say in the matter.

You can then discuss very little of a serious nature with an irrational or possessed person. As much as you might be separated from thinking and feeling folk, it will not help you to bring your story or your problem to others who are not thinking or feeling. One needs to remind oneself at times: "Wake not the brain dead," and that you are wasting your time trying to communicate immediately or in the immediate with some people. At the same time, from a longer term perspective you may be much better able to get your point across. But however things are, better to suffer your isolation than to plead with obdurate block heads; no matter how important their status and position in society might be. Christ's silence during the passion is a very helpful lesson to us in this way, both in telling us to avoid engaging in something futile (even self-defeating), and to speak more eloquently by our forbearance.

Insofar as you can help it, deal only with honest, serious people. Liars, fools, mindless slaves and possessed people are perfectly useless, and there is usually no point to try to reason with someone who is brain-washed. Only when the storm hits will you really find out who the fair-weather humanitarians are. Guard therefore against unanticipated disappointment and betrayal.

It has been an age old dilemma in dealing with the Hell problem, how to handle the "possessed." If the demons cannot be cast out, or there are just too many possessed people, what should we do? Suffer the cross? If things are really as bad as all that, the answer simply, is yes. By suffering our cross, and suffering our cross willingly, we become that much more unlike the devil, and for that reason that much more worthy of both happiness and imparting happiness to others. God blesses us, and the more we are blessed the more the devil will want to crucify us. God's purpose is the blessing. It is the devil's purpose to crucify. In the midst of these cross purposes we are to acquit ourselves as being, if only in a small way, worthy of God's blessing. Part of this is or might be suffering the cross. In passing, some might find mediating on the repentant thief crucified, rather than Christ himself, more helpful to their situation.

As pointed out earlier, deception is one of Hell's greatest strengths. If you feel they are brainwashing you, "judge not by appearances, but by the truth," not on what people say or what they say people say, but what the honest, objective, moral facts and arguments are. Similarly, do not be quick to assume that because "A" takes place it was caused by "B." Be thoughtful and consider if it wasn't perhaps "C" or "D," etc., that caused "A." More evils and mistakes are brought about through simple causal errors, logical errors, and mistaken associations than one could ever possibly imagine.

Among the very most virulent and hateful things to avoid which might be brought about through contact with these people are false ideas. For example, rape is not love, yet to a Hell person, who has been deceived by means of certain similarities between the too – such as zeal and physical contact – they will believe them to be the same thing and thus the concept and understanding of love becomes quite opposite to what it should be. Likewise, the same can happen to ideas about truth, happiness, religion, fairness, beauty.

Generally speaking, a Godly person knows truth and values first and foremost by reason and principle; an Orkonist or Goomerist defines truth according to what people think or what a particular person thinks. The former decides issues according to law and what is fair and just; the latter by whim of an authority typically based in money and or violence.

It is very rare that one can argue another into the truth. Yet where the truth is present honest, rational and empathetic disputation can be a very good and desirable thing for enhancing mutual understanding. In not a few places in the Gospels does Christ impart his message by means of a logical argument. Yet what is interesting is that as true as what He says is, in the given instance, it will be reacted to differently by people depending on whether they are of the truth or not (while allowing for some shades in between.)

For example, when he says, "Let him among you without sin cast the first stone," those of the truth will imbibe the greater meaning of forgiveness and self-examination, while those not of the truth will have taken the argument as merely a legalistic device that is a social obstacle to their selfish purposes, yet an obstacle they must, under the circumstances, respect lest they be branded a hypocrite. But if the circumstances were different, and the same truth were spoken to them in a back alley by someone with less obvious authority, such "Pharisees" would have no scruple ignoring or stifling the truth speaker.

Logical truth has only as much effect in a worldly society as that society allows it to have; that is to say the effect depends on how much as that society respects and recognizes the validity of logic. Among those of the Truth, logical truth needs no such circumstantial buttress or support. I think fear and irrationality are the biggest obstacles in dealing with the Hell problem. This need not be the case if people would only more desire and be guided in their judgment by the truth. But how to get people to love the truth and be rational, that is one of the most ancient of conundrums.

The moral law is necessary to distinguish between real life miracle and false magic, between God and Devil, between real happiness and false high. And our knowledge of the moral is made all the stronger by our ability to be honestly rational with ourselves and others. There are some people who do not know how to be very rational, or even what it means to be rational. Such people naturally make their decisions based on feelings; which in turn are based on mere images and impressions; in turn processed by overly simplistic inferences and syllogisms.

A small thing can become evil when taken to excess or inappropriateness. But what ought to be observed is that but for the excess or inappropriateness it need not be such a bad thing, or perhaps not even a bad thing at all. And what might be excess in one circumstance might not be excess in another. In addition, the presence of other factors, for example, certain kinds of people or things might make something bad or undesirable which it would not be otherwise but for these others factors present. Moral questions on some occasions then can become a question of greater versus lesser evil, and vice versa. Don't allow Hell to confuse you into thinking all wrong behavior is equally bad, and that if you have habits that need correcting, their own carrying on is therefore excusable. I am aware that sometimes in books, music and film there are things expressed which don't always reflect the best morals. In making a plea for better morals here, I would not recommend censorship or eliminating such works. Yet on the other hand I would suggest that such works be not indulged over casually; nor should they experienced or performed in a location or circumstance which unnecessarily and selfishly ignores the moral sensibilities or sensitivity of others, thinking particularly of children and animals (where unusually loud music is to be played for example.) We must be able to think and express ourselves freely, yet this does not mean we can afford to be arrogant and completely insensitive to the feelings of others. In most instances, there is really no excuse for that, and there is usually a way of accommodating both parties if the problem is approached in a reasonable and fair-minded way.

The plain life-style and solemnity of the Puritans, Amish and Orthodox Jews makes sense as it can protect against being made fool of with luxuries; which a devil will characteristically do. But it is not the luxuries themselves, necessarily, which are wrong, it is how they can be used on a person, or whether they are used extravagantly or in excess. In this the Puritan, Amish or Orthodox Jew may possibly be in error by blaming the bait rather the trap and the persons placing it. Yet sometimes Satan will abstain from using certain luxuries, clothing, certain humor, etc. as snares. These things will return to be seen as harmless, only later to be taken up by him as bait. Another argument in favor of the austere approach is that if one can't do or enjoy a thing the right way, and without some stupid devil interfering or meddling, your better off deferring it – even if that means until the next life. Such an attitude shows the most respect for whatever, in the given instance, is thought to be desirable.

If you are someone who has already been taken in by these spirit people in a serious way so as to implicate yourself, the first thing you need to do is stop kidding yourself about not being guilty because you mean well, or "pseudo-Jesus said it was ok," or "I didn't know what I was doing," or "I was forced to do it." While these kinds of childish excuses might well mitigate your offense, they do not absolve you of guilt. Start by seriously waking up to what is going on, stop cooperating any more with these people, and begin to take responsibility for yourself. Otherwise, you are not fooling anyone, least of all the people of Hell, and, as always, the more guilty you make yourself, the easier it is for Satan to take over, and control your life. Repentance of sins, being honest to others about what you did or were a part of, perhaps turning yourself in, will do wonders to help cure your circumstances. If you just go on kidding yourself, you will only make things much worse, not better.

Remember that sins can be forgiven, and there are extenuating circumstances which makes the compliance with someone raping you less culpable than it might be otherwise.

Which brings up another point. It is important to note that sin implies a literal demon,[97] because ordinarily the vast majority of the people will not have made themselves so wretched as to reach such depths of feelings of guilt as an age old demon has. Our own feelings of sin stem more from our contact with such people, and not fully disassociating ourselves from them and their master, and in this sense, sin is only known by our contact with that much darker and willfully unnatural community. Otherwise our ordinary faults, mistakes, and foibles, if not an egregious and heinous sort, would not bring up the notion of sin, except perhaps as an admonitory concept (presumably arrived at by induction.) Where else after all are you going to find the most guilty people? And how bad really are any one of us compared to them who may have been around for centuries or more, and unrepentant?

In dealing with literal Hell people, one must be prepared to deal with the most arrogant, obstinate, bullying, remorseless, violent, irrational characters in the world, and until you have actually dealt with them no words can adequately describe to you what these people are really like. Be therefore prepared for the unexpected. The worser and more crafty ones very much make it their business to surprise people.

Often Satan's or an Orkonist's "Fee Fi Fo Fum" doctrine is all he needs to get others to bend to his will. It is well then to remind one's self of the apostle's wisdom that "perfect love casts out fear." One can get through so many things, if love is present. Examine yourself for how important love is to you, and how you best are serving its call in your life. If you or someone is frightened, remember it is a person, a creation – not God – who is causing the problem, and that sooner or later this person will have to take the day off, amuse themselves elsewhere, perhaps go have lunch, etc. Even Satan or a Satan, as powerful as he is or might be, is only a creation. Seeing them in this practical and realistic way helps to take away the mystique of fear they might create for themselves. Remember as well, that when fearful, fear is being believed over reason, fairness, gratitude, and love, and this need not at all be the case if one chooses not to make it so. If one gives into the fear, bribery, strong-arming and other manipulation of and by Hell people, the claims of behaviorism become a self-fulfilling prophecy.

In combating evil then, always think "WHO is doing this?" Where was he (say Gomez or Simon Magus or Archimago) born? What was he like in high school, etc? How did he first become involved with spirits? Is this person, for all their presented importance, older than the moon? Who are his father and mother? Don't let Hell set the terms for reality. It is part of their technique to disorient you and shake you up. When dealing then with an immediately present or imminent

---

[97] See Genesis 4: 6-7

attack ask yourself what exactly is going on. Try as best you can to describe your circumstances and surroundings to yourself. What date is it? What is the temperature? How is your appetite? Is there a prior history to the Hell originating experience or experiences in questions? Asking questions like this will help put whatever is going in perspective. Use common sense. A spirit person has no more right to break into your home any more than anyone else does. Whatever grand appearance they might put on, don't let them use this as an excuse to violate common sense, let alone your God given fundamental rights.

Don't judge or worry about judging what happened to someone after they died, say an odd acquaintance or well known person. It's not for us to judge, nor is there any profit in doing so. On the contrary we could (in many instances at least) do ourselves great harm in assuming so and so went to Heaven or to Hell; when in point of fact we don't really know, and spirit people, the vast majority of the time, cannot be trusted on such matters.

There are times we feel better mentally and or physically as to be able to suffer something than at other times. It is helpful then, when under attack, to think of the kind of circumstances in which the suffering we now undergo, would not cause us as much pain if we were in that better mental or physical state. When Hell actively, physically and directly attacks, one should view what you are being assaulted with as a disease, which although it touches us, we are not a part of it, at least as long as we are making a sincere effort to behave rightly.

With respect to individual pains, try to identify and define what it is that is bothering you. Use a kind of Cartesian or analytical method of identifying the problem, and then analyze it, that is break it down into as many of its simple elements and possible causes. The intelligent understanding of what specifically ails us, or what we are specifically threatened with, go a long way toward our successfully dealing with these as problems.

Here is a list of possible pains a person being attack by Hell might go though, bearing in mind that they might be suffered in a combination or combinations.

* Ask what kind of pain it is? 1. Physical, 2. Emotional, 3. Psychological, or 4. some combination.

* After this consider as possible diagnoses:

    a. Outer darkness syndrome (perceived abandonment by God)
    b. Emotional dehydration
    c. Intellectual dehydration
    d. Despair
    e. Mental agitation, confusion: doubt, uncertainty, disorientation
    f. Loss of self identity. Disorientation of self

g. Physical pain, as say from a wound or illness
h. Jealousy, envy
i. Loneliness
j. Rejection
k. Embarrassment
l. Anger, rage
m. Extreme fatigue
n. Loss, particularly of a loved one
o. Fear
p. Thirst, starvation
q. Helplessness of thought while in physical pain
r. Defilement, degradation
s. Extreme stressfulness, impatience, agitation

Having identified what might be hurting you, you might, being honest, ask yourself questions such as:

1. Are you yourself doing anything wrong to hurt someone in a serious way?
2. Is your own attitude possibly a selfish, carnal, egotistical, faithless, ungrateful?
3. Could your attitude or feelings be something implanted on you by a demon or demons (i.e., demons as described in this writing)?
4. Is the problem really so urgent as it at that moment seems, or could actually suffer it longer if you reflected more intelligently about it?

Most of the pain Hell causes is due to demons and sprites. There are so many ways a demon or sprites can hurt you, and there are as many different demons as people that it would take a long time explaining how one might deal with them. Here, however, are a few suggestions and expedients.

1. Calm and regular prayer, including saying prayers for them perhaps. It need not be anything elaborate. A simple "God make me good. Keep me from foolishness," "Thanks be to God," "Lord have mercy on me a poor sinner," on a regular basis, I find, will work fine.

2. Try to make them feel what they are causing you to feel. It's their own problem after all.

3. In the case of some sprites, if you grab hold of them and hold on to them long enough you can drain them of their energy.

4. Sprites in certain instances will sit on a person's head. One can sometimes get rid of what is bothering them by removing such a sprite.

5. Demons and sprites commonly come into a person through the back of their neck. If you feel the back of your neck, one sometimes may be able to grab them and pull them out.

6. The five pointed star, mentioned previously, is sometimes drawn on a persons forehead. You can get rid of this by simply rubbing it off.

7. One can usually speak to a demon or sprite that has invaded them, and there are instances where you can politely persuade and argue them into leaving. On the other hand, one can't always be sure in a given instance that a given demon is not deaf, or else unable to speaks your own language.

8. Certain spirits are allergic to things like incense, tea, vinegar, tobacco. This kind of chemical remedy, even so, should never be viewed as a long-term solution, but they sometimes help.

9. Loud noises: machinery, music. Again, as with no. 8., only as temporary solutions. It is interesting to note, that various cultures around the world have traditionally used bells, horns, and small explosives to chase away demons, and these methods often still work very well.

10. In certain circumstances of justified moral outrage and indignation, one can violently expel a demon. On the other hand, there are instances when an empowered demon, or fiend, will cling on to you regardless of any amount of moral and physical pressure you place on them. A sorcerer or warlock can make the demon deaf, duped, or doped up, or there may be other distractions which make it difficult to rebuke, cast out, or repel a demon.

11. If a demon is trying to get you to think thoughts of murder or blasphemy, you can "split their head open," by turning their own thoughts on themselves. How exactly this works, I leave for a person to figure out for themselves, as it is not easy to explain.

12. Some particularly filthy demons cannot take frightening spirits, so that amusingly enough it is possible to actually scare certain spirits by talking about things like "true" life ghosts, e.g. Borley Rectory, or by reading well done ghosts stories, like those of Le Fanu, Beirce, Blackwood.

13. If the sprites you dealing with a are not too bad, you can get them to treat you better if you read fairy stories and folk tales to them: Joseph Jacobs, Crofton Croker, Andrew Lang, Brothers Grimm, etc. But as always, avoid making it seem as if you are welcoming them into your presence. It should be noted that Hell uses such people, and joking simply to give itself a benign appearance. While some sprites or little folk, might not themselves necessarily be so bad, they are usually paid, forced, or fooled into assisting Hell, and it is not usual for both demons and sprites to be acting while drunk or drugged. Such mischievous sprites, or little folk, should not be confused with "goons" or devil sprites who are typically rotten little monsters and murderers (though granted ones not undeserving of real pity when you consider who they are enslaved to.)

14. Shaking one's head a certain way, while holding your hand to your forehead, will disrupt certain tiny sprites that might be sent into one's head – bizarre or ridiculous as that sounds.

15. Witchcraft works best in stagnant air, and electric fans are frequently effective in thwarting its efforts.

16. A good, healthy sneeze can rid one of demon inside a person, and this will possibly help explain the attraction of snuff in earlier times.

17. If they incessantly bombard your head with offensive thoughts, speech, or images, repeating the word, "eternity" (or if you prefer "infinity") over and over again is helpful to offset the attack: the idea being that it reminds them that there will be plenty of time for someone to get around to punishing them for their crimes, plus, the notion of eternity, if focused on correctly, will drown out any other thoughts they are trying to engulf you with (inasmuch that, aside from God, nothing is greater). "Cast out evil," said repeatedly like "eternity' can also be a good spell breaker. By saying "cast our evil," one is assuming the greater power of God and his goodness over who or what ever forces oppress you, so it is not necessary to say "God , cast out evil;" since who but All Goodness would be implied or invoked by such a statement? At the same time, peculiar as this might sound, say "cast out evil" with a view to being purged of your own bad tendencies, faults and habits, as well as being rid of the spirit person attacking. If the devil is persistent on being present command that he say "cast out evil," if he will not, and he remains after being told to leave, then he will be defining or damning himself as evil.

18. Under some conditions a person will rather just think about what ails them without, at the same time, thinking that it is a demon causing the problem. This is only understandable and to that extent permissible as one pain can be extreme enough without the added distress of having to realize what is the cause of it.

19. I have found that I have been most vulnerable and susceptible to their tricks, and deception when just waking up from sleep, or within 15 to 30 minutes after waking up. In this period of time, I could be made more prone to anger, doubt, feelings of defilement, false reasoning, – all of which would to a not insignificant extent be caused by the presence of a demon, accompanied by the manipulation of a sorcerer. If you react badly after waking up then (say, for example, they got you to swear) don't be too hard on yourself, and wait the 15 to 30 minutes to collect yourself.

20. Say to them: "You think Satan is God." By doing so you can catch them in their contradiction.

21. If you are being physically tortured, do your best to keep balance and harmony in your body, while perhaps thinking on the infinite in his or its various known forms.

In their invisibility, swiftness and cunning, demons can often have the advantage over us, yet they are significantly weak and vulnerable in at least two ways:

1. By doing wrong, they, like anyone else, are still hurting themselves. And to the extent they are given to great evil, it manifests a wish to kill themselves.
2. By doing evil – in their extreme way – they are working for someone completely filthy (a warlock, Satan). And in turn, by doing this they degrade and harm themselves all the more by submitting to the will of someone who is even more filthy than they are.

Demons can be persuaded, say to leave a person, but a sorcerer will (typically by means of some kind of threats and if he is well enough established financially and otherwise) usually be able to keep sending the demon (or another one) back in; so that mostly, if not always, they are more attuned and answering to him (or another powerful spirit person) than to others who would (instead) tell command them to do the right thing.

If they are sufficiently near, one can chase away a spirit person by striking at them with a stick, or throwing something at them. Hell Spirit people usually don't like fire, and focusing on a flame can sometimes disrupt or break certain spells, as well as being useful in chasing a spirit off. Of course it goes without saying, care and safety are essential whenever using fire under any circumstance.

When a ghost-magician has had me surrounded by very frightening spirits or ghosts, I will sometimes have said to such spirits, "Are you doing the right thing?" (If not "be gone" or "get lost") or else "Do the right thing, you shouldn't be bothering people," or words to this effect. It is important to remind spirit people that they, like anybody else, have an obligation to "do the right thing." And if you have the courage of your moral convictions, and stick to your guns, you can sometimes get them to leave you alone. Similarly, tell people from down below to pray for their victims; "die to self" live for virtue; think of making others happy as a way to make themselves happy. Tell them things like, if they can do the right thing then God is still, after all, with them, and the more they can do the right thing the more God is still with them. Naturally, this does not mean that all will listen to you, yet it is often necessary that you have given them due notice and warning of these things.

Although some ghosts are pretty frightening, and can appear like personages from a timeless, shadowy void, don't forget that they are still people. I have ridiculed such by calling them things like the "ghost of Frankenstein," or "the inhuman fiend," and often this kind of mocking is sufficient chase off such dreadful presences. Some nights in which I was attacked I might turn to humor. One occasion had a hooded monk looking through a window right next to the bed in which I was lying. More annoyed than frightened, I asked him why he didn't make himself useful, and run up to the store to get a six-pack.[98]

Spirits are not without a sense of humor and a joke well placed can be a great way of breaking the spell of fear. Washington Irving's story "The Bold Dragoon" brings this out very nicely. This said, again be careful and joke sparingly, as such tactics can be easily used against you, by making it seem as if their tormenting of you or others is itself only a joke. This is not so true of ordinary demons however, not usually being all that bright, they are not likely to turn your jokes on you.

As a rule, do not under any conditions be seriously frightened unless you at least clearly know who is frightening you. If they say you must be afraid, say to them, "I under no circumstances will allow myself to be frightened unless and until I know clearly who and what you are." Don't assume that because someone lives "upstairs" physically and glow in radiance that they are necessarily better. There may be someone higher than them, or else you might be "higher" than them in certain significant ways (such as moral character.)

Every person is potentially special and important, no matter how many people. It all depends on whom one looks to as authority. To Hell a person in the

---

[98] One interesting thing about the film "The Exorcist" (1973) I recall is how it very much shocked and frightened people when it came out. I was effectively a child at the time and myself had no desire to see it -- yet not being necessarily for or against others doing so either. Years later, I still hadn't seen it, as I imagine others didn't for the same reason; because once I'd heard the basic synopsis it didn't seem necessary. Watching some of it in the way of clips this passed weekend only confirmed this surmise. I don't mean to say therefore that the "Exorcist" isn't a good film, but when it comes to the graphic and explicit, you know, I'd just as soon not. But more to our point, how does "The Exorcist" stack up and compare to real life? While I like dramatically how the story is resolved in the priest's sacrificing himself at the end, it is worth noting that you cannot medically prescribe for a fictional situation. So that what in the movie seemed insurmountable, in real life circumstances given the presence of more rational, intelligent and inspired people might be less so. For one thing, a given devil is a person like anyone else -- who was born and has their personal history. Consequently, consider what difference there is to seeing them in this light rather than as some religious, mythical, or cosmological abstraction. At the same time, they themselves are sick people, medically speaking, and this is another significant aspect to taken into consideration when it comes to sizing them up. Also, spirit people, like regular people, and as individuals are different. The magician, as rotten a devil as there ever was (or as I heard of or knew), for example, forbids swearing like they have in "The Exorcist;" because he himself, and in his own time, has had to hear so much of it. Things like the shaking of furniture, lights being turned off inexplicably, being held down physically by "demons," and having living apparitions enter your bedroom I have experienced and seen myself, and again if viewed a certain way can be ridiculed. However, this is not so easy to do if you are half asleep in bed; in which case you are more easily frightened. So that in all, more common sense and perhaps a little bit of humor will more likely go a much longer way than invoking "The Father, Son, and Holy Ghost" in sacred ritual. This is not to dismiss or make light of ritual, but each thing in its proper place, and just because you respect such things doesn't mean a spirit person has to. And to realize this is simply to apply common sense.

crowd doesn't matter. To Heaven the fact that they are one of a multitude does not detract from their importance as an individual. Again, it much depends on one's perspective on the matter. A person's actual meaningfulness is based on God and Truth's standard, not the world, other people's standards, least of all fallen angel's or the people of Hell. Though in individual circumstances you will feel quite otherwise, normally, one can't want to be too hard on these people since they are already, for all their cockiness, so miserable as they are. Resist then intelligently, know what part of them can actually be harmed, size up their strengths and weaknesses, and don't think that just any kind of striking back will force their retreat.

While it is well to be indignant and combative with demons and sprites, at others times one just has to bear these things patiently, suffer them, and "bear one's cross," with a mind to thinking about one's own sins. Some spirits will not be too hard on you if you show yourself tolerant and understanding of their own situation. However, be careful not to allow yourself to be *too* friendly. One can make politely, yet firmly, clear to them, that they have no right being in your home, let alone attacking you. At the same time, they can use your own rage against you. On one occasion in which they (in cooperation with "regular" people) were literally trying to murder me, I was constantly reminded by them that "they have nothing to live for;" that there's no amount of my threatening or insulting them that is going to make their situation any worse than it is, and my being angry only incited their fury the more. What I did then was to calmly tell them that I was sorry about their situation, that they were wrong, and that what they were doing wasn't going to help anything. The remainder of the time I spent grappling with truly excruciating physical pain, while meditating on Christ's passion and the Seven Last Words – all of which, for three to four hours, proved very effective and helpful in strengthening me in that hellish circumstance. They had said that if I insisted on being a Christian they would increase the torture all the more. Yet it was blatantly obvious to me that no physical pain was worse than the worse degradation of giving into them.

In most instances, it is important to be compassionate and fair with these people, and in a sincere, not phony way. In fact, be self-sacrificing, forgiving, and giving as best you can, but without being taken in by them or compromising first principles. Be more concerned with your own faults and misdeeds, and behaving yourself towards others better than they might behave toward you. This doesn't necessarily put your petty wrongs on the level of heinous evils of others. Nonetheless, your concern for your own wrong doing should be more great than your concern over the wrong doing of others – at least the vast majority of the time. For one thing, one can better help one's own sins than that of others. In addition, the worst people do us the most harm when they can get us to be more guilty.

If you prevail over Hell people in their attacking you, you are actually helping them, though needless to say they will not necessarily appreciate the fact.

Some Satanist or Orkonist might claim that if they keep to themselves, there should be no need to blame them for anything. As a matter of respecting adult choice, I (for the most part) agree except that I don't believe Satan keeps to himself, even if some of his followers are willing to do so. This said, I see no reason why those Satanists or Orkonists who can keep to themselves, don't commit crimes, or force themselves on others cannot be respected in their rights and beliefs.

The above are, for the most, part stop gap remedies, for after all, life for most people is going to involve conflict with Hell on some level or not, whether or no they are actually conscious of the fact, and its manifestation is not always of the "magical" sort. And where overt witchcraft is used, keep in mind, one's real problem is not really the demons or sprites, but Satan, Hell's agents, sorcerers, warlocks, and the regular people who order or pay for these things to go on. There is then no stock, all-purpose method of dealing with Hell. Their relentlessness at causing suffering to others is no myth. It is best to keep moving, in an orderly fashion; to take up different tasks and interests one might in an ordinary day pursue. The more different subjects and endeavors one can occupy one's self with, the harder time Hell will have in keeping pace. Hell is quick to adapt, and will have counter measures for one's own counter-measures. Do therefore be prepared to have counter-measures for their counter measures. As devils will inevitably concoct new ways to attack free people, free people must be correspondingly creative and innovative in responding.

The minds of Hell people are not in reality, yet their actions are, at least these actions affect us so. Now it is true if we were truly refined in our faith and virtue their actions would be less real to us, we could choose to make them so, and we would therefore have more power to ignore them. But few are this perfect, and that we potentially have this power for greater virtue, should be not allowed to be made excuse for the very extreme behavior of these people.

Orkon and Goomer people, for all their often phenomenal wealth and fantastic power are just low down criminals, and a person of faith, who sincerely tries to do the right thing, has authority over any of them. This is not to say that a given Satan will necessarily recognize that authority, nevertheless the more a person is strong in faith, courage, love, charity, compassion, reason, wisdom and innocence the more they will legitimately be able to put that Satan and his people in their place. If you can do nothing else, endeavor to always find virtue in your situation as best you can: avoid peevishness, cowardice, self-pity, laziness. Regular work, staying busy, getting something done, will always recommend themselves as a ways of making best use of your time and avoiding of trouble.

If things get to be extreme, it might help to think you are in Hell, i.e., things aren't as they are normally. Make the most of this contrast, keeping in mind that if things were well and good before they can be well and good again if only you will acquit yourself with faith, love, courage, virtue. Though others are brainwashed now, the day will come when they will see the truth once Hell's storm has passed, or else they will be somewhere where they cannot bother you. If what you are going through seems to be your *darkest hour* then say so. And if it actually is your darkest hour, think of how others bore theirs.

Some seem to think that if they give in to misbehavior they will only miss out on the glories or true happiness of Heaven. But what they don't realize is that it is not as simple as that. The further one is away from Heaven and doing the right thing the more easily is it for Hell to attack and possess them. Humbling one's desires for happiness will not, of itself, serve as a substitute for faith and trying one's best to do the right thing, if one wants to avoid and fend off Hell's attacks.

We invariably feel frustration and disappointment when we seek good in the wrong place or places, and to see Life through Hell's alternative reality is what causes us most to feel fear and discouragement. In air combat it is a well established rule that having higher altitude than one's opponent can be an utmost decisive factor in the pilot's defeating him. Take then the high ground in terms of defining reality by being moral and rational. The unity of all knowledge cannot be in us but only in God. Yet we must be circumspect therefore in our ideas of God, and, in addition to reason and morals, the one way we can tell God from Satan is the cross. When we see innocence suffering cruelty that is a sign of God to us. That victim should be protected, rescued and saved. Christ has already suffered for them. What a disgrace and a shame it is to us to allow extreme injustice and cruelty to take place, and be indifferent and unmoved by it! To be thus unmoved and indifferent is like not recognizing God. We know therefore that perpetrators of or accomplices to such acts cannot be of God in their behavior.

Hell is falsehood, and the road out of Hell is justice, compassion, and truth. Let us with spirit pray then, "Lord, Deliver us from all that's phony!" Infinitely better it is to be tortured a million times over, then to have "health" and "wealth" by cooperating with the ruthless murderers of children, the innocent and, yes, the admittedly not-so-innocent. Infinitely better it is also to die free in trust of God's beauty and hope then to live life a slave to worldly power: to suffer in anguish for love and truth than to continue to abide in fear and illusion.

For myself, I was born happy, and will ultimately *choose* to be happy – as I originally am – regardless of what Hell throws at me, and it is mine, as well as yours and others right to do so.

I believe very deeply that a truly mature person should see themselves as holding off Hell, like a soldier, to protect the innocent, to protect Heaven from having to be bothered (whether or not it is correct to see Heaven needing such protection.) Be a good father, son, brother or, a good mother, sister, daughter,[99] and not just to your immediate family, but to all. Help others, stand up for what's

---

[99] 1 Timothy 5.

right, and in these ways get through this life with something to be truly proud of. While wisdom speaks of deferring happiness, some Christians have the idea that this includes deferring virtue and morals as well. "Be perfect as you Father in Heaven in perfect."

~ * ~

In Conclusion

"'Then I beg you, father, send Lazarus to my father's house, for I have five brothers. Let him warn them, so that they will not also come to this place of torment.'
Abraham replied, 'They have Moses and the Prophets; let them listen to them.'
'No, father Abraham,' he said, 'but if someone from the dead goes to them, they will repent.'
He said to him, 'If they do not listen to Moses and the Prophets, they will not be convinced even if someone rises from the dead.'"
~~~~~~~Luke 16: 27-31

"...and right with that ich wakede,
And called Kitte my wif and Calote my doughter:
'Arise and go reverence Godes resurreccioun,
And crep to the Cros on knes and kusse hit for a juwel,
And rightfullokest a relic, non richer on erthe.
For Godes blesside body hit bar for our bote, (redemption,)
And hit afereth the Fende, for such is the mighte,
May no grisliche gost glide ther hit shadeweth (where it casts its shadow)."
~~~~~~~William Langland, *Piers Plowman*

There is apparently a kind of mechanism in the human condition (in its fallen state and under the influence of Hell) by which an individual on a given day will be saddled with great misfortune – misfortune goes the round to all – and it is unloaded on them without any particular reason, other than perhaps that it's "their turn." This could and does seem to happen to just about any and everyone. Notwithstanding, the more powerful, wealthy and or important a person is, materially or otherwise, the harder they are likely to "get it" if they get it. Exceptions to this, might be people protected by Hell who seem to get away with murder, but who, nevertheless, are people who are ultimately the least protected of all because of the debt they accumulate securing Hell's protection.

Pure evil does not exist, where there is evil there is some amount of good, even if only in the sense that existence itself is good.

The main causes of evil I think it is fair to say are:

1) Individual sin or guilt, and individual lack of responsibility
2) Collective sin or guilt, and collective lack of responsibility
3) Hell's influence, deception

But for these three things we would be naturally happy and good.

It is common, particularly in this day and age, to point to nature and say evil is natural. Yet if one considers the matter more closely, I personally believe that what we see as defects in nature are the results of Man's fall, its repercussions for the whole, and mutations brought about by evil. Evil does not exist in nature, but in a person's will and desire. Cruelty in nature, for example, is ultimately preventable, at least in theory, by the "direct" presence of God and or rule of man on the earth through reason and compassion. Even if, for some reason, this view proved incorrect, it is easy to see that universal belief in it as a truth could only make living conditions better for every one, including animals. Moral and Rational correctness are like correctness in architecture or engineering. If things are done "right" they will "work." If things are done wrongly, they will not. Although in our materialistic times people are lead to think human beings don't really matter in the grand scheme of things, the truth is it is the destiny for any single one of us to be as important as gods if we would but serve God, love our neighbor, have faith and do our best to do what is right. [100]

God shines his sun generously on both the good and the wicked. Sometimes we are given to wonder why this is so, but in any case it is something we must accept. Even though on occasion it might make God seem unjust and aiding evil, it would be foolish to interpret it this way. We ought even go further and never grudge others happiness, but grudge only specific, and very specific, wrong doing if applicable or called for. In this same vein, where there is injustice and wrong-doing, we should look more so to halting or removing the offense rather than blaming or punishing the person.

We are a part of the universe, and to that extent the universe is in us. God loves the whole, and parts can't simply just be eliminated that don't fit. That would not be the way of the true artist, let alone the ultimate artist. Parts must be reconciled to all other parts: hence pain. Yet by following God's will we can over time become part of his plan for harmony He has in mind, and thereby avoid worser pain, and enjoy greater happiness. The devils are our cousins, and we and

---

[100] An interesting case can be made that death itself is unnatural, and that all instances of it are the direct or indirect result of deliberate murder, in turn originating with the willful act or reckless negligence of someone. For example, say someone is struck by lightening. It could be argued that because they or someone else was careless in allowing them to be exposed, such "allowing" was based on certain free will choices made, which in turn arose from other choices. While this interpretation is perhaps reading too much into an event such as "death by lightening," cases of a person dieing from an illness are more obviously a result of people's choices. Having said this, I do not insist upon this argument, but merely offer here it for reflection.

they are each part of the greater whole of humanity. Our avoiding them does not by itself necessarily or permanently solve the problem of their presence in our lives. We must act according to God's way, with faith, love, patience, courage, charity, morals and reason, be willing to embrace suffering rather than do wrong. In this way, then can we ultimately escape the scourge and hurt caused by devils. But until the process is complete, until the tares are somehow separated from the wheat, we suffer their guilt along with our own; we suffer their pain along with our own pain.

One of the objections I have heard is, if what you are saying is true, where is God, where is Heaven, in the midst of all this?

In asking such, do they mean to say if these spirit people had not shown up they would have no reason to believe, let alone, be grateful to God?

Some of the reasons one is not likely to encounter visible spirits from Heaven is that Hell routinely sends out spirits disguised as people from Heaven, and a person is usually hard put to know which is which. At the same time, Hell can (under certain circumstances) do things to people from Heaven or Purgatory who try to interfere with their activities. This is not to say that people of Heaven could not fight the back, but there are apparently certain rules that must be adhered to; such that it would perhaps seem like bullying if Heaven intervened in an open overt way. I personally suspect that Heaven probably could get rid of these people in a particular circumstance, but does not do so because it would be giving Hell the attention it desperately seeks, and, by doing so, only encourage its bad behavior all the more; for you see as much as they regularly hide from us, not a few (or else a very powerful few) of them are fanatically obsessed with getting attention.

As well, it can well be argued that too frequent familiarity with regular people would often breed contempt by regular people of people of Heaven; so much is the world under false-Pride and Satan's influence. We see this for instance in how some genuinely wonderful and beautiful people are treated here on earth. On not a few occasions, Hell (using its own people in disguise) depicts the people of Heaven as cold, snobbish, unfeeling, self-righteous. The reality is the people of Heaven are like thoughtful and caring parents and will do what they can to help us. But the fact remains, God, in his unfathomable wisdom, does allow these things to happen, and we must ultimately avail ourselves of His wisdom, grace, and truth to get through them. On the other hand, there are times when Heaven *can* and does help us, but only insofar as we are capable of keeping up the fight ourselves, much as was required of the colonists in the American Revolutionary War till the arrival of the French.

Hell can make the most good and most beautiful look bad, without exception – and that includes God himself, and to the degree a person allows him to do this, the more successful he is. Typically, Satan will blame God for evil, but the truth is God is incapable of evil. In Job, God allows Satan to do evil, but does not will him to do so. God can and does allow evil. But He can also override it. To do evil goes against his Being, see for instance *Wisdom,* 1:13-16 and 1ˢᵗ John 1:5.

Some Church Fathers have even argued that God does not even know what evil is; as it is useless to him. God is and can only do good. No good comes from evil. Evil only comes from sin, error, irrationality, disobedience, and Satan, not from God. It is wrong therefore to say Good comes from evil – ever. Any and all good that occurs ultimately comes from God. If God brings good out of evil circumstances, it is simply his will, and goodness, bringing the good out of what good of his is already there. Evil as such has absolutely nothing to do with producing good. Evil, by definition, is the absence of good. Good occurs solely and purely through the will of God. God does not react or answer to evil, it is simply his will to bring good where He wills it. Really it is his absence that can actually be considered evil. The idea that he is the cause of this absence is one simply difficult to maintain when we see the extent to which ourselves and others fail to do what is right.

God is not frowning authority: a giant, sadistic ogre. Rather he is unassuming innocence: a child born in a barn. Act then toward God as you would, as a loving and responsible person; as you would toward the innocent; with perhaps this difference: recognition of your failings, and limitations of your wisdom, and the inferiority of your status.

God is in the upholding of truth, justice, mercy; the innocence of children and animals; the examples of heroes and martyrs; the order and beauty in nature, the harmony in true music, poetry and painting; the wisdom of sages; the joys of lasting love; and life itself. The Kingdom of God is within you, and within others, and one's doing their part to bring it about in this world, while preparing to meet it in the next.

If "down below" can influence us, we have the power to influence down below. Indeed as believers in Love and Truth *more* power. Fighting Hell is fighting slavery and tyranny. Indeed, it is fighting the most formidable of all slaveries and tyrannies, and to lose sight of this is to lose sight of our freedom. It was with such a concern in mind that St. John of Damascene wrote: "The Son of God became man [that]... by calling us to the knowledge of God [He] might redeem us from the tyranny of the devil, and might strengthens us and teach us how, by patience and humility, to overthrow the tyrant." [101] To surrender to Hell and fear is to ultimately surrender all that's good, and think how others have fought and given their lives for far lesser causes. If there are to be military expenditures, why not such as will go towards fighting these (whether physically

---

[101] *The Source of Knowledge*, Part 3, Ch. 4. Eusebius (X.4) quotes a quite beautiful hymn of the early church in which Christ is referred to as a tyrannicide.

above or below) who truly, after all are mankind's greatest enemies? If we have not as yet the technology to track down such virulent enemies and apply such military force as we are capable, then why not begin to explore, research and develop the possibility of doing such?

We all then seek good, yet our idea of good is different, hence we have conflict. Yet if the good we all sought was the same there would be less conflict. It is reason, love, and morals which allow us to find harmony and unity in diversity. And if we really knew the truth, above and beyond reason, love, and morals is God. But this is truth for us only if we see God through reason, love and morals first. "By their fruits you shall know them," that is assuming what you know is based on a clear headed, honest and rational assessment of the facts. Satan, after all, is a clever imposter. Without reason, morals, and common sense, backed up with courage and persistence, we could obviously be deceived from knowing and recognizing who and what God is. It is the absence of these things, in the final analysis, which makes us the slaves and chattel of spirit people. Rationality, honesty, and a compassionate morality then are essential to our greatest unity, and to this extent are very much part of what give us our greatest strength and protection from the despotism of powerful and arrogant spirit persons.

Finally, I think it is very important to emphasize, one should never think that devotion to God is only needed to ward off Hell and its influence. Rather God is what we most desire, the greatest happiness of others and ourselves, the infinite in others and ourselves – that is, if we think and feel morally, rationally, aesthetically. In very real, down-to-earth, practical terms, Hell has nothing to do with God. Its presence in our lives is merely incidental: a state of things or way of thinking arising out of the immoral and irrational choices of others and ourselves. If there is Hell in the world, it is only because people are freely choose and or permit its presence.

To not know and appreciate this is to risk being among the greatest of fools – and much worse besides.

~~***~~

APPENDIX A: Spirit People Primer or "Spirit People for Dummies"

1. Generally speaking, spirit people are separated from regular (flesh and blood) people by a physical and moral barrier that science has as yet been unable to properly explore and explain. Whether this division is a natural one, so to speak, or the devising of some certain person or persons is also unknown as such.

2. As known in this world, those spirit people who have power over regular people, are, in truth, nothing more than rich and powerful criminals; who acquired their status through assassination, murder, torture, robbery, terror, slavery, sophisticated manipulation. Some see such people as representing God and Heaven. Others see such people as more or less a manifestation of Hell.

3. Spirit people, like regular (i.e., flesh and blood) people, generally need money to do things in this (our) world.

4. While such as persistence is a virtue, the presence of one virtue, however so excellently possessed or displayed does not absolve the person of requirement of possessing other virtues, such as common fairness and moral decency for example.

5. No spirit person is inherently better than any regular person, at least we would not know them to be such, except insofar as they displayed both higher moral character and perhaps, but not necessarily, intelligence as well. If you encounter a spirit person then, know that you should in no wise allow them to treat you any differently than how you would let other people treat you. For example, they have no right entering your home without your permission, let alone your body – just as with an ordinary person. Know also that even the most vicious regular person is a lamb compared to the most vicious of spirit people, and as a rule spirit people who are notably vicious are exceedingly more vicious than any "regular" people are or otherwise could be.

6. Even the worst and most sickening and hideous monster of a spirit person can have people working for them who are genuinely likable, attractive and or awe inspiring (at least to the unreflective or undiscerning.)

7. Even the worst and most sickening and hideous monster of a spirit person can appear as if he or she came from Heaven, that is depending on how well a person is able to tell true from false.

8. The types and character of spirit person one might find oneself dealing with is in large part a result of the moral, and intellectual caliber of the given community they live in. So the more depraved society is the more it is going to attract the worse sort of spirit people. The more well-behaved and responsible society, conversely, will have less of the worse sort around, and more likely the good kind of spirit folk.

9. As a general rule, contact with conversant or else visibly manifesting spirit people is something that should be discouraged and turned away from. An exception to this, however, might be when a meeting is brought about in an open, unthreatening, and above board manner, between such spirit people and educated and intelligent "regular" people who are intellectually and morally competent to deal with them.

10. Spirit people respect religion no more no less than regular people do – which is to say variously and differently.

Additional Remarks

On The Power of Lying

There is this sort of understanding that times change, and the public taste changes. This assumption is true as a practical matter of course. Nonetheless, what many people fail to appreciate is that such changes do not always come about through voluntary informed consensus, but perhaps instead through manipulation, deception, disguised assassination, all of which is then foisted on us as consensus. Whole cultural eras can be formed and created by such means, and anyone one of that ill founded era's "beneficiaries" might never know exactly why they so prospered (compared to say another), though they might have a vague suspicion if they were honest with themselves and really thought about the matter. Those who go along with the aforementioned deceit and underhanded dealing will be rewarded with wealth and status by the criminal perpetrators, while those who don't will be shut out and denied their basic rights. Now if you happen to be one of the latter and go to the police or other public institutions to complain about your mistreatment, what they (in effect) might typically say is that they can't help you because you are going against consensus.

How can one help but continue to be aghast and astounded by some people's obstinate irrationality and willful blindness, even among well-educated people? "Hear no evil, see no evil, speak no evil."

Yet others seem to say "if we all lie and we force our will on others then what we say is true." The truth, such people seem to maintain, is what one says it is, no matter how much one lies.

One reason for such attitudes is, I strongly believe, the influence of spirit people, so that I think it is worthwhile to list some of the devices and methods certain spirit people have at their disposal to promote such indifference to and resentment of real and intelligent thought generally. Some of these things are somewhat hard to describe because they have no clear and evident parallel in ordinary experience, but we'll try.

1. Spirit people through their machinations can manipulate people and circumstances so that they can create the impression that they control things, and

when they can get people to believe such impressions they then actually do end up controlling things.

2. Certain spirit people seem to have knowledge of the future and can predict events. It may even be true that they do have such knowledge and can actually predict things. Yet even if one can be said to predict things, it does not necessarily follow that they are wise or know what they are talking about. The same is true with respect to very private knowledge about your past, perhaps reminding you of things you thought were known "only" to yourself.

3. Great personality cults such as that of iconified dictators and fulsomely lionized leaders and celebrities are sometimes symptomatic of evil influence. People see the success of great movements and popularity of certain figures as signs of God, and therefore authority. Yet by means of doctoring appearances false causes and incompetent leaders can be catapulted in people's minds to divine or divine like status. The status then itself becomes a sign of divine endorsement of the person, so that the people are not only deceived about the given cause or leader, but about the nature of God himself. I don't mean by this that all popular causes and popular leaders are not legitimate, but that "popularity" by itself is no proof of legitimacy.

4. Certain spirit powerful people can make their presence felt for fairly large distances. The power of such presence can be sometimes mistaken as a sign of authority.

5. Certain spirit people have great wealth and other kinds of power at their disposal. This also some take as indubitable signs of authority.

6. Certain spirit people can induce a wide variety of experiences and feelings. They can make you see in your mind and feel things like the radiance of "heaven," or the pure light of "holy" being, when really these things, coming from them, are no more necessarily credible signs of authority than a person flashing enormous sums of cash. Note also how certain secret societies are denoted "illuminati," and how the Scientologists speak of a person being "clear." The light and clarity referred to in such conceptions is, I believe, something often brought about through psychological trickery of spirit people, again manifesting extraordinary kinds of lights and visions to a person. I know in point of fact from personal experience, how some of the most muddle headed and confused people will considered themselves enlightened because they have been shown or had such things "revealed" to them. The same can be said of religious pronouncements or declarations. If one were to suddenly hear in their thoughts "I am the Lord your God who brought you out of Egypt," accompanied perhaps by a vision of distant shining lights, do know there are dishonest spirit people who can cause this to happen, and feel no qualm nor suffer any evident repercussion for doing so.

7. "My mother listen to some of these same spirit people, and she was a good hearted and devout woman. Do you mean to say that she was wrong—and about something so important? Whatever answer you give I cannot accept it."

8. "You can't be responsible for or guilty of what you don't know." People will be encouraged to not know what is going on for exactly this reason. Even more, in some instances, a spirit person can get someone to commit a crime, and then, with that person's permission, "do something to them" to cause them to forget what they did, and or give them what seem like plausible reasons to think that they were somehow "forced" to commit the crime—and therefore are not really guilty of anything.

9. As well as seeming religious, con artist spirit people can easily be irreligious, or else indifferent to religion, tailoring their approach to whomever they are trying to persuade.

A Start

If, as I have now long contended, there are spirit people and these are *the* or *a* primary cause of real (or extreme evil) in the world, what steps or measures can be taken to combat them? The following are some preliminary ideas.

1. Admit to or at least allow the possibility of their existence. Know that something can be known by it effects without it being necessarily "seen" as such, as is the case for example with molecules and atoms. This understood, possible Categories of evidence for spirit people might include:

a. *Contemporary personal testimonies*

b. *Historical personal testimonies*
If certain phenomena is encountered which is or may be ascribed to "spirit person," say reports of such as with the spirit person who spoke to the Native American tribesmen prior to the massacre at Wounded Knee, or, to use another example, spirit people who are said to have spoken to Jean d'Arc, these can be cataloged and compared.

c. *Second hand accounts*, that is spirit people as referred to otherwise in history, world literature (including mythologies.)

d. *Otherwise inexplicable phenomena,* as in certain behaviors, such as various instances of incomprehensibly vicious crimes, such as massacres, and comparing these in their similarities and differences to each other.

On the basis of these and perhaps other means of proof, a more cogent and thorough understanding of spirit people (including who they are; what kinds of them there are; where they are located, what their needs and motives are; what their strengths and weaknesses are; etc.) can be arrived at.

2. *Raising public awareness* about the existence of spirit people, and the kinds of threats they can or might pose to an individual and or community, as well as steps one might take to protect oneself against them. As part of such a program seen from a long-term view, the study of *logic* should be encouraged at an early age, and indeed made a standard part of a child's curriculum along with reading, writing and arithmetic.

3. If the existence of spirit people can be objectively and empirically established, then efforts should be made to explore *the possibility of technology* and or other evidentiary methods or tools (including medicines) which would allow of their (to some extent) being detected and put under surveillance, and, as need be, combated.

4. True evil is something unnatural, and for this reason, that which is *Natural* should be seen as of greater importance and necessity than that which is "unnatural" – at least as a general rule or principle.

5. If we look to the long term, we might consider ways of *assisting poorer spirit people* who are themselves slaves or victims of more powerful spirit people and who use them to commit crimes. If these poorer kinds of spirit people have no alternative to the way they are made to live, then they will always be there to be used by other spirit people more powerful than themselves, and for nefarious purposes. Aiding such people perhaps assumes a bit much at this point, let alone what steps (outside of prayer) might be feasible. Nevertheless, it is goal that will work both to our own protection as well as fairness and compassion toward these typically wretched and miserable people who are forced into a life of crime (say by their being made to participate in "witchcraft" activities.)

6. Similarly, with respect to the long term, there should be contemplated and considered possible laws and legislation which might be enacted to make explicit what society will and won't permit of spirit people, and what rights they do and don't have as far as society is concerned. For example, a law might be passed in a given state or community that requires that conversant spirit people who are present or guests of one of that community's citizens be publicly identified and licensed, if they are to be granted permission to be present in that state or community. Even if it is assumed enforcing such laws would be somehow impossible with respect to sprit people themselves, nevertheless, such laws could, at minimum, be used as a deterrent or incentive to regular (flesh and blood) persons who, willingly or unwillingly, possibly have potentially dangerous spirit persons in their company.

~~~***~~~

APPENDIX B: Proof for the Existence of Spirit People

"But the angel of the LORD called to him from heaven and said, 'Abraham, Abraham!' And Abraham said, 'Here I am.'"
~~Genesis 22:11

"Fan Chi asked [Kongzi, i.e., Confucius] about wisdom. The Master said, 'Devoting yourself to transforming the values of the common people, to serving the ghosts and spirits with reverence and yet keeping them at a distance – this might be called wisdom.'"
~~Analects VI. 6.22

[Samuel Johnson:] "Yes, Madam: this is a question [i.e., regarding the existence of ghosts] which, after five thousand years, is yet undecided; a question, whether in theology or philosophy, one of the most important that can come before the human understanding."
~~Boswell, *Life of Johnson* (1791), vol. 2, p. 232.

INTRODUCTION

A Scientific Challenge

I believe if one stops and considers the matter, some of the most persistent, and, at the same time, pernicious *myths* there ever were are:

1. "There are no spirit people (i.e., 'gods,' devils, angels, ghosts, etc.) or else they are remote from or irrelevant to our life circumstance."
2. "Spirit people, especially such as a 'god' or an angel, necessarily represent a higher and superior kind of life form."
3. "Even if there were spirit people, and especially the powerful kind, science could never detect, track, or identify them."
4. "Tremendous or magnificent power is necessarily a sign of goodness or benevolence. Similarly, if a spirit person can 'titillate' or cause in us feelings of religious rapture, these are an indubitable indication of their kind intentions and benevolence."
5. "Making a point to be honest and rational, even if we are talking about public affairs, is simply one among a number of options or alternatives available to us when we make important decisions and decide matters of policy. Honesty and reason then, and in most instances, are not strictly necessary in ensuring public safety or promoting public welfare."

It is in part for purposes of laying the foundation for effectively demolishing most or all of the above attitudes and assumptions, the work before you was prompted. At the same time, it is, as its title and short length imply, an informal attempt to scientifically establish the existence of "spirit" people. It also serves as an introduction to the topic of spirit people, and which I have addressed at greater length in my book, *A New Treatise on Hell*. Because not everyone is prepared to go through an entire book, this brief paper serves as more convenient means for more busy people to become acquainted with the subject. By spirit people, I mean what are traditionally referred to as angels, ghosts, gods, and demons, to name some of the more familiar kinds. Simultaneously, I want to offer a tentative framework for how such spirit people can or might be understood and classified from a modern and scientific viewpoint; using such traditional nomenclatures as the above for that purpose.

For no endeavor is more important and replete with profound and wide-ranging implications as our concluding that spirit people literally do exist. Indeed, such implications are so tremendous that it could be said to make my task far harder than easier; so much and many pre-established norms and assumptions will be upset or revision required should what I assert prove true. Consider, for instance (and but very briefly), how each of the following studies or areas of endeavor (in no special order) would be affected by such a conclusion (given here briefly and in no special order or sequence.)

a. *Morality and Law*
If spirit people exist then, and based on what else is known historically and elsewhere than they *could* affect people's behavior, and to some extent their culpability. If a person is actually infested with a literal "demon," is it possible that under certain circumstances this could mitigate their guilt? Further, if spirit people exist, and were involved in crime (thinking particularly of organized crime) then this would give a major power to criminal activity of which not only is law enforcement not prepared to deal with, but which they (presumably by and large) don't even know about!

b. *Government and Society*
If spirit people exist, do they interact or influence those who act in government and or who occupy positions of law making and law enforcement? And *if* so, than it is not hard to see how the integrity of various key governmental and societal institutions, particularly in a democratic state, could be compromised or seriously threatened.

c. *Religion*
If spirit people exist, do they interact or influence those who exert significant influence in the church or other religious community? Can spirit people take on the mantle of divine or heavenly authority, be received as such, and yet who are nothing more than imposters and confidence tricksters? Prior to the Massacre at Wounded Knee in Dec. 1890, for example, a spirit person ostensibly appeared to some of the Sioux, and professed to be or speak for Christ. According to what I am saying this spirit person (assuming him to have been real and not imagined by the Indians) was, and so I would maintain, such an imposter. Similarly, spirit persons who spoke to Jeanne d'Arc (or Joan of Arc), I have contended elsewhere were actual spirit people (not imagined) and also con artists or imposters. In

raising this point and using these and like examples, let me make clear, that I am in no way somehow challenging honest religion. Rather, I am merely pointing out that there are certain false and deceiving spirit persons who will pretend to be or represent such, and what tragedies and or calamities have or can result.

e. *Medicine*, including psychology

Various physical, emotional, and psychological disorders and states (including dreams) can be brought about by spirit persons (see my *New Treatise* for an enumeration of some of these.)

Any one or all of these areas can or do affect other branches of endeavor, including science and education themselves, and therefore the existence of spirit people can or could pose a threat to these also.

I am a historian, philosopher, and author, and am not a trained laboratory scientist, nor do I hold a science degree. In a sense, and certainly for some, I speak as a layman, so I make no claim based on my credentials. Instead, I have to rely here more on what facts and arguments I can provide than any ostensible authority in my person.

In what follows, I want to bring together a preliminary checklist of things which will help prove the existence of spirit people. I am very much aware of the limitations and inadequacies of myself as a scientific investigator, let alone as a formal scientific authority. Yet perhaps what I can offer is one starting point for which others, better situated, qualified, and suited than myself, can take the inquiry further. My background in history, literature and philosophy are of great help because often times discussion of spirit people, or at least addressing them, has only been permitted there. I want in no way to make to seem my own claims or testimony carries any kind of special weight or authority. Rather in the true spirit of scientific investigation I want to present facts and arguments which can be independently confirmed and corroborated regardless of who proposes them. With respect to what I have to offer in the way of proof, some proof I candidly admit is more persuasive and compelling than other proof I submit.

After going through this work, some might say there is still not sufficient proof to establish the existence of spirit persons. Even granting this, there is enough I contend at least to warrant *further* and more in depth *investigation*. Because of the limitations of my circumstances; some of what I offer can be better established by others elsewhere. I simply do not have the personnel or resources at my disposal to more fully substantiate all my claims or conclusions. This inadequacy will hopefully be excused if I can at least provide a rudimentary notion or idea that is not only reasonable and plausible, but also probable and better than likely. Such as I am submitting here might not qualify then as hard proof for some, but will, I would at least like to think, serve and act (or help to serve and act) as an empirically justified mandate for additional inquiry, analysis, and testing.

Here it must be understood at the outset that there are some very difficult, indeed in some cases violently difficult, obstacles to a genuinely serious inquiry into the potential existence and nature of spirit people. As I will come to address, for some individuals and groups, spirit persons have been seen as higher authorities to (ordinary) human reason and science. This might for example be said to be true of some Swedenborgians and other Christian and non-Christian denominations. Granting my premise that there are literal spirit people, if such tell or command these religious not to be open, rational, or scientific, the latter will defer to such spirit persons and not do so. Therefore if you go to them and speak on the subject they might deny or be reticent about spirit people, when in point of fact they do know of such; only the way they understand the matter, they are not supposed to discuss them. Nor is the issue merely an academic one for such religious, and your questioning could be taken by (some of) them, and whether rightly or wrongly, as something hostile and impious.

A further necessary caveat is to be consistently rational. Some, naturally, don't need to be told this. Yet it will or will not be surprising to know that others have this idea that it is possible to have an extensive knowledge and understanding of things without being rational, or without answering to reason. If you are or in any way like this, please rid yourself of this attitude, or else cease reading this immediately; as the subject of spirit people cannot be intelligently addressed without an insistence on being rational (and honest.)

Certain spirit people spend not a little part of their time literally leeching onto people and attempt to manipulate their emotions and thinking. Often times this can be done in a very vicious and arrogant way, and which is tantamount to rape. Yet some "regular," that is flesh and blood people, will accept being treated in this fashion by such malicious spirit persons; either because they don't know how to resist or fight off such abuse, or because they acquiesce to it think that's the way things are or supposed to be. If they are conscious that it is a spirit person afflicting or attempting to manipulate them, they might also acquiesce because they see or have been encouraged to see that spirit person or persons as a wiser authority than themselves or other people. This can occur not only uneducated, but also highly educated people.

This problem can be all the more difficult when it is understood that some spirit people, thinking particularly of such that are conversant, are masters of manipulating and brainwashing people. Such spirit persons do not hesitate at brushing aside ordinary human judgment and understanding, including even that of experienced and intelligent professionals, as inherently inferior to that possessed by spirit people. This attitude I would point out – while it is or may be to some extent true or justified in certain respects and with respect to certain individuals and certain areas of expertise – is simply often a kind of self-serving propaganda or posturing certain spirit people will adopt so that others will more readily believe everything they say or follow their commands.

Others will have had palpable and overt contact with spirit people, yet will suffer from a voluntary or involuntary kind of amnesia which makes them avoid thinking or being in denial about spirit people, and which ostensibly arises from sheer fear and terror of these people.

Last but not least, some spirit people are very much involved in crime and organized crime, indeed are, in some instances, actually the origin and heads of it. Consequently, one needs to understand that attempting to tackle the subject of spirit people may cause you to run into conflict with organized crime. This is one major reason why there is and has been such a reluctance to address this topic seriously; all the more so when one is reminded that in many instances law enforcement itself will not even bother to combat, or else go very far to combat, organized crime.

1. What is a Spirit Person and what are we trying to prove?

In describing a spirit person, some church Fathers, such as St. Augustine, make reference to an "aerial body." The etymological origin of the word "spirit" relates the term to gas or gaseous. We are then speaking of a kind of person whose bodily substance is, for lack of a better comparison, gas-like, yet which, unlike ordinary gases, retains an essential form and cohesion, united in a (more or less) conscious or thinking personality who both acts and reacts. Herein, I will make reference to a "regular person," in contrast to a spirit person, and meaning a flesh and blood person like ourselves.

Traditional kinds of spirit people spoken of in traditional myth, literature, and culture are:

– angels
– ghosts
– demons
– sprites, or little folk
– "gods"
or purported "gods" (the word "devil" itself comes from "deva" or god)

In Chinese lore "dragons" could be considered a kind of spirit "person" manifestation. However, here we will relegate our discussion to the more usual types most people have heard about or familiar with. Some of these types may simply be two different form of what is essentially the same kind of person. For instance, a given spirit person perhaps could be reasonably classified as being both a ghost and a demon, or perhaps both an angel and a ghost. Consequently these kinds of designations I am using on mostly tentative and practical grounds, and need not be taken as hard and fast (or irrefutable) demarcated categories or classifications.

Spirit people otherwise may be considered of two kinds: 1) those who were once "regular" people, such as in the case of a "ghost" who was at one time a regular person, and 2) spirit people who were originally and always spirit people, as is commonly believed or understood to be the case with some "angels."

In addressing this topic one might be forced to have recourse to crude analogies and comparisons. Because we understand one thing or phenomena, yet another thing or phenomena can be understood by means of something that is like it. The two may be alike in a very close or literal way (e.g., a horse and a zebra), or else in an abstract way music which flows like a river or vice versa. In dealing with spirit people one sometimes encounters things without evident parallel; so that, perhaps necessarily, one might employ an analogy to attempt to describe it.

We observed that at the outside, the substance which spirit people are in large part made up of has no better parallel than that of gas. And yet if we knew more about the subject, gas may actually be a very poor or comparison for such a substance. Obviously then there may be an extent to which the comparison is correct. At the same time there may be other respects in which the term is misleading and inappropriate. It is well nonetheless when the information valuable to us leaves much to be desired to allow the use of analogy when addressing the subject of spirit people, though with the clear understanding of the possible inadequacy of such comparisons if insisted on literally.

2. On Scientific Proof

This paper cannot be a dissertation on the scientific method, what the scientific method consists of, and what makes for valid scientific proof. Yet a few things can and should be mentioned here. For we must in some measure ask how does one prove anything? The answer is by means of criteria, or certain measures or certain tests. Both measures and tests are devised or formulated by means of rules.

Science, it could be said, is finding the best explanation for something compared to other possible explanations. To that extent a scientific theory to be valid, must be able to withstand the challenge of alternative explanations which might potentially explain more and better, and or are comparatively less liable to error or uncertainty.

What is accepted as proof depends upon the person, and a given person may accept a scientific claim with different degrees of being convinced or not. As a practical, matter, often what is accepted scientifically is simply a matter of fashion, depending on the scientific community in question. In one community value or legitimacy of a particular scientific claim may have greater weight than that same claim when viewed by another community. One community will view certain facts and reasoning as having greater weight and persuasiveness than they have with another. Cultural and political factors, for example, will make certain scientific claims more acceptable or unacceptable than they would be if those factors were not present. If for instance, people think little of lying or concealing facts which make them uncomfortable this can drastically affect whether a given scientific claim is accepted in a given community as valid or not. Such subjective biases and idiosyncrasies, and their possibility, ought very much be borne in mind when we attempt to address the question of whether or not there are spirit people.

Commonly, the best or most accepted kinds of scientific evidence or proof are:

-- *Testable proof.* Someone makes a claim that such and such an event or phenomena will or can be repeated if certain circumstances are in place. So, for example, if we say go to such and such a location at four o'clock, and you will see a ghost, and we can repeat this procedure (and then see a ghost at the location), this would be testable proof; that, among other conclusion we might draw, tells us that there is such a thing as a ghost. Nonetheless, as compelling as "smoking gun" evidence is, it is useless without some amount of deductive and inductive inference.

-- *Deductive and Inductive inference.* Often times in science we know something not because we see it, or whatever it is, but we deduce it based on its effects. Atoms and molecules are, for instance, not "seen." Rather their existence is deduced based on other observable and measurable phenomena. Such deduction is further scrutinized to see whether one explanation is more consistent with observable facts, testimony, and sound reasoning, than an alternative explanation. We have an established effect or effects, and we then ask now what caused or accounts for them? Deductive and inductive inference is then applied to analyze and weigh possible alternative explanations. With out such applications, mere sensation and or description of phenomena is inadequate to establish a scientific claim.

3. Attempting to Prove the Existence of Spirit People

What can be collected in the way of evidence for the existence of spirit people will (I think it can be easily seen) understandably vary in its persuasiveness and weight as proof, and people will tend to view various "evidence" differently. I myself have had direct, immediate and prolonged contact with spirit people on a number of separate occasions, and need no convincing. However, others of course have not seen or had direct contact with spirit people (at least in a way such that they were aware of them), and so it is necessary for me to make my case to such people. If we cannot prove the existence of spirit people based on what follows, it is again my hope that we can at least establish their plausibility, and the feasibility and worth of a more full and in depth investigation, including approaches to the subject which are not covered here.

a. Historical and Cultural Evidence

It is possible that by looking into and exploring various traditions and beliefs in myth, literature, and religion from around the world, to find some common patterns, and evidence for such phenomena and which is independently arrived at. Take for example the notion of "ghost" itself. It would be hard to imagine a culture ancient or modern anywhere around the world that does not have some notion of a ghost and what a ghost is. How is this possible if a ghost is merely the product of psychology or the imagination? The very prevalence of the idea, suggests that there is something to it, and it would be worthwhile to collect what we can on the subject as it is found around the world and compare peoples' various different notions and conceptions of what a ghost is, and what kinds there are. How are these notions similar? How are they unlike? While I won't do much of this here, I have made at a rudimentary attempt at such in my *New Treatise on Hell* (to which I would refer you.) To illustrate, the idea of there being an underworld, or nether region(s), is found in ancient Egypt, India, China, Greece, Persia, and Greece. The question then becomes how are these notions identical or alike? How is it possible for such diverse cultures to have seriously embraced an idea that is supposedly so completely fictitious? Either then it is fictitious or else there is an under or outer world in external (as opposed to imaginative) reality. It cannot be both altogether fictitious and not. Yet who yet has seriously attempted to put this question to the test?

b. *Contemporary Evidence*

In looking to contemporary or modern culture we encounter claims for ghosts (using them for simplicity sake as representative of spirit people generally) that are oftentimes backed with very specific individuals and their testimonies, such as official police reports. There are people alive to day who we can go and speak to who claim to have seen ghosts. Now granted we know some of these claims to be hoaxes or misinterpreted phenomena. Yet is not at all clear that all are such, and it would thus again be worthwhile to collect different stories and analyze and compare them for common patterns; especially if it is possible to gather stories or accounts from around the world to attempt such. The more widespread our information base, the more easy it would be (assuming for the sake of argument there are spirit people to begin with) to establish the reality of a given phenomena or occurrence on the basis of independently arrived at evidence. If we say, for example, there are *not really* any such thing as ghosts, what alternative cogent and compelling explanation can be offered can be put forth to account for people's claiming that they have seen or encountered them? To my knowledge no such specific and compelling explanation has come forward yet. Instead, based on my own experience, what we find is that those who deny the existence of ghosts say that those who say that those who believe in their existence are people who are somehow delusional. Yet if delusional, exactly when and how do such delusions arise and occur, and among people from around the world no less?

c. *My Personal Testimony*

It will not be surprising that much of my own conviction of and about the reality of spirit people, as well as in part my motive for proposing a serious attempt at proof them, is my own personal experience. I have seen, met, and conversed with them on many occasions, beginning in the Spring of the year 2000. Prior to that, while I had accepted the possibility of spirit people, I had only an abstract and vague notion of them. In both my *New Treatise of Hell* and "Narrative" I speak at some length about spirit people and my personal contact, and I would refer you to those writings, available for download at:

http://www,gunjones.com

for my more in depth testimony. Here, however, I would like bring up a couple points of interest in the way of how my personal experience can be objectively confirmed.

As I make note of in my *New Treatise*, some spirit people, like any people, are sometimes fond of gossiping, and relating old stories as well. The following are a few historical anecdotes I have heard from some of them, though without venturing to give you any exact assurance as to their veracity.

* The original Simon the Magician in some way betrayed the apostle Philip, after Philip tried to help him, and this resulted in the latter's martyrdom.

* A Simon the Magician or ghost-sorcerer of some sort or another followed Burgoyne's army. Burgoyne knew of this (evidently had seen him) and did not want him around but could not get rid of him. During the second battle of Saratoga this ghost or someone working for was able to maneuver Gen. Simon Fraser to the position where he was felled (reportedly) by Timothy Murphy's rifle ball, thus in this way indirectly causing Fraser's death. The spirit person responsible for this did not want this known however so that I tell you this against his wishes (I didn't invite his story to begin with by the way so I don't feel especially bound to oblige him.)

* George Washington was visited and bothered by spirit people shortly before he died.

* Spirit people deceived the Union commanders at Chancellorsville, and some Union officers blame (not blamed) this interference for losing what would have been a great victory.

* Phil Sheridan tried locking a (full size) demon in a closet one time. What this achieved or what came of it I don't know, but I take it the story was related to me as something intended to be funny.

* The Unknown Soldier (buried in Arlington) did not want to be the unknown soldier, as he has a name and family like anyone else.

And there are many other similar stories I could relate.

The question I raise now then is this. Did I imagine or dream these stories or did I in fact hear them from spirit people as I claim? If the former, what could possibly account for the logic and historical consistency of narrative accompanying such a hallucination or dream? My knowledge of the battle of Chancellorsville, for example, is very meager. And yet somehow in my "dream" I have attributed a certain attitude and behavior on the part of some of the Union officers there. What psychological explanation could realistically and specifically account for such a delusion? It would be easy for someone to say I imagined all this. But if so, give me a sufficient and convincing explanation as to how such imagining is possible. I defy anyone to do so. And if I did not imagine or hallucinate these anecdotes, and unless I am lying, these stories cannot but be convincing proof that spirit people really do exist.

The same is true of other strange and peculiar phenomena I describe in both my *New Treatise* and "Narrative." If what I recount there is not true as I relate it, I challenge anyone to provide a cogent and plausible alterative explanation(s).

Other sorts of evidence I would submit as proof of spirit people are as follows:

* Attempt to discuss spirit people seriously with educated people and see what kinds of reactions you get. What I can now tell you will most likely encounter is a summary dismissal, if not overt ridicule and hostility of the idea, and devoid of impartial and thoughtful reasoning.

* Try joining my "Spirit People and Science" discussion groups at either Yahoo groups or Face Book. See if it is even possible to do this. While I won't insist here that if you attempt to sign up and are then prevented or denied doing so that this is incontrovertible proof of anything, I would nevertheless suggest that the cause is related to certain regular people who are determined to prevent serious discussion of the subject. Similarly, and if it is possible for you, try coming to visit me at my home at 1604 N.W. 70th St., Seattle (Ballard), Washington. Despite innumerable appeals via e-mail, the internet, and phone, for people to do so, and for a number of years now, only one person, a police officer, has come to my house to hear or become acquainted with my story and claim. One would think that making known a purported haunted house would have been enough to have attracted some people out of curiosity. But again, and for a number of years now, not a single person, aside from the aforementioned police officer, has come to look into my situation or claims.

4. Evilology, or the Science of Understanding Evil

Another way of approaching the issue for and against the existence of spirit people is the question of evil. Put another way, can studying Evil (as in, for example, willful and unjust maliciousness of an egregious and extreme character), be a way of establishing the existence of spirit people? From whence does excessive and unnatural hatred, cruelty, and belligerence spring? Can the *effect*, in this case "Evil," be traced to an ultimate *cause*, and could spirit people be that cause? To be blunt, and allowing for some relatively small technical qualification or slight modification, I certainly do believe this to be the case. Yet since what follows is written more in the way of theorization and philosophical speculation, I hesitate to offer it as proof of spirit people as such. What instead I intend to do is to suggest or argue that if we attempt to account for why "Evil" or extreme evil (as in deliberate maliciousness and cruelty, for example) exist, spirit people is the best explanation possible. In what follows, some will perhaps understandably take exception to my assuming spirit people, while attempting to prove their existence. My excuse for this is to present a possible hypothesis that will best account for evil and why it exists, and in order to do this I find it worthwhile to offer a theoretical scenario that can then *afterward* be considered, accepted, or rejected as an explanation that applies in reality, whether as an explanation of Evil and or as additional proof of the existence of spirit people.

In *Paradise Lost*, Milton uses the figures of Satan and Death (joined by Sin), as the ultimate manifestations of evil. Going by these figures of personifications we might say that the purpose of evil is to control and or destroy. As control or destruction, of themselves, are not always thought of as evil and unnatural, when we speak of evil, we mean control and destruction that is peculiarly unjust and extreme. Though we might fine tune the argument to ask "are not all attempts at control evil," or "is not all destruction evil," here, for practical purposes, I simply mean evil as in an atrocity, a massacre, torture, or other vicious and heinous crime. Granted we might speak at length as to what constitutes evil. Yet, for brevity's sake, we will assume that by evil we mean the most extreme kinds of crimes, using "Evil," that is with a capital "E," to denote this sense of the term.

When we hear or read stories about some horrible murder the typical interpretation is "well, this person (who did it) must be crazy." Perhaps in a given instance this view might be refined by pointing to some physiological or chemical malfunction. Some times an explanation is attempt by making reference to a dark side or that man has a dark side.

Yet darkness of itself does not seem to be necessarily either bad or good. Rather it becomes bad when it becomes a place for evil to conceal itself. Speaking of darkness then as a metaphor for evil, it can and has been likened to disorder and chaos. With respect to a chemical or physiological malfunction it is easy to see how these can be taken as some form of disorder or breaking up of order.

Time and chance are often pointed to as the cause of something breaking down, but are these real or merely accidental (or else apparent) causes? Yet what is breaking up this order, chance circumstance or deliberate intention? It is true we could speak of an occurrence of real evil (in its effects) as being the result of folly or negligence. But such would, I think, be the exception rather than the rule as the more usual and predictable incipient cause of actual Evil. If we grant that there are spirit people, and such with bad or criminal intentions toward us, does it not seem to be the case that this best explains why there should be violent and malevolent disorder? Where else further could disorder be more specifically traced to?

Now natural breaking down of order, to use a few obvious examples, we see take place in geographical-scale convulsions such as a volcano, typhoon, or earthquake. With water there can be said to be an inherent and regular breaking up in as much as their molecules move and shift so easily, likewise light and fire. With things of geological nature movement is also always present, yet such usually occurs more slowly and gradually.

When life is broken up it is either to replace something wearing out (and what is wearing it out?), or by another life breaking it up, say to eat it. Now truly Evil-causing spirit people could be seen as an all consuming and or enslaving force. Yet life for such is often so very different than it is for the rest of us, even other spirit people. They seem to draw their view and strength of life from an order that is quite distinct from natural life, yet with connections to our own which permits them to invade our midst.

If this order of theirs is not natural, on what might it be based? Conjecturally speaking, it seems to spring from an evidently ancient way of thinking which sees destruction as the superior state of things. Why take such a position? Is it not contradictory?

Perhaps their response would be to say that if one were the greatest destroyer one would be the greatest person in the universe, or at worst second. On these grounds then they then understandably seek, by means of fear, bribery, and or deception, to get others to aid them. Of course we commonly find collaborators with the exponent of such a view to deny that there really is a dark side, except in story and pop culture. This I personally take as being done to avoid their or its being scientifically examined in order that they or it may be that much more free from attack.

Evil (as in willful maliciousness) requires Good, but not vice versa. Therefore it is false to say Evil is a necessary or complementary opposite to Good. Now some might think this is merely an academic point. Yet there are some people in this world who do think Evil is necessary adjunct of Good, and that the latter somehow could not exist without the former. Moreover, others not bothering to reflect on this, are mentally manipulated into adopting such a belief, with some sometimes horrendous or even fatal consequences. If science is going to effectively study and examine evil, one of the very important points it is going to have to address is how more ordinary and ostensibly respectable people are got to agree with or go along it. One reason for there doing so is unthinkingly accepting the (contradictory) idea that "evil" is somehow a necessary or positive good. Such a belief, I maintain, has its origin with certain spirit people.

One possible symptom of such a sickness, paradoxically, is that the person denies that Evil even exists. Further more, says this way of thinking, if you do not pay tribute to Evil, you will not be able to live your life. For this reason, many people pay tribute to Evil, and see it both as a high authority and the ordinary course of things, perhaps even give up their own child in sacrifice to some monster. This then is why real Evil occurs.

Chrysippus, the Greek philosopher of the 3rd century B.C., tells us that nothing is bad per se; only how it is used makes it so. Yet while this is or may be true of just about anyone and anything, logically, this cannot be true of Evil itself. We should add here then that it makes no sense to speak of Evil being a good or a potential good just like anything else in the sense Chrysippus means. We have that must less reason therefore to accept the aforementioned way of "thinking" – at least if we are rational, of course.

Going on what I have written about spirit people elsewhere, thinking especially of my *New Treatise on Hell*, if we assume there are spirit people, does this not best account for the occurrence of Evil? And if this is incorrect, what alternative theory is available that is superior to it that is more rationally coherent and more consistent with the facts? That Evil, and again I mean very powerful and extreme evil, exists and has and does occur is undeniable. What is hitherto been lacking in scientific understanding is an explanation from whence it arises. The assumption of spirit people best allows for such. In establishing that spirit people exist, we can best explain Evil, and on that best attempt a cure and remedy for it.

If we were to ask the question, "how does one acquire good" we might come up with the following:

a. By receiving it as a gift (e.g., God and Nature's gift)
b. By working for it (this would possibly include asking others for it, or also freely giving good, i.e., generously and without condition, as a means of obtaining good)
c. By stealing or otherwise acquiring it illicitly

If one were to increase the amount of good they could obtain, b. and c. would seem the only alternatives. Certain religious can get an infinite amount of good out of what God has already given. But leaving this question and this sort of person aside, most will either work or steal what's good in order to increase what they have of it. This is true of societies as well as individuals, and some societies might in fact place a greater importance on stealing good than working for it. Some who work for good will consider stealing to include taking unfair advantage of animals and the environment. Others will not, and will not think taking advantage of animals or the environment stealing.

Now one thing to observe about the stealing sort of person or stealing community is that there are some real monsters in this world when it comes to stealing or obtaining good (or goods) illicitly. And a certain custom or mentality is adopted in some realms where the one who steals the most, gets the most good. One of many drawbacks one might mention about such is that this could make the worst, most hateful criminal, materially, the most wealthy person. As a result, the most stinking and odious person could be the one who ends up possessing the most good in such an order, and possibly as a result, the power of governing itself. Hence some will think that work is the best way to obtain the most good, otherwise, even if we somehow obtain more good doing things the other way, our evils will increase dramatically as well, thus making the possession of more good nugatory, or worse, self-defeating.

In consequence of this, Evil can be made to seem as a justified, or for some even necessary, way of acquiring good. While there is nothing in this viewpoint that *requires* (as such) that spirit people be invoked to account for it, nonetheless I raise this point to suggest to you that including spirit people in the explication of it makes that explication most consistent with the facts.

~~~***~~~

ADDENDA

The physicality for spirit people can be described or characterized this way. Imagine a jar filled with paint. Now what most spirit people, if not necessarily all (since some are very weak and or tiny), can or could do is put their finger in that jar and then make a mark on a wall with the paint. In the case of certain others, much more than this; including writing or depicting a figure is possible – depending on how much energy they have available to them or at their disposal, either of themselves and or from someone else. In this sense, spirit persons such as we think of when we speak of ghosts, angels, or devils consist of a palpable and physically measurable composition and being; which, as best we can tell, largely partakes of a gaseous like substance, yet a gaseous like substance than is honed in such a way to give it a potentially remarkable force and effect on other physical bodies – again depending on the amount of energy with which it is infused.

APPENDIX C

Excerpts from *The Life of St. Anthony* by Athanasius of Alexandria,

In quoting these extracts from Athanasius' Life of St. Anthony I would like to point out that I was unacquainted with the work until after I had written "How to Fight Hell" and its Supplement. What is particularly interesting then is to have found similar observations and conclusions reached independently on these same topics.

19. [St. Anthony:] 'Wherefore, children, let us hold fast our discipline, and let us not be careless. For in it the Lord is our fellow-worker, as it is written, "to all that choose the good, God worketh with them for good." But to avoid being heedless, it is good to consider the word of the Apostle, "I die daily." For if we too live as though dying daily, we shall not sin. And the meaning of that saying is, that as we rise day by day we should think that we shall not abide till evening; and again, when about to lie down to sleep, we should think that we shall not rise up. For our life is naturally uncertain, and Providence allots it to us daily. But thus ordering our daily life, we shall neither fall into sin, nor have a lust for anything, nor cherish wrath against any, nor shall we heap up treasure upon earth. But, as though under the daily expectation of death, we shall be without wealth, and shall forgive all things to all men, nor shall we retain at all the desire of women or of any other foul pleasure. But we shall turn from it as past and gone, ever striving and looking forward to the day of Judgment. For the greater dread and danger of torment ever destroys the ease of pleasure, and sets up the soul if it is like to fall.

22. 'First, therefore, we must know this: that the demons have not been created like what we mean when we call them by that name for God made nothing evil, but even they have been made good. Having fallen, however, from the Heavenly wisdom, since then they have been groveling on earth. On the one hand they deceived the Greeks with their displays, while out of envy of us Christians they move all things in their desire to hinder us from entry into the Heavens; in order that we should not ascend up thither from whence they fell. Thus there is need of much prayer and of discipline, that when a man has received through the Spirit the gift of discerning spirits, he may have power to recognize their characteristics: which of them are less and which more evil; of what nature is the special pursuit of each, and how each of them is overthrown and cast out. For their villainies and the changes in their plots are many. The blessed Apostle and his followers knew such things when they said, "for we are not ignorant of his devices;" and we, from the temptations we have suffered at their hands, ought to correct one another under them. Wherefore I, having had proof of them, speak as to children.

23. 'The demons, therefore, if they see all Christians, and monks especially, laboring cheerfully and advancing, first make an attack by temptation and place hindrances to hamper our way, to wit, evil thoughts. But we need not fear their suggestions, for by prayer, fasting, and faith in the Lord their attack immediately fails. But even when it does they cease not, but knavishly by subtlety come on again. For when they cannot deceive the heart openly with foul pleasures they approach in different guise, and thenceforth shaping displays they attempt to

strike fear, changing their shapes, taking the forms of women, wild beasts, creeping things, gigantic bodies, and troops of soldiers. But not even then need ye fear their deceitful displays. For they are nothing and quickly disappear, especially if a man fortify himself beforehand with faith and the sign of the cross. Yet are they bold and very shameless, for if thus they are worsted they make an onslaught in another manner, and pretend to prophesy and foretell the future, and to shew themselves of a height reaching to the roof and of great breadth; that they may stealthily catch by such displays those who could not be deceived by their arguments. If here also they find the soul strengthened by faith and a hopeful mind, then they bring their leader to their aid.

26. 'Wherefore the prophet sent by the Lord declared them to be wretched, saying: "Woe is he who giveth his neighbors to drink muddy destruction." For such practices and devices are subversive of the way which leads to virtue. And the Lord Himself, even if the demons spoke the truth, – for they said truly "Thou art the Son of God" – still bridled their mouths and suffered them not to speak lest haply they should sow their evil along with the truth, and that He might accustom us never to give heed to them even though they appear to speak what is true. For it is unseemly that we, having the holy Scriptures and freedom from the Savior, should be taught by the devil who hath not kept his own order but hath gone from one mind to another. Wherefore even when he uses the language of Scripture He forbids him, saying: "But to the sinner said God, Wherefore dost thou declare My ordinances and takest My covenant in thy mouth?" For the demons do all things – they prate, they confuse, they dissemble, they confound – to deceive the simple. They din, laugh madly, and whistle; but if no heed is paid to them forthwith they weep and lament as though vanquished.

27. 'The Lord therefore, as God, stayed the mouths of the demons: and it is fitting that we, taught by the saints, should do like them and imitate their courage. For they when they saw these things used to say: "When the sinner rose against me, I was dumb and humble, and kept silence from good words." And again: "But I was as a deaf man and heard not, and as a dumb man who openeth not his mouth, and I became as a man who heareth not." So let us neither hear them as being strangers to us, nor give heed to them even through they arouse us to prayer and speak concerning fasting. But let us rather apply ourselves to our resolve of discipline, and let us not be deceived by them who do all things in deceit, even though they threaten death. For they are weak and can do nought but threaten.

35. 'When, therefore, they come by night to you and wish to tell the future, or say, "we are the angels," give no heed, for they lie. Yea even if they praise your discipline and call you blessed, hear them not, and have no dealings with them; but rather sign yourselves and your houses, and pray, and you shall see them vanish. For they are cowards, and greatly fear the sign of the Lord's Cross, since of a truth in it the Savior stripped them, and made an example of them. But if they shamelessly stand their ground, capering and changing their forms of appearance, fear them not, nor shrink, nor heed them as though they were good spirits.

37. 'And let this also be a token for you: whenever the soul remains fearful there is a presence of the enemies. For the demons do not take away the fear of their presence as the great archangel Gabriel did for Mary and Zacharias, and as he did who appeared to the women at the tomb; but rather whenever they see men afraid they increase their delusions that men may be terrified the more; and at last attacking they mock them, saying, "fall down and worship." Thus they deceived the Greeks, and thus by them they were considered gods, falsely so called. But the Lord did not suffer us to be deceived by the devil, for He rebuked him whenever he framed such delusions against Him, saying: "Get behind me, Satan: for it is written, Thou shalt worship the Lord thy God, and Him only shalt thou serve." More and more, therefore, let the deceiver be despised by us; for what the Lord hath said, this for our sakes He hath done: that the demons hearing like words from us may be put to flight through the Lord who rebuked them in those words.

43. 'And for your fearlessness against them hold this sure sign – whenever there is any apparition, be not prostrate with fear, but whatsoever it be, first boldly ask, Who art thou? And from whence comest thou? And if it should be a vision of holy ones they will assure you, and change your fear into joy. But if the vision should be from the devil, immediately it becomes feeble, beholding your firm purpose of mind. For merely to ask, Who art thou? and whence comest thou? is a proof of coolness. By thus asking, the son of Nun learned who his helper was; nor did the enemy escape the questioning of Daniel.'

~~~~~~

Copyright 2003. William Thomas Sherman
Copies of this work may be freely printed and distributed.

William Thomas Sherman
1604 NW 70th St.
Seattle, WA 98117
206-784-1132
wts@gunjones.com

~~~~~~

Though the greater intensity of these things has subsided, they do still go on. I am in fact a de facto hostage, and am being denied my right to an attorney with regard to more mundane crimes perpetrated against me. The Seattle Police, the local Catholic Church, the King County Bar association, have all refused or been unable to afford me help, and indeed have been acting in cooperation with regular people connected with Hell, indeed in particular circumstances, in the most flagrant and criminal way. And I am not talking about these witchcraft things, but regular crimes. If you don't believe me, try writing or calling me and see if you can even contact me. Most of my e-mail, regular mail, and phone have been shut down now for month;, even though I had my home broken into times numerous, pets sadistically murdered, computer vandalized, food poisoned. I have been repeatedly been denied an attorney, nor would the King County Bar

Association even recommend me one. The idea is to discredit and marginalize me, lest some of the community respected citizens be implicated in some very serious crimes, and because unlike many people I *can* explain these things. There have been so many crimes been going on these past years by these maniacs, including attempted murder, that it would take a book to write about them; which I may get around to doing one of these days, difficult as they make it for me. If you write me via the regular mail on this topic and your letter is serious I promise to send you a response. Indeed, if you write me certified mail and don't get a reply you should contact the postal authorities as it will mean your letter was never delivered and the U.S. mail interfered with.

I ask people also to write the Catholic Archdiocese of Seattle and request that they look into my claims and situation. I have been trying for some years to contact Father Michael Sweeney of Blessed Sacrament Parish, located in the University District of Seattle, but without success. Father Sweeney I know to be an intelligent and conscientious individual, and I would like, with his permission, to speak with him on these matters. Yet others have kept me from being able to reach him, though a number of years back I was able to speak with him without any difficulty.[102]

I am the author of *Mabel Normand: A Source Book to Her Life and Films* (see website at http://www.mn-hp.com), have appeared on Cable TV, "E!" Entertainment Networks "Mysteries and Scandals" program (see installment on Mabel Normand). I am a long-time student of philosophy, literature and history, and am also subsequently the author of *Christ and Truth*, book on epistemology, titled *Peithology*, and, as well, a separate work *Calendar and Record of the Revolutionary War in the South: 1780-1781*. These can all and more of my writings can be obtained at http://www.gunjones.com or http://www.scribd.com/wsherman_1

My personal, political situation unfortunately continues. Since making that paper available, I have received no regular mail or phone calls relating to it. I did receive less than half a dozen e-mails, yet only one of these e-mails was from a person willing to discuss these things in a disinterested and objective way. Some have acted as though my situation were no big deal. In truth what I have been though the past number of years is one of the most brutal and sadistic experiences a person could possibly go through. So much so, that even if you gave me the benefit of the doubt, you could not begin in your most far fetched imagination to comprehend the pains and horrors I have been forced to "sup" at the hands of these monster men and their accomplices.

---

[102] At the time this was written, Father Sweeny *was* the pastor at Blessed Sacrament, but is presently (Jan. 2014) the president of the Dominican School of Philosophy and Theology, Berkeley, CA. Nevertheless, my last attempts to contact and receive a formal reply from him he has effectively refused (on one occasion he actually hung up on me without advanced warning.) Consequently, I respectfully request, and in the interest of honest truth and fairness, he write or call me (or visit) to discuss my situation and claims; and as I have previously spoken of and raised with him.

Any help or support you might be able to lend, in whatever shape or form, would be most appreciated.

~~~~~~~*********~~~~~~~

APPENDIX 2: NARRATIVE

Originally written sometime in and by 2003. *Last updated*: 22 Jan. 2014.

I truly hate to have to write (and have spent the time) writing this because it seems self-indulgent to dwell on or take extra time going into crimes suffered and misfortunes of the past rather than leave them behind and get on with life. But if only (in my case) it were so simple! These problems and the people who brought them about continue to plague me and unless addressed and faced up to (i.e., with the help of others) I *cannot* get on with my life. At the same time, yet others need to know about these things that I will relate; so that such extreme crimes and outrages can help be prevented in future.

What is presented here is a narrative and account of what must certainly be one of the strangest and most horrifying ongoing series of events ever testified to. At the outset here, I must point to (three) obstacles I face in both telling and having this story received by others.

By commonly held beliefs the premise of my story involves spirit people, and (ultimately) if you don't accept the reality of spirit people; none of the rest of what I say can (so I aver) be properly or adequately be accounted for.

There are some who accept that there are spirit people, but who assume that such spirits have a higher jurisdiction over manmade courts and legal systems, and therefore they need not given any heed to my claims against such persons.

The people responsible for the crimes I will enumerate are some of the most vicious people you could possibly *not* imagine, and who at the same time have holdings of such wealth, power, and influence to easily intimidate those who attempt to do good or rectify evil.

The response to these above unusual claims or arguments I will leave aside for now, but will try to address and give answer to in the course of what follows.

I frankly don't know or expect that any one will necessarily listen to me. But if nothing else, I at least need some sort of record of what happened for legal purposes; i.e., such as creditors and financially related persons and matters. When they ask me why will I do not pay back this credit card or student loan you owe,[103] I will say I have been outrageously cheated, been (somewhat) physically disabled; robbed and denied both of my human and civil rights, including the right to receiving my communications (including phone, e-mail, and regular mail) and consequently am denied a fair and due opportunity and means to make a living. Now if someone says I wrote this narrative merely to avoid my student loan and credit card bill, I think those who will have taking the time to thoughtfully review this matter will see that such an allegation is completely absurd. The disgust and very real and deep dislike I have in having to write this in the first place should (I think) more than refute the charge.[104]

What is offered here is especially addressed to lawyers, doctors, clergy, educators, activists, government officials, and scientists. I would have like to make a more formal legal document of this, but then how to explain these spirit people in that wise; while at the same attempting describe the reality of things which are and were intended (by those who commit and perpetrate them) to be bizarre and unbelievable. In some instances, I provide information that would not so much help a lawyer, but yet provide a lead for a criminal and or a more close scientific investigator. Similarly, I make reference to a fact or incident which on the surface might not seem significant, by itself, but which will, I believe, be seen as having greater meaning in the greater context of my story. This account is valuable I think because it is a very rare, if not wholly unique, instance where someone is able to record and report ongoing crime that spans a decade, and which involves spirit people. Many, many others have undoubtedly suffered similarly, which is to say in secret and unheard due to the insurmountable or seemingly insurmountable difficulty faced in attempting to provide a narrative of this sort; such as remembering what happened, describing spirit people, getting others to listen to and believe them.

I have decided then to tell this story simply as I know it; which is to say in manner that can be more readily understood by most or all. Simultaneously, I am being candid about things many others refuse to discuss; having particularly in

---

[103] My being state of not paying two long term financial debts may understandably be off hand inferred by some as in some way reflecting badly on my character. While I think the inference is very understandable, I cannot, given the larger circumstances, fully agree. I forbear addressing the matter of these debts in this writing for the reason that doing so would only add unnecessary complication and material to an already complicated story.. This said, my basic position regarding my debts is this: because I was prevented from being able to work for many years now due to the regular and ongoing criminal machinations of others, and which have included debilitating physical injury to me, I have been unable to work such that I could not work to pay them off without completely unreasonable and extreme hardship being required of me – thanks to the malfeasance of the perpetrators will I will get on to discuss. In addition, I am still more than glad to pay back these, and all the mores so as some of those who owe me criminal damages be made to reimburse me for what *they* owe.

[104] As of 6 November 2008, my student loan debt that has up till this time been in the hands of at least a dozen different creditors at present stands at $4,937.63 and is held (as last I heard and after at least a dozen previous and separate handlers and debt holders) by NCO Financial systems, Inc. of Trenton, N.J. & Horsham, PA.

mind those who would claim what I discuss is somehow outside the power and scope of criminal or civil law. To restrict this exclusively to a more ordinary legal presentation would I think tend to trivialize the seriousness of what this is all about, of which the legal is only one – albeit an important – aspect.

My book, *A New Treatise on Hell*, gives an account and explanation of spirit people from a more general and scientifically oriented perspective, and I highly recommend, if I cannot require, you read that first in order to better understand this.

If some of what I get into talking about sounds crazy, I respectfully ask, as I did in *New Treatise*, that you suspend final judgment till you have fully and properly heard me. It is far too easy to take out of context what I am saying here and have it distorted or misunderstood otherwise.

This narrative has to a large extent been taken from letters I wrote years ago reporting what was going on then. In a number of instances (made more easy these days by the welcome wonder of word processing), I have slightly modified a given telling of what happened; for what seem to me purposes of better clarity, and a more properly described and or explained event. I admit I am foggy on many details, but unfortunately I simply have been through so and too much. Do know then that in between two given events I describe, there were or possibly was some other excited event or assault that also may have taken place, but which I have left out in order to keep this narrative manageable. For instance, in between the time of my returning to Seattle from Los Angeles (as I will relate) and my going to a north Seattle clinic I suffered repeatedly, and not without considerable pain, from biting my tongue (as if by accident.) This strange occurrence lasted about a week. Another sort of example might be the strange and irrational treatment and attitudes I encountered from specific individuals I might mention and describe; when I tried to discuss my situation with them.

So many things happened it would be not be possible to include them all; really thousands of assaults, tortures, and acts of maliciousness of various kinds. What is given here I would say, quantitatively (and roughly) reflects about roughly one-fifth of what happened.

Unavoidably, a number of people, and public institutions as well, will simply look bad in the course of what I relate; indeed, in a given instance, mayhap criminally so. Despite this, I am not writing this to get back or retaliate on anyone, and would myself be the first to extend leniency and pardon to any and everyone I mention here – except *possibly* those culpable of the worser and more violent of what I relate and who possess great wealth, and or who give orders to others to commit crimes.

Where it would be best to start my story is not so easy to say, As I will consider much later, there is good reason to think these things go back much earlier than what I present. But to make this all easier to understand and digest, I want to start at about early 1991 when a four part series I wrote on silent film actress Mabel Normand was published by *Classic Images* magazine (a nationally circulating film history and collectors monthly.) It was about this time that affliction in my life seem to begin taking on a darker cast than could be ascribed to ordinary setbacks and misfortune. At the time, I was living in the basement of my father's home at 6322 Woodlawn Ave. North, which is a block away (to the east) from Greenlake in Seattle.

Now though I was an author at that time, writing (and given the circumstances) was not something I could reasonably expect to make money from. Consequently, I was working at the University of Washington, mostly as a helper in mail and supplies in the Department of Education there (located in Miller Hall.) Among other things, I was also occasionally involved in protests against animal experimentation, sometimes, for example, writing "letters to the editor" to the U.W. Daily (newspaper) when the topic came up in news stories.

Sometime in (I believe 1991) I had two very unusual things happen top me. One occasion I came down with a horrible skin disease that covered my torso, which resembled small pox or something similar. I was quite shocked and surprised, if not horrified, by this completely unexpected event, and to this day, how and why I got the disease I don't know. I did research and found what I had was ptyriasis rosea, which self-diagnosis was afterward confirmed by a physician I went to see. Fortunately, the illness although it looked awful and felt uncomfortable, was not serious otherwise, and I was able to recover and my skin cleared up after a few weeks.

The second occurrence involved my Toyota Corona that was parked in front of my house, was rammed into by a pickup truck; which hit my own car from the rear.[105] The person who did this fled on foot before he could be spotted, and it was later discovered (or else believed by the Seattle police to be the case) that the person was on drugs and had stolen the pickup. My damaged car was subsequently given away and my father purchased a black Mazda pickup truck to replace it.

In June 1992, I left Seattle for Los Angeles to do research for my book on Mabel Normand. Also, and to be brief, I felt the culture and quality of living had gone seriously downhill in Seattle and I had the idea of making the move to California a more or less permanent one.

I had made some contacts in Los Angeles in the course of my Mabel Normand and silent film research, and with their help ended up moving to 1377 Lucille Ave, and which is a house on a hill overlooking a view into the Hollywood and Sunset area. My dwelling there itself was located in the basement; with my front door situated at the back of the house (and facing west.)

---

[105] I still have a photograph of the vehicle and the damage done.

For a while things went smoothly (I rather liked Los Angeles), spending my time doing research for my book. Yet as time passed a few odd events began to occur. In one peculiar instance, at 12:30 in the morning, an LA Police helicopter hovered (that is essentially remained in place) over my backyard and shined a flood light over the yard and in my house (why? I haven't the foggiest). Awoken I came outside, and gestured to them with my arms as if to ask what was going on? They continued hovering and shining the search light on me. This continued for a prolonged series of minutes; till exasperated I went inside to get my camera (that happened on the occasion to be unloaded of film.) Only when I came back out and I pointed the camera at them did they speed away (towards Sunset Blvd. – a few blocks down from where I was living.)

One time while visiting the downtown public library to do research for my book, I went downstairs to go outside and take a smoke break. As I passed through the main corridor to go out, about 3 or 4 L.A. police officers were standing in the small passage way there, and when I passed by them, one of them sort of gestured toward me. The others nonchalantly glanced where he directed, as if to recognize or size me up.

Another even stranger incident occurred when I woke up one morning to find literally thousands of ants pouring in a stream through (and under) the front door of my home in an essentially compact body. It was just like something out of a horror movie. Apparently the ants came up out of the ground in my back yard (remember my "front" door was located at the back of the house I was staying in.) In order to get rid of them I used a wet towel to collect them, basically having to wring the towel outside, or flush them down the bathtub drain after collecting them.

About or shortly after these events, which is to say in the summer months of 1992, there were signs on at least two or three occasions that it looked as though my apartment at 1377 Lucille Ave. had been broken into.

Some of my things looked as if they had been gone through, and my three cats – which is very unusual for them, – were all hiding as if they had been frightened. I called to have a police officer, Lt. Vega, come and take a report of breaking and entry. To be brief, he said that because there was no clear evidence of a break in and he could not file a report. Naturally, I was not all that pleased with this response, but what could I do? I did ask him however if he would at least write me a little note to say that he had been there to investigate. To this he agreed.

Despite all these incidents, my stay in Los Angeles was mostly pleasant, and I had no special reason to think I was being targeted by someone. For a few weeks I signed with a temporary employment agency (Kelly and or Manus services I recall), and consequently did some part time office work in the Downtown area.

Then on a Sunday in early October 1992 (I believe this was October 4), I attended a musical concert at the Roxy on Sunset Blvd. Prior to this concert I ate at the Spaghetti Factory located (as I recall) on Santa Monica Blvd. My meal there consisted of spaghetti and a glass of wine. Afterwards at the concert itself, I had a diet coke. At first I enjoyed the concert very much, the group performing was an English pop-group Shakspear's Sister. However, during the course of the show I suddenly began to feel very ill, and for no apparent reason started to sweat uncontrollably. While it is true the club atmosphere was rather stuffy, this no way seemed to explain why I was, in a gradual flash as it were, rendered inexplicably queasy and absolutely drenched with perspiration. Prior to the concert I was in the best of health and spirits and the onset of my discomfort came as a great surprise, as I was up to that moment in relatively (and for me) excellent physical shape. There is no doubt in my mind that the reason for my being ill on this occasion was because I had been poisoned. Though granted it may, taken by itself, have been unintentional poisoning; later events, as I will relate, would seem to suggest it had been otherwise.

In the following week I did not feel very well. I attributed this to simple fatigue. Then on one day in the week I ate at the Milano restaurant; on Alessandro Blvd I believe, and thought a proper meal would do me good. Instead I became dramatically worse. I began to experience very extreme constipation; a condition that I had never suffered in any way shape or form prior to this; such that it became impossible for me to go to the bathroom. As well, to my shock I discovered that my sexual organ felt as though frozen, would not function other than to urinate. My breathing became more difficult. Hoping that all this was simply some temporary malady brought on accidentally – I was not then suspicious of it being the result of someone's deliberate intent, I rested for a few days hoping it would go away. But it didn't, and by the end of the week it finally became obvious that I needed to go to a hospital or clinic. I hadn't been able to go to the bathroom, other than to urinate for almost an entire week! On Saturday morning, Oct. 10th, I called a friend and asked him if he knew a clinic or doctor he could recommend. This friend was Sydney Thompson, an elderly gentleman; who I understand is since deceased. His address was 100 W. Edgeware, Los Angeles, CA 90026. I became acquainted with him through a "Mabel Normand" contact, that is museum founder and Hollywood archivist Don Schneider (also an elderly gentleman), and who was very kind and a great and perfect help to me in my stay and getting set up in Los Angeles for the purposes of which I had originally intended.

Unfortunately, after a lengthy search, every local medical clinic we encountered on Sunset was closed! Why this was so I did not then nor now know why. In any case, Mr. Thompson at last suggested the Queen of Angels Hospital on Vermont Ave. To this I agreed.

We arrived at the emergency room of the Queen of Angels Hospital at around 10:00 a.m. I signed in and was told to wait. After sitting in the waiting area for five hours, during which time I felt fairly miserable, I was finally admitted about 3 p.m. to the Emergency room. There it was all a hub of noise and doctors, and cadets of some sort (police, medical, or police/medical I don't exactly remember, they were wearing Navy blue cover suits) were running about – the place gave the appearance of being very busy. I was assigned a bed behind some portable curtains; given a smock and told to undress. There were at least two policemen present in the room who apparently were waiting while someone in their custody was being treated. No one could see in or out of the curtained area I was located in; except very slightly through breaks where the curtains formed into corners.

After a doctor, I forget which one, briefly heard what was ailing me, he left and different staff persons came and took my blood pressure. My blood pressure was registered clumsily at least three times by different persons while I waited for a doctor proper to return. At the time, I thought the repetition of having my blood pressure taken by different persons, including at least one "cadet," was done rather carelessly and incompetently.

Finally, Dr. Elmer Eley (Note. Despite the fact that the person who ultimately signed my medical report, which I later obtained a copy of was a Dr. Phillip Fagan, I only later discovered that the physician who actually examined me was not Dr. Fagan, but rather Dr. Eley – Fagan reportedly having been at no time present), a rather muscular, middle aged, black male with a mustache, came into my spot to check my breathing with a stethoscope. Customarily, it has been my experience that when this procedure is done the patient is either standing or sitting up. Dr. Eley had me lie back on the bed/examination table and told me to breath as he applied his instrument. As he came to the area upward to the left of my heart, he made a pointed clenched fist and suddenly and with thoughtful and quick deliberation PUNCHED me just below my left shoulder! I was so shocked by it I didn't know what to say or could think what could account for his doing it. Finishing up, he asked a few questions (as if nothing had happened) and left me. I waited a while longer very much perplexed. Then a middle-aged female staff member with short, light colored hair came in by herself into my spot. And telling me to roll over, she gave me an injection Simply assuming she knew what she was doing; desperate to be rid of the "chill" that suffused my groin, and without questioning I simply acquiesced. And without informing me of what I was being given, she injected me. She then departed and I was again left to wait.

As I sat there I gradually began to feel what was apparently the "medication" taking effect. I suddenly began to have great lapses in my train of thought and suddenly found it difficult to form words. It is all somewhat difficult to describe except to say that it felt as though I had been given a very, very strong narcotic of some kind. By this time, I became very fearful, after being punched and now this apparent drugging, and didn't quite know what to do. Each time I tried to get hold of a staff member for help I was very rudely told to wait. For the *next* three or four hours I lay on my bed waiting for one of the doctor's to return, during which lengthy time feeling utmost distress at my situation. I literally felt and thought I could very well die then and there, due to the effects of being punched in or very near the heart, and the injection which had some sort of mind altering effect, causing my thoughts to be disoriented.

By eight o'clock p.m., a Dr. Eley gave me a bottle of liquid laxative and directions to buy antihistamine. Without ever telling me once what might be ailing me, or saying whether the problem with my groin was cured or not, he finally released me. Naturally, by this time I was dying to get out of there; so I didn't trouble to ask him about what my condition was. In any case, he made it so very clear that he was busy, that even if I had tried to get him to talk for more than one or two minutes he would have put me off or casually allowed himself to be distracted. He was not entirely unsympathetic when I spoke with him, only he would not stay to answer what seemed to me were very pertinent and straight forward questions. Given the actual amount of time spent seriously dealing with my case (in contrast to the time I time spent there) one would have thought I should have been there no more than ten or fifteen minutes, been on my way; while freeing the "doctors" to devote their time, and my bed, to other cases. Instead it took around four hours for me to simply wait around to get a bottle of laxative and directions to get antihistamine.

After I came back from the hospital, I found indications that someone had been in my home again. This time, the note I had Lt. Vega make (and sign), when he came to investigate, and the original copy of my birth certificate were missing.

The next day I returned to the hospital to complain and make inquiries. I asked to know what it was I had been injected with. After a lot of running around for an answer a staff person showed me a document of some kind with "Penheglian" written on it, that presumably being the medication. My current records mention "Phrenegan," but this was *not* what was originally written on the document. To add to this, the Queen of Angels staff, despite my inquiries, never explained what the Phrenegan (assuming that was what I was even injected with) was for, or exactly why it was administered to me. At that same return visit, the doctor's name then was given to me as "Herb" Fagan; though as mentioned it was actually Dr. Eley, not Dr. Fagan. This was written on this same document. I then got a senior staff person of some kind and sat down with her and told her what happened. After hearing my story, she politely told me that this was a "good" hospital and that they didn't do things like that. I then requested my medical records and was sent to the records division. Once there, they told me that my records weren't ready and that I would have to come by on another occasion.

Disgusted and frankly now a little fearful, I shortly after all this, Columbus day I believe, I left Los Angeles and came back to Seattle. Not surprisingly, and after what I'd been through, I did not feel comfortable remaining in Los Angeles. Although the laxative seemed to cure my constipation (which was hitherto something entirely unknown to and unexperienced by me), I still had difficulty breathing and was feeling the effects of Dr. Eley's having punched me. At the time, it felt as though my heart had been injured; hence my feeling that I might die; which I mentioned before. When I'd returned to Seattle, however, I went to a clinic. My injury was diagnosed as damaged muscle within my left shoulder and they prescribed Advil, which had the desired effect of alleviating the pain.

I resumed staying at my father's house, in Seattle, where I had lived prior to going to Los Angeles. Though he took me back; there was certain unaccountable hostility toward me. And after I got through telling him and my brothers what had happened to me, far from getting any sympathy, they summarily pronounced me crazy. At the same time, one of my brothers blamed me for some how putting them in danger by returning to Seattle. This unwelcome and rather inexplicably contentious reception very much surprised me. Instead of moving into my old room in the basement of the Woodlawn house; which was then occupied by one of my brothers, I was put in the attic to stay.

Sometime in the last week of October, I suffered what I felt at the time and to this day believe was poisoning. I was watching the 1992 Presidential debates one evening and upon drinking some coke from a bottle, which had already been opened, at my father's house. I began to feel the extreme effects what seemed like a street drug of some kind, possibly speed. Why or how this could happen I have no explanation. The coke was a 16 liter plastic bottle with the top off and three-fourths full resting by my father's couch; which I matter-of-factly drank out of without having any reason to think there would be something wrong with it. Whatever the cause – whether the coke or something else – I was rendered suddenly and inexplicably ill in a manner which made me feel I had ingested a foreign substance resembling speed in its effects. At the time I was taking Sudafed, and later at the hospitals directions Benadryl, both antihistamines, for some difficulty I was having in breathing. However, the last time I had taken this was the night before, and didn't see any connection between how I was feeling at the time of watching the debates and the (consciously administered) medication.

As the hours passed and I grew worse, it became clear for me as I felt to seek medical assistance, and the first place I thought of was the University of Washington Medical Center.

The following are accounts of separate visits I made to the University of Washington Hospital Emergency room as a result of my being poisoned (or, if someone insists, my feelings of being poisoned.)

"1st visit: 10/27/92. Physician: Dr. Stephen Burns. Some hours after the incident with the coke bottle I had myself admitted to the Emergency Room there and to start out with was questioned, had my blood pressure and temperature taken. I was hooked up to an EKG machine, the tapes and wires attached to my chest area. Some time during the course of my time on the examination table I blacked out. Whether this blacking out was due to fatigue or what I do not know. In any case, I was rendered unconscious for a unknown duration, at the most an hour or two. After I left the ER I went home to go to bed. Because I was so tired, it was very early in the morning by this time, I did not bother to get undressed when I went to bed. When I awoke the next day to take a shower upon undressing I found an EKG tape attached to my scrotum. The only logical explanation for how it could have got there was that someone, apparently on the ER staff placed it there during the time I was unconscious. This at least seems to be the logical conclusion. Yet because I was unconscious when it happened I cannot say that I unequivocally know that this in point of fact is what took place, let alone who the individual might have been. Nevertheless, given the staff's peculiar and ingenuous behavior and mistreatment of me in other ways and the fact that I could not have acquired such medical tape from another source, I personally am convinced that this is what happened.

"The record of this visit reports a tightness in my shoulder I was feeling as the cause of my complaint without any reference to my stating I felt I might have been poisoned. While the pain described regarding some strained muscle in my left shoulder, this pain was secondary to my being or my perceived being poisoned, and was not what I actually had myself admitted for.

"On an occasion following the coke bottle incident I again felt as though I had been poisoned after eating something from the refrigerator. (And no, it is not lost on me that it should normally seem very odd that I should suffer poisoning so close upon the first incident) Whether I was intentionally or inadvertently poisoned, I don't presume to say. It was around this time in the media that stories came out about the E coli bacteria and Sudafed tampering. All I can say is that my physical constitution was such that I felt as though I had been poisoned, nor did I have reason than or now to believe otherwise. This time, as with each such occasion, the effects of the poisoning were similar to the effects of a street drug, in this second instance pscillocibic mushrooms. Back in high school I had on at least two occasions taken these so know how these effect a person. Be this as it may I returned again to the ER. Although I did, of course, have some misgivings after what happened on my first visit, I was willing to give the hospital the benefit of the doubt since there did not then seem any ostensible reason for why such people would violate the law, let alone human decency, in order to hurt me. I also thought as well at the time that even given that wrongdoing had been done to me, i.e. the tape found on my scrotum, there would be no reason to believe that such an gross outrage could possibly be repeated. Lastly, being puzzled by the tape, I suppose a part of me wanted to see how they would react when they faced me again. Having said this, however, my ailment was genuine and my curiosity and indignation with respect to the tape itself was not itself what caused me to return to the UW Medical Center for medical attention.

"2nd visit to the UW Medical Center, 10/30/92, - Physician: Dr. McMullen. This visit, as I recall, was uneventful as far as misconduct is concerned, except that I was not given a proper prescription for the medication I was given. The problem was that no where in writing was it indicated what the dosage the medication was to be taken. At the time, I retained a good deal of evidence on this point, including the medication itself which I decided that I would not (not knowing the dosage) take. I had the original "prescription" as well. Yet for reasons unknown to me, these, as well of my Queen of Angels records, vanished from a specific storage place of mine, and I cannot explain or account for what happened to them. Due to present lack of evidence then, I will keep things simple by not making anything of this particular charge about the prescription.

"3rd visit to UW Medical Center,11/13/92, - Physician: Dr. Weaver. On a third occasion I again inexplicably felt the sudden effect of having been poisoned. As in the prior instances, I could not give an adequate explanation about why this was taking place only that it was. The nature of the third poisoning was unlike the previous `poisonings' or anything I had ever experienced before and am consequently at a loss to what I can liken it except to say it seemed like a street drug resembling speed.

"At some point early on in this third visit, I was brought into a small examination room and was told to lay flat on the examination table, which I did. A male staff member with blonde hair and glasses, after doing some routine checking, including some extensive looking into my ear of all places, proceeded to feel out my entire body. Now not surprisingly the experience of involuntarily being made to feel the effects of a drug is an extremely traumatic thing, as I would hope would be self-evident, and this made me malleable. At first, though I thought it strange, I assumed the doctor or staff person knew what he was doing so I did not protest. He never said he was giving me a massage and I presumed he was engaged in a legitimate examination. The only problem for me was the question of how slowly and deliberately feeling out my entire body is supposed to have been an appropriate method for diagnosing or dealing with a poisoning. I am no expert, so I could be wrong about this. I can say however that the experience made me feel extremely violated as much as if I had been physically molested.

"Prior to going into the ER, I had called the Seattle Police Department to make what was now clearly an overdue report about having been poisoned. Not long after the 'examination' described above, a police officer arrived at the hospital. This Officer's name was Underwood, badge no.#682. After staying to listen to me for at most two or three minutes in which I described how and why I had felt I was poisoned,[106] he told me in effect that "he didn't have time for this" and ran out without allowing me to state what had happened as far as the poisoning was concerned, let alone file a report. Months later I filed a complaint

---

[106] My complaint to him did not concern the ER's treatment of me, only the poisoning itself., and basically as I have already described it here.

against the officer with the Police Department Internal Investigations. My Contact Log File number is CL#93-227. After some letter writing I spoke with the officer's supervisor. He, in sum, stated that while the officer acted improperly in running out on my complaint it was not bad enough to require disciplinary action. The reason for this in turn was because the ER staff had discredited me with the officer without my knowledge, thus putting themselves in a position to deny me my civil rights. I only found out about this after I made my report to Internal Investigations. If the ER staff, for whatever reason, didn't want to treat me as my case required there was no reason whatsoever for them to have interfered with my simply making a report to the police, while at the same time not informing me of the fact.

"Later I spent a lot of time (in hospital examining room where I was placed) talking with Dr. Weaver who insisted I was a lunatic while at the same time refusing (except until the last minute) to give me a blood or urine test to determine if I was enduring the effect of a noxious, foreign substance. At the same time as he insults me, he effectively denies me the very means by which I could verify the nature of my complaint. It says on the hospital file for this visit that I was given a toxicological test. This I assume refers to a last minute urinalysis that was hastily done. It is true my urine was taken, but only as a last gesture on their part to show that some test had been taken. Even if the test was legitimate why hadn't it been done on the two prior visits, and why only on the third visit only after repeated requesting and finally insistence on my part that it be done? It would seem clear to me from this that having initially diagnosed my case as psychological in origin it was in their vested interest to deny the possibility of my having actually ingested a foreign substance, and in this way cover for their mistake on my first two visits. For even if I [had] been poisoned in actuality only once, let alone three times, this would not have reflected well on the ER staff's conduct refusing me a chemical test.

"Dr. Weaver, at the time of my visit, while refusing me a urine or blood test, described me as suffering from 'paranoid delusion.' Well, there are at least two things wrong with his assessment. My statements with respect to my speculation then as to what might have been the origin of my poisoning have been distorted to make it sound as though I left no room for doubt. In point of fact, I never at any time said I was certain as to the cause of why I had ostensibly been poisoned and merely offered when asked what I felt was a possible interpretation of what had occurred. The doctor's report on the other hand would seem to suggest that I had some definite and conclusive notion as to the reason for my apparently having been poisoned when in fact I had no such definite or conclusive notion. Finally, as stated before, never once could the doctor explain why I had all along been denied a urinalysis.

"A word in conclusion about the 'poisonings.' Now four separate and distinct poisonings in the course of less than two weeks would seem to stretch the credulity of some, and I am well aware of this. If someone prior to my experiencing it would have asked me what I thought was the probability of such a thing taking place, I would have said I thought it highly unlikely. Indeed to this day I cannot claim to be able to explain or understand it all. Yet just because something sounds improbable does not make it impossible. The palpable and physical sensations of separate poisonings did take place, and I don't have the least doubt in my mind about this fact. To put this another way, I am absolutely certain that my ailments were not in any way the product of delusion or hallucination, or psychological indisposition and am outraged then and now that they were treated as such without proper chemical tests having been made. If we grant, just for the sake of argument, that I had indeed been poisoned as I claim, how could I possibly have presented or handled my case differently then I did? Imagine what it would feel like if one had actually been poisoned, yet upon seeking a physician was told one was 'crazy.' Well, this is precisely what happened, and I don't know what is worse, the actual poisonings or being treated as I was by the hospital under those circumstances. [107]

"Within the subsequent months, I reported what happened to the University Hospital Administration, and my complaint was directed to Leah Kliger. My purpose in contacting her was not to cast blame upon the hospital itself but that part of the staff I did encounter. I called her in June, and after she requested it I sent her a letter providing the essential details of what took place. I waited a week or two for a response, then called her office and was told she went on vacation. I waited a few more weeks I was not able to reach her, and it soon became obvious that the reason for this is that she refused to speak with me. On a second or third call to her office I was told a letter had been sent me. This letter was brief and advised me to seek psychiatric help. Naturally, one could not presume that she would necessarily take for granted the truth of my story, yet there was no reason that she should treat my after all serious complaint in this peremptory, insulting, and frivolous manner. Simple answers to a few questions hardly seems like much to ask, again, even if I was crazy. I find it disconcertingly ironic that at least four major Washington or Seattle area medical organizations whom I inquired with prior to writing this had hardly a clue as to whom one would report a complaint of staff misconduct at a hospital. I site this example, in some detail, as typical of the cavalier and condescending attitude I met with when trying to bring my problem to attention of those who might be in a position to help remedy my situation.

"A close examination of the medical records drawn up by Dr. Burns and Dr. Weaver reveals that a deliberate and conscious effort is made in them (if one can read the handwriting) to discredit me and not in the least is there any consideration of the possibility that my supposition of being poisoned had any merit whatsoever. In each case, without there having been any chemical or urine test done to ascertain whether there was a foreign substance in my system as I

---

[107] The passage here was written not long after the events described and is not something I have added later in composing this narrative; so that yes in retrospect I could say now I have a better idea now of what happened; that is, that I was deliberately poisoned.

claimed, it is assumed that what I was suffering was merely a disillusion of some kind.

"I had no reason to think prior to my visits that I would receive anything but professional, intelligent and ethical care and treatment from the University Medical Center ER, but unfortunately in this I turned out to be much mistaken. It did not occur to me that it would be possible that something similar to what had happened at the Queen of Angeles could be repeated all the way back up here in Seattle."

In what follows I go on to describe (separately) what I experienced shortly after the U.W. Hospital visits. In retrospect it needs to be supplemented with and seen in the light of what I have stated in my *New Treatise*. I leave what I wrote here, essentially not counting minor corrections and very slight re-wordings for the sake of better clarity, as I wrote it back then to give you some of idea of the perspective I had at the time, and the very awkward position I was placed in trying to survive the regular and varied sorts of violence and injury being done to me while attempting to figure out what was going on. This said while the explanation I give may leave something to desire, the description of "the apparatus" (as I called it then) or "KGB" brain radio is an accurate one. Also, it goes without saying that in the course of what was going on I was suffering from no little amount of trauma, though I think I behaved myself as rationally and conscientiously as one could under the circumstances.

"It is at this juncture of my account that I come to what, to many, is perhaps the most extraordinary of all the events and strange occurrences of which I speak. Because it concerns something which is outside the experiential purview or expectation of the vast majority of people, it is very difficult to discuss. On top of which, if what I say is taken the wrong way, it will very likely, as has happened in the past, it could be used to discredit *anything else* I might say. Therefore, I particularly beg your patience and open-mindedness then in considering what I realize to many will sound outlandish in the extreme, yet which, nevertheless, is very real.

"Following shortly upon the aforementioned events at the University of Washington Medical Center, I found myself being made the 'guinea pig' or in otherwise victim of some sort of mind control/torture technology.

"Before scoffing as some immediately will at such a claim, let the thoughtful, objective, and honest ask two simple questions:

"1) Would it be technologically possible to come up with a device that would inflict pain, and indeed even read a person's thoughts, in our age of unprecedented technical marvels? Certainly, as has been amply documented, both the former Soviet Union and the United States have been engaged for decades in related research of this sort. With what we have seen technologically in the past decades the question has often become not 'what is possible?' but rather 'what isn't possible?'

"2) Do there exist people in this world, devoid of all real moral conscience, for whom no crime is so bad they will not think twice about perpetrating it, if they thought they could reap gain and or wreak their wrath, and did not (at least as they believed) risk getting caught doing it?

"If the answer is yes to both these questions, then one can be no question as to whether such a thing as mind control/torture technology is possible. When I first found myself a victim of it, I did not know that mine was not merely an isolated case. However, in Feb. 1996, I discovered there WERE more victims of this, and similar technologies, and that there are mounds of evidence to establish its existence to anyone willing to look at it. In Appendix [XYZ][108] of this document you will find what are only mere fragments of what is available in the way of evidence, history, and testimony concerning the inhuman application of technology to experiment of, torment and, in effect "imprison" people. For suggesting such a thing, there are obviously many who will instantly denounce me as some sort of crack pot - this is to be expected. Yet I would respond by saying, after the reader has completely read my narrative, look over all this material, and judge honestly and intelligently whether there is not after all something to what I claim. Who, for example, would otherwise have believed the non-consensual testing of tens of thousands of citizens in the fifties with radiation; the MKULTRA mind control experiments in the 50s and 60s; the non-consensual testing of LSD on U.S. service personnel during the Vietnam war, or the Tuskegee syphilis treatment which lately got into the news - could ever have taken place? Yet all of these, albeit many years later, are indisputably now part of the public record. Since those experiences, research and technology has only become more sophisticated, and certainly there has been no dramatic increase in public and private ethics and morals – to say the least.

"Essentially, the technology is used on me as follows:

"1) As best as I can tell it involves some sort of implant in my brain. Someone later told me that mind control/torture technology can be used without an implant as such being necessary. Be this as it may, as best as I have been able to tell these past five years of enduring it, what I have is an implant of some kind.[109]

"2) In some way I can't claim to completely fathom, this technology can be used (as preposterous as I know this sounds) to read my thoughts. By thoughts I mean not my 'mind' as such, but rather my thought perceptions, i.e. images or worded memories.

---

[108] Not included here. Someone interested in finding out about technological mind control, such as in the work of Dr. Jose Delgado, can easily find the sort of things I am referring to on the internet.
[109] This was written in 1998.

"3) Initially it was used to send 'signals,' usually in response to my thoughts. Later this expanded into what is in effect a two-way radio.

"Signals refers to a kind of transmitted communication into my brain. One might liken it to a radio communication in which my brain, or something implanted in it, serves as a transmitter and receiver. These signals usually take the form of one or two word communications; or snippets from popular songs. They are not audible forces as such, but thoughts distinct from my own to the extent that I it has been possible for me to carry on a dialogue with them. These are not 'voices,' but more like thoughts - only thoughts that are not my own. What happens sometimes is that I can have a memory of these 'alien' thoughts and can usually tell the difference between these memories and the signals themselves.

"In late November and December of 1992, the intensity of the 'auditory/thought' signals was at their highest pitch, and musical snippets were very frequently resorted to respond to my thoughts. Since about January of 92, however, the intensity has been toned down and the musical snippets made much more infrequent. At the present time (Feb. 1998), the occasional song snippets have long since ceased, most of the signals take the form of one and two word communications. As time has gone on this technology has been used as a full bore radio, and it is common for me to carry out extended conversations with people using it (who, incidentally, claim to be from Microsoft.)

"4) Aside from thought reading, perhaps the most unusual feature of this mind manipulation technology is the ability to have my sleeping dreams invaded, in which I am forced to sit through what (for lack of a better term) I will call a 'dream production.' As best as I can tell this aspect of the technology works in either or both of two ways:

"a) Images are somehow 'suggested' to me while I am asleep, which images are then interspersed then in a given dream which is otherwise my own.[110]

---

[110] Having been put through literally thousands of "dream productions;" either done by the magician or his (evidently) "sorcerer's apprentice," I've come to the conclusion that the subconscious is essentially and nothing more than much the equivalent of the cognitive mind running on low batteries; making allowance for "heart" or emotional consciousness also to some degree present but that is and can be aroused differently and with much, much greater difficulty. What dream productions are, in effect, are a kind of hypnotism carried out on you in your sleeping state; so that your subconscious lacking sufficient waking energy cannot think properly. In them you are an observer, sometimes someone who reacts, but who also guided or in effect "told"" by some subliminal method used, what to think and feel. Sometimes you think or feel what you are told, but depending on the kind of person you are, not always. In my case, they can sometimes impel me to a feeling or belief, but as or more often not, my response is one of " I don't know what you are talking about." Or similarly, they try getting me to act or feel a certain way, and when I do not, they find themselves having to change scenes; since a dream production is virtual reality that is much like a movie, but in which you participate. Someone of these are intended be satanic and terrifying in the extreme; with extremely violent, obscene, and sadistic situations created for one to go through.

"b) Images are 'broadcast' by means of the device via some sort of electromagnetic wavelength into my brain (while asleep).

"One can tell they're not ordinary dreams because, unlike ordinary dreams, they are so frequent, and carefully orchestrated. They usually take the form of propaganda, sometimes trying to shame me about drugs,[111] or something I did wrong in the past. In other instances the dreams are used to attempt to degrade me, to shock me, mock me and my values, intimidate, or even attempt to flatter or be friendly to me. True, all of us have had nightmares, yet these 'nightmares' have the unusual distinction of resembling some of the distasteful and obnoxious rubbish that often comes out of Hollywood movie studios and television of recent memory. Prior to being subject to this 'torture,' I rarely could ever recall even one out of ten of the dreams I have had while sleeping. Yet after the introduction of this device into my life, I can recall (whether I want to or not) the vast majorities of these. And these dreams, which I endure DAILY, are so pronounced and often of considerable duration that if I did actually want to, I could probably recall the contents of even more of them. Contrast this with before 1992 when I would perhaps, on average, remember a dream I had once every one or two months.

"Other aspects which have, on various occasions, marked these dreams as being of artificial origin are presence of luminous phosphorescent colors; distorted and warped facial images (e.g. a very long nose, or bulging forehead – I later saw a computer program that i sable to do this with photographic images); 3-D images which make objects looks like they are 'coming at me.' None of these things has ever occurred to me in a dream prior to 1992.

"Another very telling aspect of these 'productions' is that they have often included appearances by celebrities (or at least images which give the appearances of being such, for example computer generated images:)

"Some of those persons who have 'appeared' in these 'dream productions' are:
Ted Kennedy
Tom Hanks
Tom Brokaw
Clint Eastwood
Dan Rather
Warren Beatty
Woody Allen

It further comes as small wonder or surprise then that such current day masters of mind control, as they themselves have many times freely admitted to me, should have ties with film making, television, and that world of make believe; and which is confirmed as well by the film industry's affectionate and overt devotion to subjects like sorcery and the occult; and by which and related means all competition can be completely snuffed out and obliterated; hence the egregiously monopolistic character of big budget movies and decrepitly sordid television in recent decades. If the war against Hell is anything, it is a war against mind (and heart) control.

[111] Marijuana, though I have never been much of a drinker.

Katie Couric (twice)
John F. Kennedy
Lyndon B. Johnson
Paul Newman
Bill Cosby
George Hamilton
Kathie Lee Gifford (three times)
Justin Heyward
John Lodge (twice)
Ray Manzarek
Elizabeth Taylor
O. J. Simpson (prior to famous crime incident)
David Letterman (three or four times)
Bill Clinton
Hilary Clinton
Jerry Seinfeld
Al Yankovic
Alec Guiness
Mike Meyers
John Candy (prior to his death)
Daryl Hannah
Gene Siskel
Robert Duvall
Alan Dershowitz
Martin Landau
Luke Perry
Tom Hanks
Geena Davis
Demi Moore
Jim Morrison lookalike
Bob Dylan lookalike
William F. Buckley, Jr. lookalike

"There have been more well known people and celebrities than even this extended list, yet I hope this catalog will suffice for the purpose at hand.

"Now it would simply be far too incredible for someone to have naturally occurring dreams in which such a list of celebrities appeared – even if they WERE mentally ill!

"Lastly, another feature of this technology is to inflict physical pain on any and every part of my head above the level of the ears, which I could best liken to having one's head stuck in a microwave oven..

"Assuming that I am not lying, I submit therefore this example of celebrity 'appearances' as convincing proof of the reality this technology. If on the other hand, somebody thinks I am lying, then let them give me a lie detector test. I am more than happy to oblige them, indeed pay for such a test myself

"Keep in mind that there are such methods known as 'psychological warfare,' and the obsession among certain segments of our society (particularly here locally) with Hi-tech and 'futuristic' gizmos. There is no reason to assume that some of the more very well to do among us would use such on their opposition particularly if that opposition happens to be poor people like myself. In other words, for some of society's despots an approach of this kind works well because it is so hard to believe, so difficult to prove that in many ways it is the ideal criminal method. 'Psychological warfare' using technology if being used on a poor person is not something that poor person is going to be able to prove – hence its obvious effectiveness as a criminal method.

"At present I am still hostage to this technology and various sadistic, often disgusting and perverted, assaults. I have had CAT Scans done to locate the brain technology, but either because the doctor's are crooked, or the technology is of such a nature it is impossible to detect unless under the naked eye (as someone once told me), there has been no way as yet for me to establish its existence with others."

At the present time (of this writing), it is my sense that, with respect to "brain radio" (such as described above) in the course of what I have been through, both technology and spirit people were used. But it is not always easy to say which was which in a given instance. Some of what I wrote in retrospect seems rather (though justifiably and understandably) naïve. Yet despite what has developed since (in the way of spirit people being made manifest), there have been and are technologies, including radios of some kind, which are used on a person's brain. This sounds both utterly horrible and ridiculous. Nevertheless, I maintain this with conviction as much as I could maintain anything else, having gone through so many experiences; which I could describe in close detail, having had ample opportunity and evidence (admittedly in subjective circumstances) to test my findings and conclusions. The circumstances of such an investigation are admittedly subjective. Yet I believe I have made an earnest effort to be both impartial and reasonably scientific (if not methodologically thorough) in my investigation and analysis of the phenomena – that is an effect – of or in the brain; which I am certain (as mortal can empirically be) is brought about by technology.

I returned to the dept of Education (Miller Hall) where I worked with Jack Thiem in the supply and mail room, and just tried to get on with my life normally. Yet with each passing day of abuse and indifference from others I was afraid I might be killed. For this reason, and in fearing for their safety, I took two of my cats, Timmy (or Timmina, a small gray striped tabby) and Hindman (a large size mostly gray with white Persian) to the Animal Shelter on 15th N.W. and had them put to sleep (October 14, 1992.)

Basically at that time I felt I was being set up for bad or perhaps for good, and I thought that somehow others who knew of my predicament would take care of and shelter the cats for me. Or if this was not the case these (at the time) unknown people were trying to kill me, and therefore it was just as well the cats be spared the situation I found myself in.

Jebo or Jeebo the third cat (a golden and light brown-striped tabby) was not with me at that time, as she had run away when I was packing to leave Los Angeles, and in my (very foolish) haste and anger at her running away, I decided to leave her (a choice I later deeply regretted.) Later, I was going to return to LA myself to get her, but my father flew down instead, was able to find her in the house (I had left the door open) and he brought her back up. Afterwards, as I will relate, I lost Jebo too, and I must say that these cats were in a sense the only close family I really had, and losing them, and losing them the way I did, was by far and is among my greatest personal griefs to me in all this; despite the countless painful and truly agonizing things I have and still continue to go through.[112]

Although I was still suffering at that time (and since), I tried as much as possible to live my life normally as I could. Some people from the E! Entertainment (cable tv) network contacted me to do an interview for their "Mysteries and Scandals," regarding the William Desmond murder case, which I had covered in my *Mabel Normand Source Book*. I agreed, and they came up from Los Angeles and did an interview with me. There was, of course, nothing wrong or untoward about this interview, but I mention it as one of the things of significance that took place in late 1992 or early 1993.[113]

Other events which took place at that time, and of which my general recollection of these is somewhat vague. But basically I went to have Cat Scans done of my brain for Dr. John Chapman. To make a long story short he did not find anything in the Cat Scans to show there was anything of a foreign nature detectable in my brain, and I believe that was the upshot of my visits with him.

Sometime in 1992 or early 1993, I left living at 6322 Woodlawn Ave. N. and moved to 3014 NW 75th, which my father had bought. Sometime after moving in, I went to work for Gray Top Cab (then independently owned and located in Magnolia area of Seattle) various bizarre assaults and occurrences continued to afflict me.

At some point, I don't exactly recall when but probably shortly after I moved to 3014 NW 75th St., I seriously contemplated suicide after Jebo became

---

[112] My sense here is that the suffering of the helpless and the innocent is worse than anything we ourselves suffer.

[113] Later note, Jan. 2014. Upon looking it up the later, the show is dated as broadcast in 1998; so it is very possible my original narrative text is incorrect in recording exactly when I was interviewed. As far as a reason for the discrepancy, as best as I can guess I may have simply remembered wrongly – though admittedly and in retrospect it seems rather strange that I should have, and so materially on such point.

frightened so as to in one instance run up the chimney. She was now all alone from her sisters and there was something frightening her. I adopted a kitten, who I called Joseph Skatey (a tabby also) hoping to give her someone as companion, I myself feeling very much the loss of Timmy and Hindman. But she continued to be sad and frightened. They one morning I awoke and found Jebo had vanished and I haven't seen or heard or her since. It was at this juncture that I seriously contemplated suicide. As ridiculous as this sounds, I drove out far off into the country one time (on the route from 99 to Darrington) only to find that despite miles upon miles of forest I could not find a tree with a branch both low enough and strong enough to support me hanging myself on!

I spoke of wanting to commit suicide confidentially to someone I knew from high school, Stuart Greene (as last I know of Medina and Seattle.) He told my parents about this, and they or somebody called the King County Crisis Clinic. To make a very painful to write story short, they had me committed, and while at Harborview Hospital (at which I stayed a day or two) I was handcuffed to a bed – this formally justified on the grounds on this simple basis that I had verbally remonstrated with and objected to the proceedings as entirely unnecessary (which they were, as seen by me then and now.) Simply put, if there were people there to make a fuss or prevent me from committing suicide, than I did not (at that time) feel the circumstances were right to do it (i.e., contemplate suicide) in the first place. This was my thinking.

At my father's or the court's behest or insistence (I don't recall exactly which) I saw Jenny Becker, at Case Management Services, located at Capitol Hill in Seattle. I forgot exactly what the purpose of these visits was; except that I do remember the subject of obtaining social disability payments for my "illness;" though at the time I was dead set against the idea (being a strong believer in the work ethic.)

Again my situation at that time I perhaps found more trying than at any other because one of the things that most baffled me was how and why these things were still going on. As a result, much of my general memory of this period is particularly sketchy; so that in describing what follows I may easily be mistaken as to sequence of which took place. Otherwise the following is a list (*) of *some* (and only some) of what I went through or continued to go through as best as at present I can recall.

* Two toxic injections (administered to me by someone when I was asleep at home); one of which caused my eyes to turn blood shot red, my tongue stark yellow, and induced a feeling as though my liver were being eaten away. Basically, I woke up from a nap with the feeling of a pin prick in my arm when the first of these took place; followed then shortly after by the above described symptoms. I went to the Ballard Swedish Hospital, and despite the fact that the place was all but deserted of patients, they were very cavalier, told me they were

taking care of someone with "a broken leg" and had me waiting so long that there was no purpose of my remaining.

With regard to the second 'injection,' similarly I woke up with a pain like that of a needle inside my thigh, near my groin. This was similar to the chill in my groin I had felt in Los Angeles. This time I went to Virginia Mason Hospital ER on Capitol Hill. There the "physician" (I only dealt with one person there) sat me down and talked down to me like I was crazy, and, as at the UW Medical Center, *refused* to give me a blood or urine test to detect whether I had taken in some unhealthy, foreign substance. These two events occurred roughly 3 to 4 years ago, and unfortunately I do not have the exact dates of each; however both took place while I was staying at 3014 NW 75th (which was a basement apartment, while the house above was being rented to others.)

* Food poisonings of, what as best as I could discern, of street drugs, such as speed, also laxatives. Most of these occurred between 1992 and 1994.

* Given diseases of various kinds, from large number of out breaks of miniature moles, relentless ear-ache, appearance of acne on strange places such as the middle center of my eye-lid, to an unbelievably severe cold (which last 2 to 3 months) unlike any I had ever experienced. Again, these occurred during this 1992 to 1994 period also. It might be thought by some that these might simply have been brought about naturally. While I cannot categorically deny such a possibility, specific circumstances and the frequency of these things leaves me little doubt as to most (if not strictly all) of the actual origin of these as being criminal in origin.

* The appearance of tiny, bloodless, streaks of scars on my abdomen and inner thigh – which I still have, and which were later shown to Dr. Robert Aigner when I was examined by him. As well I discovered various tiny moles appearing at various parts of my body including my arm pits, back and eye lids.

About (as best as I now recall) early 1996. My father decided to rent the house at 3014 NW 75th completely (including the basement where I was staying) and bought a very small house for me to live in at 1604 NW 70th.

At the time of my moving, the number of brain torture dream productions, radio seemed to increase, and I remember at the time feeling great depths of loneliness and despondency. I now had two cats with me, Joseph Skatey and his sister Neffy. The latter I adopted after I lost Jebo (naturally from the same person from whom I got Joseph Skatey, and who was someone I knew personally.)

In addition, I was daily (and nightly) put through literally thousands of hours of dream productions made all the more painful by my being alone as I was.

To be frank, I really was fed up with all that was going on and others indifference (including my family), and would have abandoned everything become homeless, gone to live in the woods, die, rather than put up with the abuse I was being put through. The simple reason I didn't was there was no one who I could give my cats to take care of them for me, and I felt (and still feel) an obligation to them; such that if the truth is told I would not still be in Seattle (or perhaps even alive) but for them. This may sound ridiculous to some – that is, that I would endure such as I described for the sake of some cats – but it's the honest to God truth.

The following are some incidents or occurrences that took place after moving to NW 70th and are taken from notes or letters I had written about that time.

"On about Nov. 8, 1996(?), I found in my office an envelope of photographs of two young men carousing - drinking, smoking pot, and acting up. I never saw these two before in my life. With these photographs was a note which read 'Hey Dickweeed, you better keep your door locked or some Mexican might come in and steal your stuff.' I call the Seattle Police. An officer, one P. Fox came and took a report, and the photographs. He gave me a case number, namely 96-499535, and I have not heard back from SPD since…

"My pickup truck suffered three flat tires with a nail driven into the center of the tire within a three month period. Again, just a coincidence someone might argue, only in prior to these events I might get a flat tire every few YEARS or so. Also on at least one if not more occasions I have had a tire on my taxi cab slashed by a knife (also about this same time.)

"By means of some technology I do not claim to comprehend, these people have been able to inflict physical pain by some remote means. Aside from what I have mentioned previously with respect to head pains, I refer here to what I will call 'zapping.' On one occasion I was inexplicably struck with an atrocious pain in my spine that literally paralyzed me for 45 minutes, so that I could do little more than lie still in agony for that period. A more frequent occurrence however is throat 'zapping,' by which they cause a pronounced harshness in my larynx (or thereabouts)."

"In the past few months [i.e., early 2000] I have had my home broken into, keys and wallet stolen. One instance will give an illustration. My wallet was stolen from my coat pocket where I always keep it. A few days later my house keys were missing, and when I went to the location where I keep my back-up keys, there was my missing wallet there, with the back-up keys missing! This is just one instance and is typical of the kind of monkey-shines I have had to put up with.

"Needless to say, my life is made extremely painful, and I live day after day with this ongoing torture and agony, and I consider it nothing short of a sheer miracle that I have survived as long as I have. I please then beg your help in this, both to rescue me, and at the same time put these criminals, who have acted across state borders, behind bars where they belong. Local officials and medical people are apparently easily manipulated by these people so that despite my efforts on a number of occasions, I have been able to get no help whatsoever. I have in effect been sold into a kind of slavery, albeit at the cost of apparently a great deal of money to these criminals – though money does not seem to be at all any big hardship for them."

On one occasion I came home and found one of my two cats, Joseph Skatey, severely traumatized, despite the fact that he had been locked inside – as if someone had been thrashing him; such that when I tried to pet him he at first snarled back at me as if in fear and distrust. Joe was a very sociable sort of cat and I remember he would without hesitation friendlily approach dogs, cats, squirrels, possums, raccoons and people (as they might on a given occasion visit or otherwise show up around my home); so that I suspect when his attackers came in he was probably at first friendly toward them, not expecting or, based on previous experience, any reason to suspect any harm.

Some different odd things happened while I was at 3014 NW 75th, among them a silver Timex (hand-wound) watch my mother gave me, and rosary I had since my first Holy Communion at St. Agnes in Rockville Centre NY were stolen or went missing. It was also while at the same address that I had my first experience of being held down physically by a demon (while lying in bed); so that for a few moments I could not get up. At the time, however, I did not quite know that that is what it was, and did not spend much time thinking about it.

As mentioned I was driving a cab for Gray top. At the time one of the dispatchers, Bob Crouch, had started taking care of a stray tabby who came around the lot. To make a long story short, Gray Top was sold to Yellow cab, and the lot was abandoned. I quit the company and agreed to my father's proposal to live on disability (this was in September 1996.) I just couldn't both regularly work and stand the strain and regular harassment of what these people were putting me through – though prior to this the idea of living on disability was utterly repugnant to me. Some time before the company left the lot the cat (who Bob Crouch called "Ducket") had kitten and I returned to the empty lot, every other day to feed them (this for about 3 years.) Moreover, and again to make a long story short, I ended up taking some home and adopted them. These[114] were:

---

[114] These were actually Ducket's grandchildren, and their mother was another cat, Kitty Kates, who I recall ultimately had about fourteen children. I tried to get them spayed and neutered by a charitable group which does that sort of thing but they required I catch them all and bring them in which was simply too much for me what with all else going on. I have photographs of all the ones I adopted.

Daniel Snugby
Peanut Berry
Pete Jubilee
Huggin Coat
Spivey or as I also called him Jenkish Jenkington.
Covey Cub

I want to speak about these cats a little bit. It might seem that I needed pets, but this really wasn't true. The fact is I felt sorry for their helpless situation, and that was why I adopt any and all of the cats I did in the first place. Not that they didn't benefit me and give me company, it's just that for practical purposes they would have been too much a burden and inconvenience for me to have taken care of otherwise than out of pity and compassion for them.

This said they really were all great kids. Each had a distinctive personality, that quite simply, made them lovable. And despite all the excruciating, agonizing, and unspeakable pain I went through it was their suffering – that is, they who were completely innocent – that has and does outrage me most about what has happened, and for me personally is the saddest part of all my story; though I fully understand that for others, they might think me silly to think so. But more on these cats later.

During much of this period (1998-2000) I completed my Mabel Normand Book, and took up making wooden ship models from kits including Corel's Half Moon, Mamoli's Golden Hind, and Blue Jacket's U.S.S. Constitution – all of which I completed. I also spent a lot of time playing Red Baron 3-D on the computer, and for a while headed my own online combat squadron and we participated in a few campaigns with other squadrons. As well, I spent much time reading various books, but my focus was especially on philosophy; as I was at that time making plans to write *Peithology*.

Up to this time I had two spirit people experiences, but which left me mystified as to what they were. One (that is, of seeing a floating angel) I won't discuss here as it is rather involved, but which I can mention occurred New Years Eve of 1992-1993. But of the other I will give a brief description. In 1999 I think (though that's about as exact at the moment as I can give you), I was sitting in my chair one night and it was violently windy outside I felt a very pronounced iciness in a part of my chair arm to my left, and not just in the chair arm but to the area of my left, and then suddenly smelled fruit that smelled like flowers or flowers that smelled like fruit. Now I had read books, such as Peter Underwood's, about "real" ghost before so that my initial reaction was scientific and I wanted to check to discern if what was caused was not being brought about by some technical or technological trick. Without going into the specifics, I attempted to see what might be the source of the cold spot or the smell by checking a nearby vent and whether the wind from outside was not carrying something in, but didn't discover anything of the sort in either instance.

Otherwise my first most vivid and palpable manifestation of a spirit person or a "spirit world" (outside a sleeping dream) was in April 2000; sometime, as I recollect, about mid-morning (10am?).[115] Basically what happened was this, I was sitting at home one late morning and this ghost came to me looking like Jesus; that is, as Jesus is commonly represented to have looked in surface physical appearance; while at the same time he presumed an air of authority with me, yet authority accompanied by kindness and understanding. Some of the details of the basis of his coming to converse with me I'd rather not get into because they concern some personal feelings I had for someone. But essentially this ghost told me that my wish toward this woman would be fulfilled; only I had to show my faith., and in order to do this he told me to follow certain steps. His speaking was calm,[116] yet his manner was such that, looking at the experience in retrospect, he wanted to hurry or shake me up, while again appearing friendly and calm. The steps he ended up telling me were things like "go to your car" (in my case my truck). Then I would go to the car, and then he said go back and get your Bible. Essentially giving me instructions or suggestions to do this odd little thing or other.

Once more, to make a longer story short, he told me to drive my car from 1604 NW 70th St. east up 70th.[117] I was to follow what he was to tell me; else if I did not I would not be showing my faith. He did not speak of Jesus or things Christian, but it was implied by his manner and appearance (or so I at the time thought.) As I began up east on 70th, he told me to speed up till I was driving something 50 or more, in a 20 to 25 zone. Strangely and of course very fortunately there were not any cars along the route; so I otherwise made it to the entrance to southbound Aurora Ave. N. at about N. 65th.

Here, again in what I described I followed his instructions, I slowly drove out onto Aurora, the while chatting with him about personal things in my life. He told me he wanted me now to drive (south) down Aurora to the Aurora bridge and drive off. To not do so would suggest my faith was less than it should be, or words to this effect. Though I naturally didn't like the idea, if he was going to insist on it, it seemed something I had better do, and was to some extent prepared to do, but when he saw I would do it, he had me just stop the car on Aurora about a couple blocks south of the south bound exit about and near N. 65 and had me

---

[115] I had court records which contained the specific date of this incident, but all the records I had regarding this event, formal and otherwise, have been stolen or else have disappeared. The following notes I took at the time are as much as any original record I have left:
"Judge Brady Johnson, hearing on 8/7 [2000]
B. Lamendola
Next hearing 9/11 at 1:30"
[116] He sort of half appeared over my shoulder and could be seen in my mind's eye, though he may at first appeared in full length so to speak, I can't now exactly recall.
[117] This first part of the drive may have been down 65th rather than 70th, I frankly don't now recall which; though I am fairly certain on the drive back towards Ballard we took 65th.

park the car in the middle of Aurora (and as I recall giving me specific instructions as to the angle of my tires, or something like this.).

After lingering there a few minutes talking with him, I don't remember exactly about what other that I was glad I didn't have to (drive down to and) drive off the bridge. He had me drive back up 65th, telling me to (in effect) to floor it. I did what he said but naturally expressed my discomfort at doing so. He told me it was all right, and that if I had faith there was nothing to worry about. As I sped in the truck, going I guess about 70-90 mph (which was after all a ridiculous speed under any circumstances along that residential road.) Although I did pass a moving car or two, that is moving in front of and crosswise to me, mostly the way was – again very fortunately – free from traffic, and pedestrians though I had one or two close calls of hitting something stationary.

After we got back in the neighborhood of my house, he had me drive down from 70th to NW 65th from 17th NW and from NW 65th NW, going west to 19th, he told me to drive down 19th, and basically crash my truck into a tree (located along 19th at about the 6400 block and on the west side of the road.) The tree was a small one, and I plowed right over it; while at the same time doing major damage to the front of my truck.

After the truck had come to a halt in this fashion, he told me to get out of the truck, and run back and forth up and down the block, like a running coach would, and that in future I would remember this day as something of importance, indeed of Biblical like importance. I did as he instructed; after which he told me bring foam or spit to my mouth. He then instructed me to go to a specific house (I don't now recall the exact address but it was on the east side of 19$^{th.}$) and ask for a towel to wipe myself; as well as had me say something silly to the person there (who did give me a towel as asked.) Again I did as the ghost said, but by this time, I was just beginning to realize what trouble I had just got myself into. As well, by this time people were coming out to inspect what happened, and before long police and ambulance arrived. Some of the details that happened are a bit foggy to me now, but basically they took me down to Swedish hospital (in Ballard), and a police officer talked to me there, but I didn't mention anything to him about the ghost or pseudo-Jesus. I do remember that the police and ambulance people seemed to act somewhat strangely as if (it seemed to me) they knew in advance of what was going on, and it occurred to me about then that I had been set up. This said, for some strange reason, I did not really think to condemn the ghost out of hand; as I still wasn't quite sure who he was, or if he was or represented who he appeared to be and or represent (i.e., Jesus.) Not surprisingly, I felt very stupid As I recall I think I said to the officer I just went crazy or something like this as my explanation for what happened, but that I had no previous history of such a thing (which I didn't.) [118]

---

[118] It was claimed by someone that I hit them along 19th, on the basis of which they filed a claim with Geico and received damages. This claim is a complete fraud, I hit no moving (or stationary)

Again some of my memory is obscure on some (though not all) of what happened. But as I recollect, I was uninjured and was released from the hospital not long afterward. The ghost had left me about the same time the police arrived. The thing was so outrageous and hurtful to me that at first I didn't think I would actually be prosecuted; since I had been criminally set up (as I then and do now maintain) and I thought for them to do more to me was unnecessary aggravation. But sure enough I did in the following weeks receive notice to go to a hearing; where a date was set for trial.

I will just observe in passing that it is typical of Hell people to make light of their own wrong doing to insist on the letter of the law with their enemies, and this is what I now found myself dealing with.

Without going into any great detail here (at least at present); after some preliminary hearings I had a trial a few months later.

Each time they held a hearing and finally and including the trial, my own case was reviewed last before all the other cases present; so that by the time it came up the court room was empty of all but only the most necessary people. To make a long story short, my trial was a literal farce and a pre-arranged stage show. I remember the lawyers and judge saying silly things back and forth which seemed like legal-ease, but which was really a lot of nonsense intended to make me look foolish, and as if my hearing or trial (in the given instance) was joke – which it was. My court appointed lawyer, Benny LaMandola, I felt then and believe now, and I am inclined to believe as well the Judge Holyfield (off hand I believe that was his name though I would have to check) had some idea of my being (at least in some way) set up, and were participating in a charade. When at one hearing I said I wanted to get my own attorney, I was told by Judge Holyfield I had to stick with LaMandola. At my actual trial I protested this, and accused LaMandola of being crooked; saying I didn't trust him for my attorney.[119] I tried to get the names of the court reports and prosecution attorneys, for reference to them as witness of what was going on, but the prosecution attorneys were never the same (though they did give me their names), and at the actual trial only one of the court reporters or "stenographers" (of the three or four there), named Yvonne, would give me his/her name.[120]

I suppose one way I would describe the attitude of LaMandola and the judge was that they wanted to get me out of there but on the formal condition I see psychiatric people who could say I was not a threat to anyone. Some people

---

vehicle, and I expect the recipients of this insurance money were the same people (or connected to the same people) who set me up.

[119] On a couple of occasions Mr. LaMandola conferred with my father, against my own wishes; as if I were incompetent to address my case, really in such a way that I thought was needlessly insulting and condescending, though ostensibly well meaning in his motive.

[120] I think I have her last name written down some where in my mound of papers on these things.

from King Country Crisis Management eventually saw me; said I was ok, and that was essentially the end of the matter.[121]

A week just before my final trial, I slept for four whole days (or at least given to think I had.) In other words I went to sleep on Thursday and woke up on Sunday, something that has never happened to me before or since.

Meanwhile, the ghost had started making regular visits to me. Among the kind of things he was telling me was that I was to be the new Jesus (as if to replace "him," i.e., the ghost.) It didn't take much resistance to attempted mind control, even under such circumstances, for me to realize this was ridiculous, and I told him so, but still he persisted.

It was about this time that all kinds of different spirit people came to visit me, but mostly this ghost who turned out to be "Simon the magician" or one of the "Simon the Magicians" referred to in my "New Treatise on Hell." The events of this period are so bizarre I will forbear attempting to describe them at present; other than to mention that Magus, Gyro and some other spirit people were regularly engaged in mocking, making fun of and harassing me, while at the same time appearing to be friendly.

There are possibly some reading this who might think that these spirit people were somehow divinely sent. Let me be emphatic then, yet without addressing too closely the question, that I think these spirit people, generally speaking, and those I have dealt with are basically hoodlums; who want attention and who mooch or leech on people, emotionally, psychologically and socially. Though sometimes they will pretend (to me, rather unconvincingly) to possess some higher wisdom and understanding of things, really they are just con-artists who use people for their own selfish purposes, and no spirit person is deserving more respect than anyone else unless it is clear they are conducting themselves honestly and morally. At the same time, it is wrong to assume that those employed by Hell all look like devils; when in truth certain of them can make themselves look or appear as if they were Heavenly. Another thing that makes them persuasive to people is their *persistence*. If Hell people could be said to possess excellence, it might be said to be in this category. However, I myself attribute this to their arrogance, desperation, and hopelessness. Though such persistence is bound to make a strong impression on a person, one must never see it as justification or excuse for how these sorts of spirit people act.

As I stated, the many of the things I went through during this period of meeting spirit people and conversing with them was extremely weird, and I don't know that it is especially worth anyone's while at this juncture to go all that much

---

[121] This was Steve Woolley, and Saskia Von Michalofski. I also spoke at some point to Mekka Robinson a probation officer, and to Bob Powers of Western State Hospital a psychiatric examiner, both of whom I did mention the ghost to. At my trial itself I didn't mention him, however, but without lying or distorting the truth otherwise.

into them; though perhaps later I will think differently or else some will be curious and inquire with me. I will mention though a few occurrences that happened. I continued to be plagued by various kinds of assaults, including (in so special order):[122]

* Making a mess of my house, particularly with witchcraft dirt.
* Shrinking clothes so that clothes became un-wearable.
* Soiled a living room chair by (apparently) placing a drunk transient in it with bad hygiene.
* Did things physiologically to prevent me from going to bathroom.
* My computer was regularly and incessantly assaulted with hacking, viruses, and both obscene and gibberish e-mail. I have logs from my firewall program listening some of the ISP phone number location my computer has been attacked from. I have had prank e-mails sent out using my e-mail address, and have had e-mails bounced back to my mail box (i.e., as undeliverable) which I hadn't even sent, but which had my e-mail address as the sender.
* Removed Amnesty International sticker from my vehicle.[123]
* A full bag of groceries was stolen out of my kitchen.
* Dirty coat covered over with witchcraft dirt left in my house.
* My communications continued to deteriorate, receiving even less e-mail, phone calls and regular mail.
* House plumbing interfered with.
* Wallet and keys stolen a number of times; had to get my credit card replaced 3 or 4 times within a single six month period.
* Cat brought in a large bug, such as you might find in a Central American rain forest or remote southern desert. It was an inch to two inches long, and looked very strange. Being so out of place as it was, I remember I felt sorry for it.
* Forks – all but one – stolen. This has happened twice.
* I would find the lace in my sneaker (tennis shoe) given over to one side too much, so that one end of the lace was much longer than the end opposite to it. But more than this a knot was made at a point in the lace just outside the upper lace hole. In other words the problem could not be fixed unless you unlaced the knot someone had tied there
* Locks broken on my pickup.
* "transmogrifier" (as I called it at the time), or a feeling like a tight rope around my heart, and later other similar internal contractions and stranglings of various kinds carried out by sprites.
* Papers, records stolen.

---

[122] Some might object to a given one of these that it occurred in isolation, and had no special relation to what else I recount. My response is perhaps that very well is so, but that it won't hurt to have listed it along with the others as *possibly* connected to my main story.

[123] On a similar note, I might mention that not long before this juncture I had had a bumper sticker made which quoted John 16:11: "The ruler of this world has been condemned. " This was before becoming directly acquainted with spirit people.

Now one story involving spirit people I will tell here in order to provide information on a certain phenomena or experience that some people may or might have done to them and be taken surprise by.

My present home is on a little hillock above the street; so that the garage is below the house. On sunny days I will sometimes lie down and taken in the sun on the roof of this garage. At the time of which I am speaking, Gyro and Magus were telling me I had to make a deal and work with "Microsoft." On that same occasion, I was waiting to face an inquiry (really as it turned out inquisition) by spirit persons from the "House of Israel." (a sort of spirit person manned examination board.) As these matters were being variously discussed between these people in myself, I think I refused something they demanded of me at one point, and afterward while lying on the garage roof sunning myself, and after talking casually with some of these people, some "demons" shouted in my headed for about 5 to 10 minutes, berating me, Their voices sounded like Arthur Brown and really were incredibly belligerent and hostile. Because at that time I was still somewhat (though not entirely) believing of these people, it made the experience more painful to me because I wasn't sure if *I* wasn't possibly doing something wrong (by refusing what I had been asked to agree to.)

Now being without a car, the truck basically being given up on as too damaged, I could not go to feed the homeless cats at the Grey Top lot in Magnolia as I had for the prior 3-4 years. I tried one very early morning riding a bike up to Magnolia (from Ballard) for that purpose, but found this too onerous. I then tried the bus, but even this was too much for me; especially since the Hell attacks involving spirit people were reaching their height. Thereafter I could go to feed them no more, and what became of Mother Kitty Kates[124] and all her kids (plus some other cats) I don't know, but naturally I felt very sad about it; as I had tried to help them.

Yet, as mentioned I managed to adopt six cats from the lot by that time, and, again, they were:

Daniel Snugby
Peanut Berry
Pete Jubilee
Huggin Coat
Spivey or as I also called him Jenkish Jenkington.
Covey Cub

These were in addition to Jospeh Skatey (who had been with me most of the while since moving to Ballard in the early nineties) and his sister Neffy McKee. It is easier for me to tell what happened to the cats (who I call "the kids")

---

[124] She was a tough cat and a survivor. A more furry than usual brown tabby who she would sometimes run (up to) four whole blocks at a time to meet me when I drove down to the lot, which was funny.

each one at a time, and therefore I will do so, though what I give here is for practical purposes very brief, and a much longer story could be told about all of them. They are given in the order I lost them.

* Daniel Snugby (kidnapped or disappeared)
Daniel was the first I took home, and lost him before adopting any of the others. I was in real tears after that happened. He was like a gray and white cuddly doll.

* Peanut Berry (orange brown striped tabby)
Was a fragile runt sort of cat, who was violently poisoned, and had to be put to be put to sleep after a prolonged seizure too great for his tiny frame. Emerald City Emergency clinic, on Stone Way in Fremont, did a test when I brought him in and found he had a high toxicity level; confirming that he had in fact suffered from poisoning of some kind. I remember I used to pick burrs out of his fur; as he was rather a helpless sort, and couldn't or didn't do so himself. I buried him out near Sultan in the woods.[125]

* Pete Jubilee (furry, Black, brown mix, with white in his mouth and feet)
A heroic and extremely cute little cat[126] who tried to protect some kittens he had. I say he had because these Hell people made him pregnant by means of a sort of sex change operation. I checked his sex when he was very small, so I know he was male. He ultimately had to be put to sleep because the goon sprites kept attacking him, and the kittens also, who were finally taken down to the Seattle Animal Shelter. At first Pete took the kittens by the neck from outside the house to under a neighbor's shed next door. One time I went to check them and the poor kittens had sprites in them making them look snake like, and possessed, like something out of Stephen King. Of course, it was not the poor kittens fault, and they were made to suffer in various ways, including not being able to go the bathroom. (Ballard Animal Hospital on Leary Way)

* Spivey (gray striped tabby)
He was cripple in one leg and had to hobble around. Because he had difficulty moving he seemed like a cat that thought and reflected, and there was a certain intelligence, if goofiness somewhat in his eyes and look, and behavior, and would sort of yowl somewhat expressively rather than meow as such.
For years he had survived out on the lot like the others, but his case he had difficulty walking, and there were a numbered of times he would come out for his food even if there was searing north winter wind blowing in, and I would see him fall on the rocks or in some icy puddle because his legs couldn't properly carry

---

[125] The remains of the others who were put to sleep or passed away, were disposed of by the facility where they were put to sleep, whether it was the Seattle Animal Shelter or a veterinarian.; except for Covey Cub who I buried in a garden bed on my property. I would have preferred to have buried them all out in some far off, quiet place in the woods rather than have their remains disposed of institutionally but after Peanut Berry was put to sleep I didn't have my truck and the grounds of my property are very small; as well as my living circumstances very distressed and violent.

[126] He used to sort of chirp rather than meow, and was a pure innocent.

him. He had lived outside for so long that when I finally captured and brought him into my home, he seemed beside himself to be able now to live on a carpet, and have regular feeding. Because he was odd the other cats did not always take to him, though he was always friendly to them, if sometimes somewhat obstreperous, and sometimes I would see them play wrestling together.

Like Joseph Skatey he too could not go to the bathroom and his stomach accordingly bulged. In his pain from the sprites he sometimes seemed to specially appeal to me for attention and I would try to pet him and console him, but really they were so wearing him down that I had to have him put to sleep. The Hell people did a particularly vicious prank on Spivey, which for convenience I will refrain from telling it or else save for another occasion.[127] (Crown Hill Veterinary Hospital)

---

[127] The following is my summer 2008 account written for my website at gunjones.com of the malicious prank relating to Spivey alluded to here. This described "prank" occurred in about July 2000:

"There was this one cat of mine, "Spivey" (a nickname I gave him), who was a former feral cat crippled in one leg whom these witchcraft people delighted in tormenting, and when he came to me, brushing up against me for help, but there was nothing I could do -- all those great police, lawyers, professors, priests, government people, activists out there could or would not help me after many years trying to get someone to come forward and do so. As a result, I was left fighting literal Hell alone -- while being subject to brain torture radios, and ongoing harassments of various kinds, including vandalism, robbery, poisoning and being given diseases to name just some.

"Now in Spivey's case (and this was true of some of the others cats as well), they did thing like have it be struck or crawled over by sprites and prevented him from going to the bathroom (so that his stomach swelled -- as happened also when the other cats were subject to this treatment.)

"On a number of occasions I have had my cats abducted from me during what was going on then (and two of them never saw again as a result), and it so happened that one day I found Spivey missing even though I had left him locked in the house. He was gone for a few days and then suddenly turned up. It disturbed me, but because this kind of thing had up till that time been going on a long while I was forced to take it in stride. About the same time or a little time before my younger brother (who without getting into it here would assist these people) for no apparent reason gave me a hand fire extinguisher. Although I already had one, I took it but did not understand his meaning or purpose in giving it to me (though suspected something fishy.)

"To return to Spivey, he on the surface and under the circumstances seemed all right; when suddenly one early morning something very strange occurred. He went under the bed in my bedroom, and after a while I heard him make his short yowls (as he sometimes did.) I could not get him to come out from under the bed and in order then to reach him I had to move the bed and pull it from over him. When I did, I found him lying there with three new born kittens, a day or two old, with their afterbirth -- and the atmosphere of the room, incidentally, suffocated and acrid with sprites and some dirty spirit people. Now I knew him to be a male so of course what i was seeing didn't make any natural sense. And yet he was treating them as if they were his own.

"I don't exactly what my very first reactions were, but before long I decided I would have to do away with the kittens since Spivey nor any of the other cats could possibly nurse them. Again they were about a day or two old and I concluded that if I didn't do it they would starve to death anyway. I took them outside and one at a time bashed in each one of their heads. Interestingly, as I did this, and for what reason you can speculate yourself, I had a male angel standing behind me while I was so engaged -- perhaps as if to give me moral support -- though whether his intentions were sincere or feigned I could not tell. In any case they did nothing to console me. Following all this, I buried the kittens in one of the small garden beds of my house.

"I was then left with cleaning up the blood with some rags. But then the question arose, what was I going to do with the bloody rags? What but burn them, since if they should be found it would only

* Joseph Skatey. (gray striped tabby) and * Huggin Coat (a dark brown yet almost blonde very furry tabby)
Was regularly vomiting, could not go to the bathroom, had sprites in him; so to spare further unnecessary suffering I had him put to sleep. Because he had been with me so long, Joe was just about my best friend in the world. The same with Huggin Coat, he too was put to sleep to spare him further Hell people attacks (after about 2 or 3 months worth that is) Both of whom were extremely loving and friendly, Joe being a sort of the chief cat of the whole house, and Huggin Coat being most serene and affectionate. (Seattle Animal Shelter in the case of Joseph Skatey, and Ballard-Greenwood Veterinary Clinic on 15th NW with Huggin Coat)

This left me finally with

* Covey Cub
(a mostly black tabby with white mouth an d feet) and Neffy (small brown tabby.) Joseph Skatey's sister. Part of the reason they survived is that they kept to themselves, and went out for long stretches; whereas the others were either homebodies or just babies.
I still to this day have Neffy.[128] Covey Cub was assaulted for weeks by the sprites till he couldn't eat, and finally they violently assaulted and literally choked him to death. Like Pete Jubilee, he was sort of a hero. In his case, he actually fought the sprites back to some extent, on one occasion trying to pull them with his teeth from Jenkish Jenkington's (Spivey) back (of his neck.) Before Covey was specifically targeted these people had stolen some photographs I had of him, and afterward I was repeatedly told "Covey Cub is not someone you want to know;" which basically is an ominous declaration that whoever is referred to (in this case Covey Cub) doesn't have long to live.

For most of 2000, I attended weekly mass (which I sometimes attended on a week day rather than a Sunday) at St. Alphonsus Ligouri, Ballard's Catholic church. Earlier on I went to Blessed Sacrament in the University District in the early and mid nineties; where at one point I spoke about my problems with Father

---

possibly create complications. This I did. However, I could not put out the flames with water and found that I needed to use the fire extinguisher.
"The next day I was sitting in my chair on my lap top as I usually am, when I heard a small, sharp cry emanating from the front outside of my house. I arose, and going outside to look tracked the sound to a narrow cleft where the house overhangs the foundation. Upon examining more closely, I found there yet another day old kitten; this one yelping, helplessly alone, and which (with no other choice) I was forced to kill and then bury in the same manner as I had the others.
"In conclusion then, you (including the City of Seattle who lent them their court room) who have facilitated or made easier the activities of this gang of spirit and moneyed people I write of, perhaps have even knowingly aided and abetted them, get a good idea of who you are getting mixed up with before you get so mixed up."
[128] Later Note Jan. 2014. I finally had to have Neffy put to sleep on Friday, Nov. 12, 2010; owing in no small part to the continued harassments I was still being made subject to. She was about 17 years old by that time. If a cat could be a saint, I would think of her as such a one. Tiny thing that she was; she was so very good and long suffering during all that happened.

Michael Sweeney there. Bt the commute (i.e., without a car at that time) was too costly and fatiguing (remember I continued to be subject to physical assaults, including brain torture radio) going to Blessed Sacrament from Ballard; so I naturally went to St. Alphonsus which was more local.[129]

Father James Gandrau, then pastor of St. Alphonsus' at that time, refused to come to visit me and speak with me for over eight months despite regular monthly entreaties. He simply acted like he didn't care and when I tried to ask why, he gave some casual and not very convincing, or very well disguised excuses. I had some other strange experiences relating to going to St. Alphonsus' but which I won't go into here, except to mention that a number of times while going to receive communion I have been handed Eucharist that in one instance was heavily perfumed, and in a another part of it was obviously bitten off of (though this wafer wasn't something that came from the priest's altar bread.) At another time, when sipping the communion wine I was the last in line to receive it and when I did I tasted an illness, like a very bad cold, in pronounced, concentrated form. Of course, one might infer the cold or disease might or would have come accidentally from another parishioner. But at the time I was not inclined to think so, because of what seemed the circumstances of the service, and because the taste of the disease was so obvious and concentrated that again (at the time) I felt I was deliberately poisoned. If I was mistaken in the conclusion this was only after I sipped, and prior to that I had no thought whatsoever of being poisoned, nor would such have occurred to me as possible in church.

I have also attempted to reach Archbishop Alexander Brunett of the Archdiocese of Seattle a number of times however I believe my mail to him was deliberately intercepted and treated as something he would *not* find of interest; such that I received an indifferent response from his office, for which I still have copies of "his" letters.

In late October 2000 in an effort to find someone who would help me and to also satisfy my conscience that I did something to try and tell people what was going on, I went to the corners of 2nd and Union and 2nd and University in Downtown Seattle, stood on corners there and (after three separate occasions)

---

[129] For the last few *years* now I have been trying to contact Father Sweeney, but for reasons not completely clear he has not contacted me back. I know him to be an intelligent and conscientious person and would not think he would himself deliberately avoid contacting or seeing me. Yet there is evidently an effort on the part of someone to interfere with my reaching him (or his reaching me.) Someone who would like to know the truth behind this can contact Father Sweeney (last I heard he was at Blessed Sacrament, though often away on travel), and see if he will not come to visit me, or if not why not. If he just tries calling me by phone, it is very likely he will not get through as my phone line for a long time has been tampered with -- or so from circumstances one could reasonably infer. In the course of seeking assistance, I have tried calling certain people a few times, and the person who answered on the phone claimed to be them, but I have good reason to suspect, based on their strange attitude and way of speaking, was actually an imposter, who presumably was able to interfere with the phone line – or again so it would seem.. [Later note, 6 Nov. 2008. While it was subsequently possible to talk with Father Sweeney, and to make a long story short, he showed a pronounced disposition to avoid conversation with me.]

handed out at least 200 or more of the following leaflets. I received no response on this other than a couple sarcastic e-mails which at the time and in retrospect and in my opinion seemed to have come from some people working for or allied to "the gang" that was harassing, torturing and tormenting me. To this day it baffles me why after all I have written and sent out about my situation that no one has come to my home to see me as I have repeatedly requested.

Below is the leaflet I handed out on the streets of Seattle in late October 2000. I'll grant you it sounds rather brash and sensationalistic but for its purpose, and given the time it was written, this was I think pardonable. 'Gomez" followed me on these occasions; as if to offer advice or support, and I recall my attitude toward him at that time was becoming more and more skeptical of all of these spirit people (with the understanding that the underling sort of people could in no way compare in culpability to the chiefs.)

"Saving Seattle: Truth is the Solution

"Why am I wearing this badge – a yellow star made of cloth – the symbol of betrayal, brutality and deception? While we think of such problems pertain to people of others countries and other eras, the truth of the matter is in the past decade or so, Seattle has fallen under a strain of the same moral and spiritual illness which helped bring about the tragedy and suffering which this yellow star represents.

"If it sounds strange to say, it is only because much of what has been going on is carried out in the darkness, unknown to most of the general public. While our problems cannot be literally compared with those happenings of over 50 years ago, our troubles are in their way very malignant.

"Bribery, blackmail, extortion, propaganda, thefts, techno-lunacy, interference with communications, put-on jobs, and hocus pocus have become the primary weapons used to make a mockery of our police and judicial system and rob us of our God given rights and liberties. Our persons, let alone our property, cannot even be protected from violence, thanks to the machinations of corporately sponsored organized crime acting in concert with the both willing and unwilling cooperation of city, county and state government. Using tactics of secrecy and deception, corporate hoodlums have ruined people's lives, and families, and brought about a generally degradation of Seattle's quality of life in general. These people, reportedly connected with Microsoft and Dreamworks, among some others, have rendered well-meaning civic leaders and citizens helpless to cure these problems that plague the city, and people are left at the mercy of a ruinous despotism, that without mincing words, is completely mad and out of its mind.. Mind control, druggings, deliberately giving diseases, intrusive high technology are some of the tactics used, their very outrageousness is what helps to keep them secret from people. People who would tell the truth about these things are discredited, threatened, framed, and blackmailed thus leaving the rest of us at the mercy of corporate gangsters who want to live the good life off the backs of hardworking decent people.

"I myself am a victim of very vicious violence and torture carried out secretly for the past 7 years. Druggings, burglaries, poisonings, deliberately giving diseases, psychological warfare techniques of various kinds have been used on me. My pets have been murdered. I have tried before contacting numerous authorities and others to tell them about what is going on, but have been effectively ignored and denied any assistance. The purpose of this campaign of harassment and bullying I have been subject to has been designed to discredit and debilitate me. Although a scholar and a published author and historian, I am poor and have been systematically isolated by these people, and thus have been hard put to get any help.

"Recently I was railroaded for a crime of which I was really the victim. It is admittedly a strange story to tell,, but suffice to say I was set up to look crazy, with the obvious intention of discrediting me should I attempt to report or bring a law suit relating to what has been going on. The legal proceedings have been such a mockery and farce that they can more likened to theater than to anything remotely legitimate. It is so bad, that the persons behind this savaging of the law have absolutely no morals or scruples, and don't hardly even bother making their lies sound believable. The case has been dismissed, as far as I know, but it does not change the fact that our legal system is and has been used to perpetrate crimes in order to assist corporate thieves and murders.

"Those people who have sold out our city, and cowardly given in, are a complete shame and a disgrace, and are all the more foolish because the money and seeming security they obtain really only debases themselves. There is said to be some internal effort to solve these problems, but if so it is done secretly and what is really needed is an open and candid community addressing of the problem. It is unlikely that we can expect dramatic change anytime soon, it is a very, very difficult kind of quandary. Yet in the meanwhile it is well that people speak frankly about what they know, become stronger in their virtue and keep a resolute and honest faith in God.

"Please, if possible, keep in touch with me if you can. I have been made isolated, am not wealthy, and it is not impossible that something could "happen to me" by my speaking out like this."

[This was accompanied with my name, address, phone number and e-mail.]

In composing and going over what I have here, I have thus far *largely* refrained from speaking about the spirit people, yet at the time their incessant harassment of me was very much a part of what else I've recounted or described. In fact, they could really and justly be consider the cause of any and all of these things, whether in inflicting pain on me themselves or having someone else do it for them. For this reason, I have normally in recent years taken or have tried to take an attitude of greater leniency towards others' wrong doing in all this because I realize and appreciate how difficult it would be for some people (particularly the immature and irrational) to fend off and overcome the (often) sophisticated manipulation and bullying of ruthless and high powered spirit people, such as (the so-called) "Simon the Magician."

As yet I have not describe much of what they did and how they acted. Part of the reason for this is that I still have to deal with these people, and as a result try as much as possible to forget them. Not because I find it impossible to think about or recollect them, but there is such a thing as *too much*, and some of these people (thinking particularly of Magus") are notorious for seeking attention and wanting you to think about them and or what they are doing (as in, say, their gossip or scheming.) Consequently, I don't plan on writing about the spirit people at this time unless necessary or otherwise somehow prompted to. But, as I said previously, I may change or modify my position on this in future and decide to tell more.

Yet speaking generally they were led by some main warlock ghost or other; who is typically referred to as "Simon the Magician" but in truth it is not always clear who the given ringleader of the moment is. Typically, they have and do subject me to brain radios while regularly sending over demons or sprites to annoy or defile me in some way (say, by just their presence in or next to me.) For odd "seasons" these people will be extremely cruel and do the most sadistic things. At other seasons, they will be cruel but in a routine way without malice. Other times they are merely annoying. At yet other times they are very cruel and annoying (both.) But above all, and in all, I find the group I have been dealing with tiresome and a glorified nuisance (aside from a number of them being truly heinous and monstrous criminals.)

Some spirits that have worked for "The Monster Maker" (one of my names for their greatest leader) and "Simon" are not so bad and in many ways likable. And while I will often curse the head ghosts, I'm usually indulgent of their slaves and dupes (even though they did some awful thing or things themselves) because they are exactly that slaves; are often quite wretched, and obviously do not have much say in much of anything, and are often themselves "possessed" or infested by other spirits and demons.

The "Baby Jane" Hudson character of the film with that name in the title, is a very good example of a pronouncedly "Hell" influenced person (whether spirit or a regular person), and could actually be studied as such (though obviously a fictional character.) She is a good illustration of someone who had spent years listening and following directions or advice from a Simon the Magician who followed her around who explained to her how things "really were" and she believing him, and he sending various demons in her to better take charge of her thoughts, feelings, and reactions. She meanwhile would keep to herself any knowledge of the Simon in question or spirit people generally claiming ignorance. It is not difficult on this basis I think to construct a useful character sketch of the sort of spirit person who would do what he did to her and why he did it; bearing in mind that someone down the road very likely did something similar to him. He could inform her of any number of secret or difficult to find out things and she (as hideous as she might appear to people) might be able to obtain great power over others by this information. Not necessarily because she by herself desired great power, but possibly because the ghost did, or some combination.

One thing that is also important to emphasize is that these high-powered spirit people need money to conduct their schemes, operations, and run their organizations just as regular people do, and they ultimately obtain this money and consequent ability to interact with regular people through and by means of regular people.[130] Persons have to be paid when it comes to spirit people just as with regular people. Also understand that if, for example, a philanthropic group has an organization to aid homeless youth, the Hell people might just as likely actually have one to ruin them, and with the same corresponding staff and bureaucracy, or at least for practical purposes one is justified thinking so. Further worth mentioning, these people can take the side of any argument, and pose as any party – often times convincingly, depending upon their audience.

Despite all I have written (both here and elsewhere), there are still people who would rather believe these spirit people than my own explanation of these things. For instance, some people I have dealt with no matter how appalling or atrocious the things I describe are will brush them aside as "no big deal." One of the reasons such people take this attitude (at least for some of them) is that some people can't help but see spirit people as superior and naturally more wise than regular people, and (in a given instance) possibly have been told that I don't understand how things really are. For example, I might come to someone relating what I have been through, say poisoning, and they will react either to make a joke of it or else take an attitude that I am somehow to blame for what happened. They might not express these attitudes explicitly, and instead on the surface show concern, while just below the surface expressing indifference or else disdain. Now some will react this way normally. But I know in point of fact that for more than a few such as I have dealt with in this way, it is because they listen to one of these ghosts or "Simon the Magicians" (or perhaps someone else acting on his behalf.) Typically these people, if not entirely irrational, are capable of reasoning of only the most superficial and limited sort. They don't reason; they just "know." And as far as the kind of people I am describing and have dealt with are, they "know" better (than me) because they listen to these spirit people – perhaps have been listening to one of them for many years.[131] Certainly, and any rate, that is my strong and distinct impression.

---

[130] Naturally and in the same vein diseases need a regular food source, and it can be that by locating this and depriving the disease of it, that a disease can be reduced or eliminated.

[131] On a related historical note regarding persons who listen to spirit people, it seems more than probable that a large part of the problem with idols and graven images, as referred to in the Bible, related or listened to spirit people speaking "through" those idols to people; with the person then interacting with the idol (actually the spirit person.)

Were it possible to actually reason with such people (which with most of them it frankly isn't, including some highly educated people I have spoken with), I would respond to what I see or understand to be their arguments this way.

1. "Although you have been put through a painful ordeal, you are more honored (than are hurt or have suffered) by such attention from such important people (i.e., the spirit people.)"

For all I have written on this subject I frankly wish I rather had not met any of these people (though granted; as I have stated, some of them aren't so bad, so for them taken *by themselves* this remark does not apply.) As a matter of fact, I feel cheated having had to spend so much time with them; while at the same time having had to miss seeing and dealing with other (regular) people who, by far, I would rather have seen instead. These spirit people when they obviously manipulate and or force themselves on others are in truth, and at best, parasites who, as they see it, have nothing better to do with their time than leech on, use, and rob others. Yet they might do this while pretending to be divinely authorized to do so, and support this claim with amazing heavenly visions and or displays of extraordinary powers (such as predictions that come true after many years, deja vu or causing or seeming to cause an earthquake or the weather to change.)

As far as I am concerned these visions and displays of which I have seen many are phony copies, con-artists tricks, or the results of "technology" as yet unknown to us. "By their fruits ye shall know them." And this is what I go by – not these visions, shows or manifestations of say cherubic faces, glowing angels and heavenly lights, golden clouds of heaven, and more. Not only ages-old spirit people and sorcerers, but regular magicians and Hollywood special effect artists can achieve many of these same effects, through technology and other sorts of perceptual manipulation. Power, by itself, says nothing. It is character or moral character that says everything. Unfortunately, and in my experience, such a stand or perspective is apparently in the minority. Ordinarily I take the view that organized Hell (including pretend God and or Heaven) is made up of phenomenally wealthy spirit people, who obtained such wealth from assorted crime, murder, terror, and deception, and who use such wealth and power to persuade people as to their legitimacy as divinity, or else legitimacy as supreme power.

2. "You will receive a great reward for your suffering, and therefore it (the suffering and injustice) is of no great consequence." This by the way is a great "Simon the Magician" kind of argument.

While as a Christian one would of course see a certain truth to this, coming from devils or devil influenced people it is not at all the same thing. For one, I don't need to suffer and pay for what is already mine. Yet these spirit people think nothing of charging others for what already belong to the victim or target of their robbery, etc.

My attitude toward these spirit people is quite simply "who needs them?" The reality is I didn't need them (at least those that afflicted me) for anything, and the fact is and as far as I can reasonably surmise, I was basically sold out (by others) to feed some powerful or influential spirit person who others see as someone of great importance, but who in reality is really nothing more than a monster or vampire, or a spirit person "Montgomery Burns," yet one who can dress in himself (in the eyes of some) as if he were divine, or if not divine, as such, he is seen as representing a higher order or power.

Someone might say I only have seen the bad spirit people and not the good ones (the assumption being that the person making this argument *has* seen the good spirit people), and therefore my view is distorted. The fact is I have seen all kinds of "heavenly" visions and beings, some of them actually somewhat convincing. But glowing ethereal figures, large winged angels, saints, bearded prophets, "divine" lights, rapturous religious feelings – these don't prove anything, and I know better than full well that the devil can re-produce any of them. What he has a harder time with is justice, honest rationality, and long term, consistent moral character and integrity. Understand that Hell employed spirit people can be made to seem kind, friendly, understanding, benevolent, wise, generous and supportive, and only a thinking person is going to see past the charade. For others, they simply do not think, and take such shows or impostures as the real thing, though in fairness it should be kept in mind that mind control of some sort might be being used on them.

These spirit people are supposed by some to be so great and unchallengeable, yet:

a.) they operate in secret and have others do so as well. Why the need for secrecy, indeed outright fraud and willful deception? To teach some higher lesson? In all my years dealing with them, I have taken no especially great lesson from these people other than to discover that people can be a lot more rotten than one would otherwise have imagined possible.

b) all this past decade I have been combating them and their hench-people (or "Renfields") they have been the ones with money all this time, and people to assist them – not me. I had no money (worth speaking of), and had no one to talk to let alone assist me. Yet I have managed to survive "survivor island" now for over a decade despite innumerable and incessant attacks of various kinds.

I have received good advice or a useful tip from a spirit person (including "Simon the Magician"), but such as was no better as any I could have received from a regular person, and in any case the benefit of it, all in all, was relatively negligible. No, if I possessed any wisdom in all these years it came from books and my upbringing (which included the church.) Someone who gets most or a large part their understanding of things from conversant spirit people I would say is someone who is, or is not much better than, a lunatic.[132]

Yet observe, many corners and quarters of society are run or managed by such people, that is "regular" people who listen to spirit people, hence many of the problems, often times appalling and atrocious, we have.

Not least is the point in all thus that others have and are victimized by such people and such experiences, both people and animals. They in many instances could not possibly have the say or understanding I have in relating these things; so that in telling these things about my own experience I hope it is very clear that many others could say similar or perhaps relate even worse done to them or those they've known. But how on earth will they explain and then get people to listen to them about such people and such things? It can be so bad that at times it seems one needs a certificate admitting guilt signed by the criminal himself and publicly notarized before law enforcement or others will take evidence from you on the crimes involving these criminals, spirit or otherwise.

Assaults of various kinds continued on me and or my home of which the following are some that I had recorded at the time and taken from a letter I sent out in March 2001:

"For the past year I have been subject to quite a number of instances of ongoing harassment, burglary, and on a few occasions food tampering. It is not possible to get into the why and what this is all about here just now. But suffice to say the purpose of the persons harassing me is essentially political, with the aim of intimidating and discrediting me, lest I bring law suits and get started criminal investigations against them. These persons are very wealthy and powerful, and are extremely ruthless and amoral. They are so 'far gone,' that for them doing the 'wrong thing' is doing the 'right thing.' That there are aspects of my story which, are quite bizarre and incredible there's no denying. Yet at the same time, these are not so beyond evidence or explanations as my opponents like to casually and lazily claim. More, however, on the overall scenario later.

"In the meantime, what follows are, essentially, the most recent events relating to this ongoing nightmare.

"Somehow or other the perpetrators found ways of getting into my house, 1604 NW 70th St., and got hold of or made copies of my house keys. This was not all that impossible for on a few earlier occasions I had left the back window open

---

[132] Worth noting is the fact that the more clever Hell people can take and represent any interest or any side in any argument, sometimes convincingly so. Thus, for instance, if we all came together and agreed to fight "Hell," it probably wouldn't be long before a Hell person infiltrated the ranks and used the cause to further his own ulterior interests, perhaps even becoming a respected and recognized leader in the process. This in one reason why elsewhere in my writings I strongly urgeand recommend that proficiency in logic needs to be made an integral part of *basic* education, along with reading, writing and arithmetic.

when I had gone out, and it would have been a relatively easy matter of their climbing through there to get in, and thus get hold of a spare key from a drawer. At one point I did get new locks installed (a number of years ago), nevertheless their ability to get to my keys was somehow repeated again (probably from my absent mindedly leaving that back bathroom window open on a subsequent occasion.)

"In about the middle of last summer, I discovered that my wallet was missing. I called the Seattle Police. They came, took a report, and gave me a case number. This case number incidentally was later deleted from my computer files, and when I made inquiry with the police about it some months afterwards, I was told their was no report of any such case. A short time following this incident, I found that my house keys were missing. When I looked at a place behind some books on a bookshelf were I hid my spare keys, I found the missing wallet! The main set of keys themselves turned up – open to view – on a table (or couch) a number of days later. Needless to say, I did not somehow negligently put my wallet in that location. Invariably I keep my wallet in my coat pocket or on a specific spot on a book shelf where I know I will find it.

"Oct. 20 or 21, 2000, on the lawn of the east side of my house, I found a mutilated squirrel. What was especially unusual about this was that he was surgically mutilated and deliberately skinned with his bowels turned inside out. I called the police to report this. Officer D. W. Umpleby, badge #4852, came in response. At first he surmised a dog might have done such a thing, but upon inspection concurred with me that it some person had deliberately done it, given the way the squirrel was sliced up. Even so, he said he would not take a report as he did not consider it as a crime as such. He did say however he would enter it in his regular log book.

"Oct. 27, 2000, Upon using some food items, bottles of ketchup and mustard, I found them poisoned with a disease, like a concentrated "cold," (like something from the Center for Disease Control.) You could unmistakably taste, even smell it in the ketchup and mustard. This had actually happened before, though I did not call in about it. I contacted the police, having the bottles as evidence. Officer Dolan came to take my account. He said he saw no evidence of burglary, and refused to take a report. When I offered the ketchup and mustard bottles as evidence, he said I would have to go somewhere like the University of Washington laboratories to have it tested as the police had no facilities or means to do it.

"Oct. 30-31, 2000. Between 12 am to about 3 am, I was poisoned with something, causing indescribable and excruciating pains in my chest and ribs. During the day I had only some water and some coke to drink. As best as I can tell, it was the coke that had been tampered with. During the course of this seizure, I vomited and up came some yellow power, like barbiturates or something. The experience of it all was so awful, I literally thought I was going to die.[133] Later in the morning I called the police, and an Officer E. R. Haggerty, #6413, arrived. After telling him what happened he said there was no evidence and would not take a report. I requested he at least log it in his book, while at the same time I gave him a written account I had of what had been going with respect to the harassment and assaults against my person and home.

"My home had been got into on other occasions which I did not report. Among other instances I have had my computer tampered with, and whole programs deleted. Documents tampered with or removed. I build ship models and have had small pieces of the ship I was working on removed, in a vandalistic way. Even more absurdly, on an occasion in early past summer, I had an entire recycling bin stolen from me, with its contents! About 3 months later, this same bin turned up in my garage, with the exact same contents still in it. Needless to say, I had been in the garage any number of times in the intervening period and the garbage can size container was not in there.

"Just recently, some unknown party infected my computer with a virus which my Norton Anti-virus detected and quarantined: Bloodhound.W32.EP (memory resident, full stealth, triggered event, encrypting, polymorphic). I still have the file if any one is interested.

"Lastly, here worthy of note, the traffic sign in the middle isle of 16th NW and NW 70th right near my home (a few times it was the sign at 17th and NW 70th), has been repeatedly knocked down by someone driving a four wheeler or something over it. This has happened at least a half dozen times these past six months. The police say they don't know who has been doing this. Although I cannot categorically state this vandal is connected with the people who have been attacking me, it is perfectly in keeping with their arrogant, lunatic and intimidating ways to have been responsible for such a thing."

In the following described series of events, which I reproduce here also as I essentially sent it out to various people in early 2001. One of my main purpose in attempting to seek legal addressing of the case was not so much to seek justice or recompense but by doing so initiate a formal investigation and thereby bring out the criminals. Unfortunately, to make the long story short, as earlier no one could or would help me.

---

[133] This same sort of poisoning as later repeated 2 or 3 times later and on one occasion (Jan. 2004) the pain lasted more like 8 hours. These incidents were, I think, the most *physically* painful of all my ordeal. On another occasion they were bashing in my head with radio for hours. This was not unusual only in this instance they did something to make it more physical. It literally felt like wearing an old army helmet and it being struck by a steel headed mallet so excruciating was the kind of headache caused. This also lasted for a more than brief duration, say an hour or two. By footnoting these "attack" incidents, I don't mean to trivialize their seriousness. They do constitute attempted murder as far as I am concerned. Yet having recounted already what I have thus far, it seems unnecessary to highlight or make a special case about them.

"On Monday Jan. 15, 2001 Martin Luther King Day, 6:45 am, I was cooking olive oil, and managed to spill some on my hand, thus seriously burning it. I called 911 and they sent out Fire people, three men with a fire truck. It was my idea to have it merely treated with some spray or powder. The fire person ostensibly in charge, a young to middle age blonde male (I did not get his name), upon examining it said I needed to go have the hand treated at a local Ballard medical facility. I told him that, under the circumstances, I would rather not and just wanted them to put something on it if they would, my reason being I simply did not feel the injury warranted it. At this point, he started accusing me of being crazy, asking if I was on medication – merely because I preferred not to go down to get my hand treated. I told them thanks any way, and would just call Walgreens and get something. I at no time acted improperly, frantically or irrationally. Part of my concern was that I, not being a person with much money, I wanted to avoid any unnecessary expense, plus I frankly did not care for the derisive tone in which he was speaking to me.

"I then went back inside my house, after which they proceeded to call the police and a Harborview ambulance. Five police officers showed up, lead by officer Sgt. Jerry Harris. I told these men, as I told the firemen, that I preferred not to make such a fuss, and instead would rather purchase something from the drug store for it if the fire people themselves could or would not treat it. Sgt. Harris insisted on the other hand that I go to have it treated as I was told. I didn't like the idea but seeing they were going to force the matter I submitted, while formally protesting, to go along. At no time was I physically resisting, or acting over-emotionally about. I merely expressed my disagreement with there being made trouble over the matter. Nevertheless, I went along with what they said. I was taken down to the ambulance waiting outside and ushered inside it. Before, and while the ambulance was being made ready, I told officer Harris I wanted to make a formal report of what had been going on with respect to the aforementioned burglaries etc., (since the subject came up. He told me to contact him later and he would talk with me.

"As I went into the vehicle, the ambulance people of American Medical Response (AMR), a young woman and man, after a few words with the fire person in charge, told me they would have to tie me up. Naturally I objected to this a quite unnecessary. Again I was in no wise at any time physically, raising my voice or verbally resisting (other than some laconic remarks) about my being forced to go to the hospital. When I asked why they needed to tying me to the bed, i.e. my right ankle and wrist, and on whose authority they were directed to this measure, they said it was merely a precaution for my own safety, the male ambulance driver at the time remarking he 'didn't want to work at McDonalds.' On whose authority I was 'tied down' has yet to be determined. Again, though I formally protested, I peaceably acquiesced to their demand. At present, it might however, be surmised that the ambulance people took this measure upon their own responsibility. In passing, it should be noted that sometime later I was told by someone who knew about these things that unless a person was being taken to be committed, there was no justifiable reason for tying someone down in an ambulance like that. My being tied down like that by the way, only increased greatly the physical pain – from my hand – I was already in, as I could not use my one hand to help hold the injured one. After a leisurely drive – no sirens – down to Harborview (not, you will note, the Ballard medical facility mentioned by the fire dept. person). I was brought into the emergency wing. A supervising woman doctor was attending some people there and she asked when we came in what this was about. They told her I had burned. I stated I was formally there against my will, and was desirous of not incurring any medical expense. She said, "if its only a burned hand, he doesn't have to stay here – he can go." Someone then whispered something to her, and after a pause she inexplicably directed to them to have me taken to a hallway for me to wait to have to be treated. At the time, I politely reminded those people that I was their against my wishes, and did not want to be there, though was no in any way making a scene about it.

"I was wheeled in the ambulance stretcher to another bed lying in a hospital corridor. They transferred me to this bed and again tied my ankle and wrist to the bed. Again I politely but firmly protested, without improperly resisting. Sitting in my bed I took from my pocket a rosary I had with to pray while I waited. A security person, (upon being asked said his name was "Sinclair") took my rosary from me saying it could be used to physically attack and injure someone. (He was referring to a 2 cm medal it contained depicting on one side the Virgin Mary and on the other Jesus.) I protested this silly measure aloud to the people there, but did not have my rosary returned to me until some 10 minutes or so later.

"After lying there for about a half hour, my hand was subsequently treated, and the job well-done by a doctor. The hospital people themselves behaved rationally, were very kind and did not see why they thought I needed to be treated that way.

"Some days after the incident I called Sgt. Harris to talk with him about what had been going on, as per our mutual remarks the day of the ambulance business. him. He agreed to set a time. There was some phone tag getting in touch with him and back, but after about a week or so had passed he agreed to arrange a time to come see me. It was arranged that I call him a certain day. I did, he wasn't available, and so I left a message. He did not, however, call me back. A few days later I was contacted by a King County Crisis Commitment Services person, a Bill Bruzas. They said Sgt. Harris had called them to evaluate me While Sgt. Harris later said he talked to these people, he denied actually directing them to come speak with me – contrary to what Mr. Bruzas claimed. It should be noted in this regard that even if I was crazy I could still be subject to crime, and it was only right that Sgt. Harris, or other police officer, fairly hear my story, but this he was refusing to do. But he presumed based on his "looking into the matter" that I was in effect, some person with mental problems, and that he therefore not allow me to make a police report of what had been going on. Originally, Mr. Bruzas contacted me by phone and said I was not forced to speak with him. On that basis, I told him, I did not think it was necessary and asked that he drop the matter, understandably indignant at what I considered Sgt. Harris uncalled-for and rude treatment of me. Some hours later in the afternoon, contrary to what he had said., Mr. Bruzas, along with a woman, S. Megorden [sic}showed up at my house. It was my understanding based on what he said that I was not obligated to speak with him, and that I had told him I just as soon not have him come over. My feeling, of course, was that his was all quite irrelevant. Mr. Bruzas in effect lied, apparently with the purpose of having it go on record that I "refused" to speak with him, thus making it seem there was something wrong with me by not cooperating, when in point of fact he earlier told me I was not obligated to speak with him. Even so, rather than turn him away, and thereby make it seem as if I were being unduly recalcitrant, I invited he and his partner in to the discuss the matter. They were there for about 15 minutes during which time I politely answered their questions, explaining what had happened, including mentioning that Sgt. Harris by doing this was reneging on his earlier promise, denying my right to make a report, and hear my side of the story. It was Mr. Bruzas, by the way, he who told me they did not have a right to tie me up in that ambulance unless it was for purposes of commitment.

"Within a few days following, I called officer Harris about this, asking why he had called those people, and refused to come let me tell my full story. Again in the course of two or so phone call conversations, he refused, albeit politely to discuss with me my situation and told me to go talk to the "Harborview people." His attitude was one of, he had looked into the matter, ostensibly deciding on that basis that their was something wrong with me, and therefore refused to take a report from me about all that had been going on. Without any disrespect to Sgt. Harris, who otherwise seemed like a decent individual, his attitude and that of the police has been frankly irrational. They say I have no evidence, or the evidence is insufficient, yet they refuse to hear my story. As anyone should know, especially some involved in collecting evidence, you can't always justify the relevance of given evidence without a certain amount of analysis and thoughtful consideration, as not all evidence is of a blatant "smoking gun" character, especially when, with respect to a police matter, one is dealing with criminals possessing a penchant for cunning, and covering things up.

"About Myself:
"I am 39 years old, and am originally from Long Island, New York, and my family moved out to Bellevue (Medina) about 1973. I was raised, and still am, a Roman Catholic Christian. In High School I was a Boys State delegate in 1978 at a convening in Spokane. I attended the University of Washington where I received a Bachelors Degree in English. Later I attended Gonzaga Law School in Spokane for two years.[134] I am a published author and have done work for Classic

---

[134] I also was in the Army (National Guard) for a while and successfully completed Basic Training and then Armor school training at Fort Knox, Kentucky, was part of the 1st Brigade of the 803rd Armor Division, and afterward attended ROTC school briefly at Fort Lewis here locally (my

Images magazine, a nationwide publication and Novastar computer Game company. On many occasions I have been a social activist in the past, often writing and speaking out on behalf of animal rights. Having composed a number essays, poems and short stories my first published book in book form was *Mabel Normand: A Source Book to Her Life and Films*. Author, film reviewer and columnist Anthony Slide wrote of this book

"'*Mabel Normand: A Source Book to Her Life and Films* deserves wide readership. No reference library should be without it. It is a gallant and eminently worthwhile attempt to resurrect Mabel Normand to her rightful place in film history.'

"Bruce Long, author of *William Desmond Taylor: A Dossier* wrote:
'MABEL NORMAND: A SOURCE BOOK TO HER LIFE AND FILMS, by William Thomas Sherman, has just been published. Compared to most other books on silent film stars, this is truly a great book – more than a great book, because it stands as a prototype of the way such books should be. If only there were similar books available for dozens of other silent film stars!'

"On the basis of this book I was interviewed and appeared on television on the E! Network's "Mysteries and Scandals" program in 1998. Part of its subject involved an investigation into the unsolved 1922 murder of film director William Desmond Taylor, which involved some police of a kind of my own. I am including it with this letter as a sort of testimony as to my character and intelligence. Presently, the revised and updated edition of the Source Book is coming out and will soon be available.

"In my leisure time, I am an amateur violinist[135] and build historical wooden ship models.

"Nature of the original harassment and violence crimes.

"It is not really possible to get into any real detail here as to what my claims about my being subject to crimes, such as assault and burglary are about, given their extremely bizarre nature and the deranged character of the person (s) behind them. This business has been going on actually for about 9 years now, and these people, reportedly connected with Microsoft and Dreamworks have made my life a living hell. Part of what they have succeeded in doing is isolating me from people, by bribing them blackmail them, threatening them, and smearing my character with them. I have contacted the Seattle Police Dept., King County

---

entire service lasting 9/88 to 1/90). I was however discharged, though honorably, from the army on the grounds of false enlistment. I failed, when I enlisted, to mention a (property) trespass offence from back around my time in high school – or such was the reason I was given. My not finishing law school (once more to make a long story short) related to a dispute I had with my father.

[135] I had taken lessons for a few months from a fine elderly gentleman who lived on Capitol Hill, a block or so away from Bischofberger's, the violin maker there, and got to where I could play some sonatas by Bach and Corelli. tolerably well. However, "they" would grease my bow hair or have sprites hang on the tip my bow, or pull on strings while I played so that it has been some time now since I have tried playing.

Sheriff, FBI (some of the earlier events took place in Los Angeles), and Senator Patty Murray this past year about my overall ongoing case, and have not received even ONE single response from any of them – which speaks either to the influence of these offenders, or their ability to interfere with my mail. The whole thing is so extreme and absurd, I must admit I am at a loss to understand what these people's (i.e., the perpetrator's) problem is. But essentially it seems to be something political in nature, inasmuch as I am one particularly given to speaking the truth who they want to debilitate and or discredit. As of recent years, another obvious motive of there's is to drive me crazy or at least discredit me so that I will not be able to bring criminal charges against them or sue them. In the very near future I will be composing a detailed account of all that has been occurring with respect to my case. When that is finished you are certainly welcome to a copy. Even so, the overall story is not strictly necessary for purposes of addressing my claims with respect to the incidences of Jan. 15, and the incidences immediately leading up to.

"Conclusion
"The validity of my case regarding the prior burglaries, or my alleged being 'crazy' is not strictly speaking in question here – as such. What then is at issue is my treatment at the hands of the Seattle Police Dept., Seattle Fire Dept., AMR, and Harborview security. For even if I am 'crazy,' as they will claim, that does NOT justify:

"the peremptory assumption that I could not be victim of a crime(s), and that protection as a citizen from violence, burglary, theft be denied me by the police.
"my being denied my right as a citizen to make a report of a crime to the police.'
"my being restrained in an ambulance when the purpose of transporting me was not for commitment, nor for reasons of my physically resisting or being verbally abusive to the ambulance people.
"my being denied my religious right to have access to my rosary.

"Of course, it is my contention that I am not crazy or mentally disabled, but rather that this is something that has been falsely impugned to me. Granting me this, it further follows that there is no reason:

"why I should be denied refusing medical assistance for my burned hand. Clearly, the woman doctor on duty at Harborview, when we initially arrived, was of the same opinion as myself..
"why I should suffer the indignity, humiliation, and distress of being tied up, least of all when I was already in great pain from my hand.

"A certain smugness, sarcasm, haughty disingenuousness, hasty generalizations, dissemblings, circular reasoning one can often detect in perjurers and people given to legal obfuscation. Such behavior one will find aplenty if one talks or inquires with many of the officials involved in these incidences. A decided indifference to the truth, at times reaching to an antipathy, is not in the least uncommon. As long as the observer or interviewer is honest themselves, it doesn't require a person with great sensitivity or insight into human psychology to spot a liar when they speak with one of these people. There are things about my case which, admittedly, are not readily establishable or easily determined. Nevertheless, the credibility and honesty of my opponents, however, – to a detached intelligent observer – is certainly not one of those points of dispute."

Though there is much more to relate, I will before closing and giving my conclusions give some description of the difficulty I had trying to get an attorney to take my case of being involuntarily taken to Harborview in the Martin Luther King Day incident. For my own convenience, I will here mostly reproduce notes I took at that time.

"Wrote to Robert Wayne, Luvera, and other attorneys 2/16/01"

In a letter to Seattle attorney Charles Hamilton of Mar. 14, 2001, I list the following attorneys, law firms, or others as having been written to in February.

"Below is the letter I sent out to you previously, Feb. 16, but, for whatever reason, you had not received. I might note that I also sent copies that same day (addressed and introduced differently depending on whom I was addressing) to: Kargianis, Watkins, Werner; Luvera, Barnett, Brindley, et. al,; Paul Kirschner and Assoc., John Walsh, atty.; Robert Wayne, atty; Ron Perry, atty; Levinson, Friedman, Vhugen, Duggan and Bland; Mayor Schell, City Councilman Jim Compton (head of committee on Public Safety). So far I received one letter back from the Luvera firm (I am including the email response), and one from a clerk, Lance palmer, of Levinson, et. al, both briefly saying their firms cannot take the case. Otherwise, I have not, for what it is worth, received any word yet, from any of the others."

"Wrote Mayor Schell Feb. 16.

"Last week of March (or first week of April) wrote KCBA, ACLU (either Tues, Feb. 27)

"Received reply from ACLU, dated April 11, Claire Younker Moe, said they could not take case.

"Contacted Patty Fraser [i.e. at the King County Bar Association.} , Friday April 13, inquired about my case said she had not received it in mail. I emailed it to her, and she said she would have Joan Anderson look at it. pattyf@kcba.org

"Jeffrey Needle – April 16

"Send off letter to Mindensbergs, Gautschi, Thurs. Apr. 17

James Lobsense April 19 (no time, not within his 'expertise'), secretary 'Jenn' on Apr. 19 told me that he could not respond in writing to a case unless he was going to take it. She had asked if she could respond to me on his behalf but was told she could not do so.

"King County Bar Assn. Patty Fraser called Apr. 13 said she did not receive letter, I [had] emailed it to her, she said she would have it looked into Mon. 16th. No answer called 2:30 pm Thurs. 19th they said Joan Anderson left at 2:P00 would not be in Friday would respond Monday 23rd.

"April 20 (had spoke with her a week and a half earlier, no response after Stephani Cirkovich said Mayor did not know about business had sent report to SPD internal investigations

"April 20 Julian Saucedo of Councilman Compton's office had sent report to SPD internal investigations, upon inquiring was told they did not have report, I emailed him a 2nd copy.

"Monday, April 23
Called KCBA, was told Joan Anderson would mail me within a few days.

"Fri., April 27
No response from Joan Anderson, called KCBA again was told by a person 'who worked there a long time' there was no Sally Fraser, this by Josie Bell, Julie, Edna.
Called three times, and could not get through for like 20 minutes each time.
Joan Anderson was gone till the 14th, will mail report Monday as directed by Ms. Bell to A Tanya W. to follow up.

"Contacted by SPD about my case a Lieut. Mark Olson 684-0850 referred to matter by his supervisor, who spoke to and about case to Julian Saucedo, said he did not have my case narrative. I asked he contact Mr. Saucedo and obtain narrative form him if possible.

"Monday April 30
Called KCBA, Tanya not in.

"Tues. May 1
Called Susan Mindensbergs said she could not take case
Prior to that I was put on hold 1120 to 1200 left message for Tanya W.
Finally made contact with Tanya W. Called here earlier about noon, she wasn't there left note for her to call me. Called kcba back about 4 pm, she said she got the note but 'didn't have the number, it was on someone else's desk.' I asked her if I could start from scratch and email my case to her. She agreed. I called her back to make sure he had received them. She said she found out, apparently from a note on Joan Anderson's desk that Ms. Anderson had spoken to 7 or more

attorneys and could not find anyone to take the case. I asked Tanya if she would send me a formal letter explaining that fact. She said she would."

"Called Rick Gautschi, whom I had mailed my case to, asked that he call me back" [but he, for whatever reason, and like most of the rest, did not do so or was prevented from doing so.]

There are numerable and many other strange occurrences and happenings which took place these past 12 years, but which I didn't even mention here, endeavoring instead to relate mostly and only events both specially memorable and or representative of others. To be frank, I feel like an idiot having to live the way I do; isolated, abused, often treated rudely by people despite all I have already gone through. But the fact is, I am not to blame for what I suffer; while at the same time I am happy beyond expression to know that I am an enemy to the kind of people and forces, who to my mind reflect the worst of either Earth or Hell (whether Hell from above or Hell below); indeed, and as far as I am concerned the most horrifying people in the world today. I take great pride in the fact that, with God's grace, I have been able to fight off these people, alone, with relatively no money, no assistance; while being daily and around the clock assaulted with brain radios and other sorts of harassment, while those whom I fight and have fought are ages old, have plenty of money, plenty of people working for them, and various kinds of both spirit people and advanced technology at their disposal. To call these spirit people "losers" as I have done on many occasions seems to me more than warranted, and it is a great shame – to say the least – as well as being completely ridiculous that there are or would be some who would see such people as I have spoken of as either divine or representing higher wisdom. Yet there are such people who do, and it is very odd to think that one of the main reasons this world of ours goes to Hell (as it does) is due in no small part to people's inability to distinguish real from false Heaven, looking more to show and feelings, rather than substance and truth.

My final recommendation to others who do or might encounter such as I have described: Do the right thing, be rational, use common sense, and know that no spirit person is inherently better than you are, nor is any under less obligation to be moral, decent, fair, and reasonable than you yourself – regardless of all fantastic power and sorts of power they might display or possess.

Now a few extra comments I think are in order concerning "The Heroes of Might and Magic." Time and again they have tried to marginalize me with people, yet the fact is I was never interested in them. Yet they were and are very interested in me, and have spent enormous amounts of money in the course of hounding me and getting others to go along with them. They will mock me and say I missed out on life (so far) – yes but that's because (and since the time they started attacking and setting me up) I wasn't fairly allowed to participate or compete to begin with! Many times I have received prank e-mail messages deriding me for not having a major college degree, implying therefore I should not be listen to with respect to academic or scientific matters. My response is that I don't ask anyone to listen to me, but to look at the facts, consider reasonable arguments I have to make, and judge for themselves the truth. Even is I am correct in my conclusions on only a small percentage of what I report, something very seriously and obviously wrong is going on here.

If I am wrong why the need for the people I (have tried to) describe to bother me? I must think with respect to the wealthy person(s) who has acted as (regular person) chief of staff for the ghost or else :Faustus" that he is a sort of dictator-type (like Adolph Hitler or Joseph Stalin), responsible for the most atrocious acts, yet will go on believing that he is someone how helping people.

What at this point can be done? My problems as they stand are:

The terms of my Social Security disability payments include the stipulation that I cannot have more than $2,000 dollars at a time – not all that much to live on for many people, and certainly a preposterous amount to have to combat literal "Hell" with. These people still regularly run brain radios on me, thus severely limiting what I otherwise can do physically and mentally, and routinely hit me up with sprites and demons; though these days I rarely if ever "see" apparitions, ghosts, and spirit people in physical manifestation, palpable to the naked-eye form.

Being without a car now,[136] it is difficult for me to move around. But much worse than being without a car is if I go out to walk or take the bus they will assault me with a demon and brain radio; possibly have something set up waiting for me in advance of where I am going. At the same time, if I am away from my house for too long it gives them that much more opportunity to break into my home, and go into my things (including my papers and documents on these matters, some of which they have stolen.) On top of all this the abuse I have and still to suffer continues to take its tool on me – non-stop now for 12 years – and I am not sure how much longer I can go on like this before becoming more severely debilitated.

My regular, mail, e-mail, and phone continue to be interfered with[137] so that what gets out or what I receive is by approval of (whom I would call) "the

---

[136] Later Note Jan. 2014. I have, since this was written, had a car now as of Jan. 2006; purchased for me with funds from an uncle and with the help of my father.

[137] To give just one example, I sent a (to be signed for) certified mail to Jerry Brown's "We the People" group in Oakland and it was returned unsigned, as if they refused to receive it. The U.S. Postal Service, or certain importantly placed people in the U.S. Postal service, I do believe are definitely involved in some way as an accomplice with these people. It is worth remarking in regard the series of post office shootings that have occurred in the past decades: disgruntled employee guns down supervisor or fellow worker. This sort of occurrence (to make a long explanation short) suggests invasion of the postal service by Hell people, to obviously better control that most fundamental of societal institutions.

commandant." This means I am prevented, especially as an author, from being able to make a living, or to do so without absurd and obviously unwelcome interference and oversight by my enemies. I therefore most urgently need to have my communications properly restored to me, including postal mail, e-mail, and phone.

There are a number of ways this problem might be addressed, but for starters I would like to establish or re-establish contact with the following people. If possible, I would like them to have copies of this Narrative in advance so that (if not already) they can be made familiar or more familiar with my situation.

There are others I might list but do not do so thinking that this list here should, for starters, be sufficient and more convenient for the purpose. Addresses and numbers give are the most recent I have but are still possibly not current.

* Peter Underwood
(or someone representing)
The Ghost Club Society
The Hon. Secretary,
Mr Trevor Kenward,
Pine Trees
26 Dewlands Rd
Verwood. Dorset. BH31 6PL.
Reason: spirit people study and research.

* Local, regional, national Law enforcement generally who might have questions on these matters as they pertain to combating crime.[138]

finally and also…

any and all who have been trying to reach me but who for whatever reason have been unable to contact me.

Now there have been some who have been and are reluctant to speak to me, and this of course I fully respect. Being someone who hates being bothered, I don't want to be an unnecessary bother to anyone else. There is no question that some won't talked to me because they are, in one way or another, threatened, blackmailed, perhaps frightened as well, including perhaps one of those I have named above. Whatever a person's attitude toward me is, I respect that only I want to it from them if possible, and not from someone else who possibly has falsely taken it upon themselves to represent them.

---

[138] I have contacted the FBI a number of times, at least 3 or 4, including submitting complaints about hacking on the internet to them, or personally submitting one of the drafts of my story to their Seattle office, and have never once received a response back from them.

Again, keep in mind, unless you are talking to the person face to face, it *may* be possible that someone might impersonate them (I at least have had this happen.) So in a given instance, if you call or write one of the above named people, or myself for that matter, it is not entirely impossible that you might encounter someone impersonating them. I don't say this is likely, merely possible, and something to be on one's guard for.

~~~****~~~

Appendix A

Correspondence and Miscellaneous

a. Correspondence

(I also have copies of the most of the responses, if I received any, from the letters given below. In addition there were other letters I sent but what is contained here is sufficiently representative of them.)

March 26, 1993

Dear University [of Washington] Hospital Collection Dept.,

In response to request for payment of certain charges for visiting the University Hospital Emergency Room, I must inform you that I neither can nor will not pay it. The reason I can't pay it is that, quite simply, I'm unemployed and broke. However, were this the only reason, I would still pay you back in increments as I was able to earn some money. My credit has always been good, and I have always paid my debts when I am in a position to do so. The primary reason then that I won't pay it is because the `care' and treatment I received in the ER was not only negligent, but criminal, as I will explain here shortly. I thought that by just ignoring your initial collection notices that I was doing you a favor by not suing you instead. Were I not poor and preoccupied with getting on with my life as I am, I would bring an action against you. As this is not the case, I thought it best to ignore the bills. Yet since, however, as indicated by a phone call I made to you regarding your notices you've said you would like the truth, I'll give it to you.

What brought me to the ER was at least four separate cases of my having been poisoned. Now it is true that I had taken Sudafed and Benadryl (as prescribed by an ER staff member) for a respiratory ailment, however, the effects of the poisonings were similar to street drugs. When I was in high school, I experimented at one time or another with cocaine, speed, and hallucinogenic mushrooms. I was not a repeat user, but I had at least tried them once to three times each. Consequently, I am familiar with the effects they cause. The effects of the poisonings then were very much like the effects of these drugs, if not identical. It is my personal belief then that the poisonings and the treatment I received in the ER was some kind of twisted extension of the war on drugs of which I was made a victim. I do not know this is the motive for what happened to me in point of fact, but this is my best guess. Why me? I have been smoking marijuana for the last five years and had only recently stopped in August. It may have been that the poisoning and treatment I received was done with the approval of one or both of my parents. My parents, of course, as would be the case if it were true, deny any such involvement. Nevertheless, I offer this here as a possible explanation for the poisonings which brought me to the emergency room. In any case, it is not the poisonings themselves which are my cause of complaint against you, only the reason for my being in your emergency room. When it comes I right down to it, I honestly can't say that I know exactly why I was treated in the outrageous, negligent and criminal manner I am about describe. I offer this business about "the war on drugs" as being a possible motive, not a known and established one. After all it, one would be hard put to describe any sane or rational explanation for why people, especially "physicians," would treat another person so.

What happened to me in the emergency room:

Staff was, for the most part, callous and surly, thereby contributing to the stress I was enduring as a result of the poisoning. On numerous prolonged instances, I was made to wait for a doctor to attend me. When I was attended to it would be different doctor's or nurses, practically none of whom appeared at once together. I can understand ER staff being busy, but the nights I was there the place was practically vacant of patients. But what's more, when the doctors would leave me, I would hear them in another room laughing and making friendly chit chat with each other as if they had nothing better to do. This was after I was told they needed to seriously discuss my problem amongst themselves. I would think it would be common sense that a person who was poisoned would be suffering trauma, why did not those in charge of the staff, at the very least, take this into consideration?

I was hooked up to an EKG machine and tape was used. The tapes and wires were placed on my chest. When I went home afterwards I went straight to bed without undressing. When I woke up next morning and did get undressed I found an EKG tape attached to my scrotum. When I was in the ER I began to feel very drowsy after a while, it is my belief that what happened on this occasion was that I was put to sleep, or I fell asleep naturally thus making this act of assault possible.

The treatment I received for the poisoning was different each time, and much of it seemed to have nothing to do with what was ailing me. For example, I was made to listen to a psychiatrist rather than have a blood or urine test which would identify whether I had ingested a foreign substance. I feel, in retrospect, that these efforts in the way of psychiatric counseling were done to discredit my credibility should I protest what was done to me.

On one of the occasions, I was placed on a table and the 'doctor' felt out my entire body, from head to foot. Because I was scared and drowsy from the poisoning, and assumed he knew what he was doing, I didn't not resist. However, the feeling I came out afterwards of was of my having been sexually molested.
To this day, I don't see how 'feeling me out' (i.e. my entire body - from head to foot) would serve anything in the way of diagnosis or treatment for a case of poisoning.

On one occasion I was given a medication to take with some 'instructions.' Later when I went home to take it, I discovered that these instructions neither named what the medication was, nor described in what dosage it should be taken. I was staying at my father's house at the time of these visits and for some strange reason this documentation mysteriously disappeared from my room.

Due to the traumatic and debasing nature of what happened to me, I've tried to put these events behind me such that there are certain things which I simply do not remember, e.g. the name of the physician who molested me or the specific visit at which it occurred. It is possible that were I to think hard and piece together my recollections that I could come up with some of these absent facts. At this time though, I can only tell you what at this time I do recall. Just having to relive what happened by writing this letter is painful and degrading enough for me.

If you insist on battling this matter out in court, I will do so. However, I would hope that at least on the basis of my indigency that the charges could be waived.

(note. a copy of this letter was sent to your Downtown physician center with the relevant financial information).

                                      Sincerely,
                                      William Thomas Sherman

~~~*~~~

July 25, 1993

Dear Mr. Bernard, (Queen of Angels Hospital)

Recently I sent your hospital a copy of a complaint I filed with you charging staff misconduct that occured August 10, 1992. It is with regard to that complaint that I am writing this letter.

The other day I obtained a copy of my medical records, and no where in them do I find reference to the nature or name of the medication that I was injected (in the buttocks) with by a middle-aged, female member of the ER staff. Is the reference to it in the record contained in hospital code/short-hand or is there just no reference? What ever the case, I am writing here to request to know what the name and nature of that medication was that I was administered.

Please write and inform me as to what the answer is.

Yours Sincerely,
William T. Sherman

~~~*~~~

August 16, 1993

Medical Board Of California
Complaint Unit
1426 Howe Ave., Suite #54
Sacramento, CA   95825-3236

Dear Medical Board,

I am writing this letter to report to you a blatantly unethical and criminal incident which occurred at Queen of Angels Hospital in Los Angeles on a Saturday in August of 1992. Because my rights as a citizen visiting a public hospital were violated in more than one manner, it is best to state what my charges are by providing a full account of the sequence of events. I hope you will pardon its length, but I feel a detailed account is necessary. Why I have waited till now to write this letter will be explained in the course of my story.

I am an author/historian from Seattle who was in Los Angeles last summer doing research for a book project. I left Seattle moving to the Echo Park area of Los Angeles in late April 1992. On a Sunday in August I attended a musical concert at the Roxy on Sunset Blvd. Prior to the concert I ate at the Spaghetti Factory on (I believe) Santa Monica Blvd. My meal there consisted of spaghetti and a glass of wine. Afterwards at the concert itself, I had a diet coke. While I enjoyed the concert very much, the group performing was an English pop-group Shakspear's Sister, during the course of it I suddenly began to feel very ill, and for no reason started to sweat uncontrollably. While it is true the club atmosphere was rather stuffy, this no way seemed to explain why I was, in a gradual flash as it were, render inexplicably queasy and drenched with perspiration. Prior to the concert I was in the best of health and spirits and the onset of my discomfort came as a great surprise. Indeed, in all my life up unto this point I have almost never had need for a physician other than on one occasion when I had sprained my ankle.

In the following week I did not feel very well. I attributed this to simple fatigue. Then on one day in the week I ate at the Milano restaurant (on Alessandro?) and thought a proper meal would do me good. Instead I became dramatically worse. I began to experience extreme constipation, a condition which I had never suffered in any way shape or form prior to this, such that it became impossible for me to go to the bathroom. As well, to my shock I discovered that my sexual organ would not function other than to urinate. My breathing became more difficult. Hoping that all this was simply some temporary malady brought on accidentally, I rested for a few days hoping it would go away. But it didn't, and by the end of the week it finally became obvious that I needed to go to a hospital or clinic. On Saturday morning I called a friend and asked him if he knew a clinic or doctor he could recommend. I will omit his name here out of courtesy - he is a rather old gentleman, but will provide it if requested by legal authorities. He arrived to pick me up at my address at 1377 Lucile Ave. off Sunset and we looked around for the nearest clinics. Unfortunately, after a lengthy search, every medical clinic we encountered on Sunset was closed. Why this was so I did not then nor now know why. In any case, my friend suggested the Queen of Angels Hospital on Vermont. To this I agreed.

We arrived at the emergency room of the Queen of Angels Hospital at around 10:00 a.m. I signed in and was told to wait. After sitting in the waiting area for five hours, during which time I felt fairly miserable, I was finally admitted about 3 p.m. to the Emergency room. There it was all a hub of noise and doctors, and cadets of some sort (police, medical, or police/medical I don't exactly remember, they were wearing Navy blue cover suits) were running about - the place gave the appearance of being busy. I was assigned a bed behind some portable curtains, give a smock and told to undress. There were at least two policemen present in the room who apparently were waiting while someone in their custody was being treated. No one could see in or out of the curtained area I was located, except very slightly through breaks where the curtains formed into corners.

It was at this juncture that I began to suffer the mistreatment for which I am writing this letter.

After a doctor, I forget which one, briefly heard what was ailing me, he (she?) left and different staff persons came and took my blood pressure. My blood pressure was registered clumsily at least three times by different persons while I waited for a doctor proper to return. At the time, I thought the repetition of this procedure by different persons, including at least one "cadet," was rather incompetent.

Finally, Dr. Phillip Fagan, a rather muscular, middle aged, black male with a moustache, came into my spot to check my breathing with a stethoscope. Customarily, it has been my experience that when this procedure is done the patient is either standing or sitting up. Dr. Fagan had me lie back on the bed/examination table and told me to breath as he applied his instrument. As he came to the area upward to the left of my heart, he made a pointed clenched fist and suddenly and with thoughtful and quick deliberation punched me below my left shoulder. I was so shocked by it I didn't know what to say or could think what could account for his doing it. Finishing up, he asked a few questions and left me. I waited a while longer very much perplexed. Then a middle-aged female staff member with short (brown?) hair came in by herself into my spot. Holding a hypodermic needle filled with a brown/yellowish liquid in her hand, she told me to roll over as she administered its contents to me. Simply assuming she knew what she was doing, desperate to be rid of the "chill" that suffused my groin, and without questioning I simply acquiesced and without informing me of what I was being given, she injected me. She then departed and I was again left to wait.

As I sat there I gradually began to feel what was apparently the "medication" taking effect. I suddenly began to have great lapses in my train of thought and suddenly found it difficult to form words. It is all somewhat difficult to describe except to say that it felt as though I had been given a very, very strong narcotic of some kind. By this time, I became very fearful, after being punched and now this apparent drugging, and didn't quite know what to do. Each time I tried to get hold of a staff member for help I was very rudely told to wait. For the next three or four hours I lay on my bed waiting for one of the doctor's to return, during which lengthy time feeling utmost distress at my situation. I literally felt and thought I could very well die then and there, due to the effect of being punched, as I thought, in the heart and the injection.

Now let me say at this point, that the emergency room all this time was in utter pandemonium. It was a literal madhouse which included someone farting loudly and repeatedly, and directly at me from an adjacent bed, some patients screaming and yelling in apparent delirium. How on earth anyone could expect to regain their health in such an environment, even if it is an emergency room, is beyond me. I fully understand and appreciate how chaotic such a place can get, but this went beyond outrageous or ridiculous.

By eight o'clock, a Dr. Fagan gave me a bottle of liquid laxative and directions to buy antihistamine. Without ever telling me once what might be ailing me, or saying whether the problem with my groin was cured or not, he finally released me. Naturally, by this time I was dying to get out of there, so I didn't to trouble to ask him about what condition was. In any case, he made it so very clear that he was busy, that even if I had tried to get him to talk for more than one or two minutes he would have put me off or casually allowed himself to be distracted. He was not entirely unsympathetic when I spoke with him, only he would not stay to answer what seemed to me were very pertinent and straight forward questions. Given the inordinately prolonged amount of time spent seriously dealing with my case one would have thought I should have been there no more than ten or fifteen minutes, been on my way, while freeing the "doctors" to devote their time, and my bed, to other cases. Instead it took around four hours for me to simply wait around to get a bottle of laxative and directions to get antihistamine.

The next day I returned to the hospital to complain and make inquiries. I asked to know what it was I had been injected with. After a lot of running around for an answer a staff person showed me a document of some kind with "Penheglian" written on it, that presumably being the medication. My current records mention "Phrenegan," but this was not what was written on the document. The doctor's name then was given to me as "Herb" Fagan. This was written on this same document. I then got a senior staff person of some kind and sat down with her and told her what happened. After hearing my story, she politely told me that this was a "good" hospital and that they didn't do things like that. I then requested my medical records and was sent to the records division. Once there, they told me that my records weren't ready and that I would have to come by on another occasion.

Disgusted and frankly now a little fearful, I shortly after all this, I think Columbus day itself, I left Los Angeles and came back to Seattle. Not surprisingly, after what I'd been through, I did not feel comfortable remaining in Los Angeles. Although the laxative seemed to cure my constipation, I still had difficulty breathing and was feeling the effects of Dr. Fagan's having punched me. At the time, it felt as though my heart had been injured, hence my feeling that I might die, which I mentioned before. When I'd returned here, however, I went to a clinic. My injury was diagnosed as damaged muscle within my left shoulder and they prescribed Advil, which had the desired effect of alleviating the pain.

Quite obviously, I had absolutely no idea of expecting anything remotely like this to happen when I went to a hospital. I had always hitherto thought ordinary doctors were generally responsible professionals who one could put their trust in. Imagine than my inexpressible horror and dismay to have underwent what I've described. If these things were done to me deliberately, which I am inclined to think is the case - though I can't say that I know, perhaps this kind of shock and intimidation was these person's apparent intention.

I subsequently obtained a copy of my records for my stay in the emergency room copies of which are included here. Whether the date on my records, August 10, 1992, refers to when my they were processed or the actual date of my visit I don't know. However, it is absolutely impossible that the day I was admitted to the ER was the 10th since the 10th of August of last year was not a Saturday. What specific Saturday my visit did take place I honestly don't recall except to say that it most definitely was on a Saturday following the Shakspear's Sister concert at the Roxy which in turn was on the Sunday previous.

Why have I waited till now to make this report? I am not rich, and just making ends meet, so it is with great difficulty that I can find the spare time and energy to relive these awful events by writing them down. I cannot afford to litigate, and I've been told after enquiring that even if I could afford it, it would be nearly impossible to prove my allegations in court, since I am practically my only witness. Not very long ago I have complained via letter to the hospital itself, but have, I suppose not surprisingly, got no response other than that they said they would look into it.

What would be persons motive to do such things to me? I frankly don't know. In any case, I do not feel it necessarily incumbent on me to provide a motive since any persons who would do such a thing in the first place could hardly be considered rational. I am a writer and of the things of mine could very well be considered controversial. The historical research project I was involved in Los Angeles had implications which, I suppose, some might consider undesirable. Yet never in my wildest dreams did I possibly imagine to incur someone's ire in this malicious kind of way.

What do I hope to gain at this point by reporting all this? I am under no delusion, realize how incredible my story must sound and am more than aware of the inherent difficulties of a relatively indigent individual challenging a major institution with comparatively limitless financial resources to legally defend itself. Believe me it took a lot of initiative and energy to merely write this letter knowing full well what I am up against. However, if what I assert is true, the persons responsible will likely find themselves committing some wrong in the future. The reasoning here is simply this, if someone would stoop as low as I have described by what absurd moral standard do they scruple between right and wrong in governing their actions? If nothing else then my report then will serve as a warning and caution to those who oversee their conduct. I do this more out of civic duty at this point then anything else. I could not live with myself knowing that I did nothing whatsoever to combat such intolerable misconduct. You may do with this letter as you please, for my part in my conscience I can say that I have done all I could under the circumstances.

If you have any questions, please feel free to write or call me.

Note. Copies of this letter have been sent to Los Angeles City and County legal and medical departments who would have jurisdiction, in one manner or other, over this matter. It was from one of them that I actually got your address.

Yours Sincerely,
William Thomas Sherman

~~~*~~~

August 16, 1993

Commanding Officer of Detectives

Los Angeles Police Department, N.E. Division
33353 San Fernando Road
Los Angeles, CA 90065

Dear Commanding Officer of Detectives,

  I am writing this letter to report to you a blatantly unethical and criminal incident which occurred at Queen of Angels Hospital in Los Angeles on a Saturday in August of 1992. Because my rights as a citizen visiting a public hospital were violated in more than one manner, it is best to state what my charges are by providing a full account of the sequence of events. I hope you will pardon its length, but I feel a detailed account is necessary. Why I have waited till now to write this letter will be explained in the course of my story.

  I am an author/historian from Seattle who was in Los Angeles last summer doing research for a book project. I left Seattle moving to the Echo Park area of Los Angeles in late April 1992. On a Sunday in August I attended a musical concert at the Roxy on Sunset Blvd. Prior to the concert I ate at the Spaghetti Factory on (I believe) Santa Monica Blvd. My meal there consisted of spaghetti and a glass of wine. Afterwards at the concert itself, I had a diet coke. While I enjoyed the concert very much, the group performing was an English pop-group Shakspear's Sister, during the course of it I suddenly began to feel very ill, and for no reason started to sweat uncontrollably. While it is true the club atmosphere was rather stuffy, this no way seemed to explain why I was, in a gradual flash as it were, render inexplicably queasy and drenched with perspiration. Prior to the concert I was in the best of health and spirits and the onset of my discomfort came as a great surprise. Indeed, in all my life up unto this point I have almost never had need for a physician other than on one occasion when I had sprained my ankle.

  In the following week I did not feel very well. I attributed this to simple fatigue. Then on one day in the week I ate at the Milano restaurant (on Alessandro?) and thought a proper meal would do me good. Instead I became dramatically worse. I began to experience extreme constipation, a condition which I had never suffered in any way shape or form prior to this, such that it became impossible for me to go to the bathroom. As well, to my shock I discovered that my sexual organ would not function other than to urinate. My breathing became more difficult. Hoping that all this was simply some temporary malady brought on accidentally, I rested for a few days hoping it would go away. But it didn't, and by the end of the week it finally became obvious that I needed to go to a hospital or clinic. On Saturday morning I called a friend and asked him if he knew a clinic or doctor he could recommend. I will omit his name here out of courtesy – he is a rather old gentleman, but will provide it if requested by legal authorities. He arrived to pick me up at my address at 1377 Lucile Ave. off Sunset and we looked around for the nearest clinics. Unfortunately, after a lengthy search, every medical clinic we encountered on Sunset was closed. Why this was so I did not then nor now know why. In any case, my friend suggested the Queen of Angels Hospital on Vermont. To this I agreed.

We arrived at the emergency room of the Queen of Angels Hospital at around 10:00 a.m. I signed in and was told to wait. After sitting in the waiting area for five hours, during which time I felt fairly miserable, I was finally admitted about 3 p.m. to the Emergency room. There it was all a hub of noise and doctors, and cadets of some sort (police, medical, or police/medical I don't exactly remember, they were wearing Navy blue cover suits) were running about – the place gave the appearance of being busy. I was assigned a bed behind some portable curtains, give a smock and told to undress. There were at least two policemen present in the room who apparently were waiting while someone in their custody was being treated. No one could see in or out of the curtained area I was located, except very slightly through breaks where the curtains formed into corners.

It was at this juncture that I began to suffer the mistreatment for which I am writing this letter.

After a doctor, I forget which one, briefly heard what was ailing me, he (she?) left and different staff persons came and took my blood pressure. My blood pressure was registered clumsily at least three times by different persons while I waited for a doctor proper to return. At the time, I thought the repetition of this procedure by different persons, including at least one "cadet," was rather incompetent.

Finally, Dr. Phillip Fagan, a rather muscular, middle aged, black male with a moustache, came into my spot to check my breathing with a stethoscope. Customarily, it has been my experience that when this procedure is done the patient is either standing or sitting up. Dr. Fagan had me lie back on the bed/examination table and told me to breath as he applied his instrument. As he came to the area upward to the left of my heart, he made a pointed clenched fist and suddenly and with thoughtful and quick deliberation punched me below my left shoulder. I was so shocked by it I didn't know what to say or could think what could account for his doing it. Finishing up, he asked a few questions and left me. I waited a while longer very much perplexed. Then a middle-aged female staff member with short (brown?) hair came in by herself into my spot. Holding a hypodermic needle filled with a brown/yellowish liquid in her hand, she told me to roll over as she administered its contents to me. Simply assuming she knew what she was doing, desperate to be rid of the "chill" that suffused my groin, and without questioning I simply acquiesced and without informing me of what I was being given, she injected me. She then departed and I was again left to wait.

As I sat there I gradually began to feel what was apparently the "medication" taking effect I suddenly began to have great lapses in my train of thought and suddenly found it difficult to form words. It is all somewhat difficult to describe except to say that it felt as though I had been given a very, very strong narcotic of some kind. By this time, I became very fearful, after being punched and now this apparent drugging, and didn't quite know what to do. Each time I tried to get hold of a staff member for help I was very rudely told to wait. For the next three or four hours I lay on my bed waiting for one of the doctor's to return, during which lengthy time feeling utmost distress at my situation. I literally felt and thought I could very well die then and there, due to the effect of being punched, as I thought, in the heart and the injection.

Now let me say at this point, that the emergency room all this time was in utter pandemonium. It was a literal madhouse which included someone farting loudly and repeatedly, and directly at me from an adjacent bed, some patients screaming and yelling in apparent delirium. How on earth anyone could expect to regain their health in such an environment, even if it is an emergency room, is beyond me. I fully understand and appreciate how chaotic such a place can get, but this went beyond outrageous or ridiculous.

Yours Sincerely,
William Thomas Sherman

~~~*~~~

August 23, 1993

[Addressed to?]

Sorry about Ray Moore.[139]

Enclosed is a
a) copy of my letter to the Board stating my complaint
b) Letter from Leah Kilger, of Hospital administration after I had sent my report to her.
c) copy of Recent response from Asst. Chief of Police Roger Serra to my request to investigate the possible criminal aspects of this case. Originally, at the advice of an attorney's office, brought my criminal complaint to the Seattle Police, they said to take it to the University Police. I then spoke with a Det. Roberts of the University Polioce on the phone and sent him a copy of the report of my complaint. After he read it, I spoke with him a few days later on the phone and he simply refused to investigate. Following this refusal I brought a complaint to the Asst. Police Chief Roger Serra about Det. Roberts refusal to investigate this matter in any way shape or form. This letter from Asst. Police Chief Sera then is his response to my complaint regarding Det. Roberts.

---

[139] Washington State assemblyman Ray Moore who responded to my initial appeal for help sympathetically, but who for one reason or other left or was compelled to leave office.

1) I never spoke with a Mr. Jim Smith, Chief Investigator. The only person with whom I had person to person contact with (and that was over the phone) was Glenn C. Hay-Roe, Intake Coordinator - the individual I referred to as the Board's public relations person.

2) At the advice of an attorney's office, I attempted to bring matter up with University Police, there response was that they could not do anything to investigate because it was a medical matter. There response was to take my complaint to the hospital administration, which I have already done when I wrote miss Kliger. Although Asst. Police Chief Serra mentions Leana Osterman as the person I should contact, Miss Kliger of the hospital administration gave me no such recommendation and presented herself as the person who should be contacted with such complaints and allegations as I raised. The problem then in dealing with the hospital itself is a) I have tried then to raise my complaint then with the hospital administration, but they have refused to deal candidly. The reason for this, one can easily surmise, is that the hospital obviously has little reason to subject themselves to possible liability or to risk hurting their reputation by dealing truthfully with my allegations.

          Sincerely,
          William Thomas Sherman

~~~*~~~

August 23, 1993

Secretary of the Senate Marty Brown
Legislative Building
P. O. Box 40482
Olympia, WA 98504-0482

Dear Mr. Brown,

In response to your letter on behalf of Senator Ray Moore's office, yes, I would very much like the matter I raised with Senator Ray Moore pursued further and brought to the attention of Senator Phil Talmadge and the Senate Committee on Health and Human Services.

My response to Bruce Miyahara's letter is this.

1) I never spoke with a Mr. Jim Smith, Chief Investigator. The only person with whom I had person to person contact with (and that was over the phone) was Glenn C. Hay-Roe, Intake Coordinator - the individual I referred to as the Board's public relations person. When, after receiving the Board's terse verdict, I asked Mr. Hay-Roe if I could speak to someone about my case, he very politely told me that no one other than himself was willing to talk about it, and that for his part he

had nothing to say except that the Board cannot disclose the basis of its findings other than that they have found for or against the patients complaint.

2) Mr. Miyahara's response in sum is that the Board cannot reveal what happened during the course of its investigations, except to give the verdict. This goes, once again, to heart of my original complaint against the Board which I raised with Senator Moore. What on earth is this secrecy for? Can we expect, in future, clandestine trials to establish a persons legal guilt or innocence? Is it not possible than such a system could be corrupted by doctors and hospitals trying to protect their own?

Either the Board handled my case in a negligent manner, or is assisting the hospital in a cover-up, or there needs to be some drastic changes made to the Medical Board's review process and procedures.

Enclosed for purpose of closer examination of my case are the following:

a) Copy of my letter to Senator Ray Moore
a) Copy of my letter to the Board stating my complaint
b) Copy of letter sent from Leah Kliger, of Hospital administration, in response to my report.
c) Copies of letter from Asst. Police Chief Serra, of the University Police.

Explanation: Originally, at the advice of an attorney's office, brought my criminal complaint (alleging possible sexual molestation) to the Seattle Police, they said to take it to the University Police. I then spoke with a Detective Roberts of the University Police on the phone and sent him a copy of the report of my complaint. After he read it, I spoke with him a few days later on the phone and he simply refused to investigate. Following this refusal I brought a complaint to the Asst. Police Chief Roger Serra about his refusal. This enclosed copy of the letter from Asst. Police Chief Serra then is his response to my complaint regarding Det. Roberts. Asst. Chief Serra states I must take my complaint to the hospital administration, which I have already done in my contact and correspondence with Leah Kilger. The problem then in dealing with the hospital itself is that they refuse to discuss the case. The reason for this, one can reasonably, is that the hospital obviously has little reason to subject themselves to possible liability or to risk hurting their reputation by dealing truthfully with my allegations.

If you or Senator Talmadge or his committe have any questions or require further information about this matter, by all means feel free to contact me and I will be more than galad to oblige you.

Thank you for your attention to this.

Sincerely,
William Thomas Sherman

September 14, 1993

Dept. of Health Facilities and Services Licensing
P.O. Box 47852
Olympia, WA 98504-7852
Attn: Hazel

Dear Dept. of Health Licensing Staff member,

     This letter is being written to you to bring to your attention several separate instances of staff misconduct, negligence and patient abuse which occurred during the course of three visits to the University of Washington Medical Center Emergency room in October and November of 1992. Because what took place happened over the course of more than a week over nine months ago I cannot recollect every single detail of all that transpired. I have had in my possession records and documents corroborating or recording what I report here, yet for reasons admittedly unknown to myself, these have unfortunately 'disappeared,' including notes I kept on my computer which were entirely erased. This said, the following is the account of my charges based on what I do clearly remember or can independently confirm. I regret the length of this report, but due to the unusual nature of the events and the fact that what took place occurred over an extended period of more than a week, I feel my story requires thoroughness in examination given its concededly unusual nature. I will conclude this report with a summary of my charges and the important unanswered questions related to the case. The allegations I have to make refer to conduct which is either criminal, unethical or negligent. Some of the allegations are easily confirmed, others are less so. Yet however you credit a specific charge I make I would hope at the very least that you bear in mind that even if there is not sufficient evidence to indict an individual or the hospital on one count, particularly a criminal charge, this is not grounds in and of itself to necessarily absolve them of another allegation.

                                           Yours Sincerely,
                                           William Thomas Sherman

September 20, 1993

Medical Board Of California
Complaint Unit
1426 Howe Ave., Suite #54
Sacramento, CA   95825-3236

Dear Board Member,

I am writing this letter to amend slightly a report I submitted to you concerning staff misconduct that took place at the Queen of Angels Hospital on Vermont Ave. in Los Angeles in the latter part of 1992.

In my report, I affirmatively state that my visit took place in August. On this specific point my memory, as evinced by some records recently located, is incorrect, and my visit in fact took place in October. The reason for my confusing the two months was due to the fact that the pop group Shakspear's Sister, mentioned in my report, originally had a concert scheduled for August which was subsequently canceled and rescheduled for October. It was this memory in reference to the concert which caused me to confuse the month.

This noteworthy correction, however, is all of substance that needs to be amended, and my report otherwise properly stands as sent. I would though place added emphasis on the fact that I was given an injection, allegedly of Phrenegan, without being told what it was or what it was being given to me for, which had a horrendous side-effect. It was told me later that Phrenegan is given to treat nausea. The trouble with this is that my explicitly stated symptoms to the doctor were severe constipation, a "frozen-up" groin, and some difficulty breathing. While these naturally made me feel bad, I would not equate the feeling with nausea. It is extremely odd and anomalous to me that no where in my medical records for the hospital are my specifically reported symptoms even mentioned, despite the fact that the Doctor examined my penis when I brought up the matter about the groin. All this of course is aside form the outrage of being punched.

If you have any questions please feel free to write or call me.

Sincerely,
William Thomas Sherman

~~~*~~~

The American Civil Liberties Union
705 2d Ave., Room 300
Seattle, WA 98104

October 19, 1993

Dear American Civil Liberties Union staff person,

Please accept enclosed here copies of letters I wrote to the Los Angeles Police concerning some rather (to put it mildly) unusual events involving that department and myself. The letters speak for themselves, and in making yourself familiar with my case it is necessary that you read them. My purpose in writing you is to report to you what happened while I was in Los Angeles – via these letters, and to ask your suggestions as to how I might deal with my case. I would also be interested in knowing if there are any Federal agencies who might have jurisdiction or other interest in the events my letters describe whom I might at least send my report - do with it what they will.

Your prompt acknowledgment of receipt of this letter would be very much appreciated.

Thank you for your time and attention to this matter. If you for your part have any questions, of course, feel free to call or write me. I would appreciate it if you would promptly acknowledge the receipt of this letter.

Sincerely,
William Thomas Sherman

~~~*~~~

Oct. 21, 1993

Medical Board Of California
Complaint Unit
1426 Howe Ave., Suite #54
Sacramento, CA   95825-3236

Dear California Medical Board,

A few months back I submitted a complaint you regarding an incident involving a Dr. Fagan which occurred at the Queen of Angels Hospital in Los Angeles in 1992. My case number is 17-93-30020.

Since sending that letter some developments have come to light in the case which I think it necessary to bring to your attention:

* My original report is incorrect as regards the date. My visit to the Queen of Angels Hospital took place in October rather than August as it states. The reason for the error is due to my associating the "Shakspear's Sister" concert it mentions with August due to the fact that the concert, as advertised, was originally slated for August, but was rescheduled to October. In trying to recollect exactly what happened, I connected the concert date with the date of the incident, which was the right thing to do except that my memory of the date of the concert was wrong.

* The doctor who allegedly signed my signed my medical records, Philip Fagan, was not the physician who treated me. The latter was instead one Elmer

Eley. "Dr." Eley apparently had been investigated on an earlier occasion by Health Licensing and had been reprimanded for faulty medical records.

* Not long after the incident Eley was discharged from the hospital.

* My medical records contained no reference to my groin ailment, despite the facts that I explicitly voiced my complaint to him and that he physically examined that part of my anatomy in response.

* The medical records state that I was suffering from nausea, when on the contrary I was suffering from constipation. It was on this premise that I was injected with a substance alleged to be Phrenergan, such that even if what I was injected with was Phrenergan I was given a medication for something I wasn't suffering from.

Thank you for your time and attention to this matter.

Sincerely,
William Thomas Sherman

~~~*~~~

Nov. 15, 1993

ACLU of Southern California
1616 Beverly Blvd.
Los Angeles, CA 90026
Fax: 213-250-3980

Dear ACLU staff person,

I am sending here to you copy of a report involving misconduct on the part of officers of the Los Angeles Police Department. Because my account is rather involved, it is better that you read my reports instead of my attempting to summarize briefly what happened here. Copies of these reports to the Los Angeles Police were supposed to have been sent to you earlier by the Seattle office of the ACLU, but as I have not receive confirmation of their having arrived at your office, I am faxing them to you.

The reports speak for themselves, yet I would add that I am more than willing and happy to be subject to a lie detector test should there be any question as to the veracity of my allegations.

Your prompt acknowledgment of receipt of these reports would be greatly appreciated.

Yours Sincerely,
William Thomas Sherman

~~~*~~~

January 3?, 1994

United States Attorney
312 N. Spring St.
Los Angeles, CA   90012

Dear U. S. Attorney's office staff person,

(*Note*. A copy of this letter was sent to your office on December 9th. However, upon calling your office on December 29th, I learned that you had not received my original letter. I am therefore sending you this letter a second time.)

This is letter is being written to you to bring to your attention a case involving misconduct and corruption on the part of the Los Angeles Police Department. The technical basis of my complaint is the failure of the police to investigate a charge of assault and battery, inflicted by a hospital doctor, that occurred to me while I was staying in Los Angeles during the summer of 1992. My case, however, has far deeper implications beyond this failure to act on a citizen's complaint and concerns as well what would suggest to be criminal behavior and criminal involvement on the part of the Los Angeles Police Department.

Enclosed with this cover letter are copies of 1) my initial letter to the Commanding Officer of Detectives reporting the assault and battery that took place at the Los Angeles hospital, and 2) a later letter sent, at the recommendation of the Los Angeles City Attorney's office, to Chief of Police Willie Williams complaining of the department's unjustified and unexplained failure to investigate my complaint. These two letters describe in detail what took place when I was in Los Angeles; the reading of which will provide you with the specific nature and background of my allegations.

My case is admittedly some what involved, yet I will briefly sum it up here to state that there is good reason to believe that certain members of the LAPD have seriously broken the law, participated in vicious terrorism and violated my civil rights in the interest of serving political ends. These, as I am sure you will agree, are very, very serious charges, charges which I do not make lightly. And while it might be understandable that some one reading or hearing my report and allegations might question their veracity, let me unequivocally state that I am more than happy and willing to submit to a polygraph or lie detector examination.

I am not an attorney, so technically I cannot specifically recommend to you how you ought to deal with my charges. I would say, however, that if what I allege is true that this is a matter of extreme seriousness inasmuch as the wrongs alleged to have been done or participated in were of a calculated and ostensibly organized nature; and that to let something of this nature slide or be swept under the rug would be to condone activities which bespeak the very worst kind of depravity and corruption. I would hope at the very least then, you would take the trouble to investigate my story.

Thank you for your attention to this matter and please feel free to call or write me with your questions.

Yours Sincerely,
William Thomas Sherman

~~~*~~~

July 14, 1994

Jeanne Kohl
House of Representatives, #402
John L'OBrien Bldg.
Olympia, WA 98504

Dear Representative Kohl,

I am writing to you to lodge a complaint and possibly get your assistance in an issue regarding the state medical board.

In the Autumn of 1992 I was subject to negligence and mistreatment at the University of Washington Medical, in consequence of which I filed a report with the state medical board. Due to the somewhat involved nature of my complaint against the hospitals physicians, I cannot go into the details of my case here. Having said this, I will provide you with as much information about my case if you are interested or need to know. However my purpose in writing you is to address the procedures of the medical board and how my case was handled. Please inform if you want the details and I will send them to you.

Needless to say, my case, after many many months, which include an appealing and additional review by the board, was deemed by the medical board to be without merit and my allegations against the physicians dismissed. While I can respect the board's position to decide such matters as they think proper, I object strongly to the bureaucratic, insensitive and unfair way in they handled my case.

My objections to their handling of my case and then deciding against me is based on the following:

(a) Though my complaint in writing was received, never once was I given an opportunity to speak with an investigator.

(b) Though the nature of my charges were serious and then decided against, I was refused having any questions about my case answered, despite repeated requests.

(c) The members of the board who were specifically responsible for addressing my case were kept anonymous, such that I do not know who specifically investigated the case or who decided it.

In sum, the response I received from the board's public relations persons was that medical matters are too complicated for ordinary people to understand and that I must simply trust their judgment without desiring any specific explanation. Quite frankly, I find this kind of response coming from a department entrusted with overseeing a profession which deals with people's lives on the most intimate kind of level arrogant and irresponsible. Either my particular case was treated unfairly or else something must be done to make the medical board process of review more accessible to the public and physicians made more accountable for what they do. I am not rich, nor can I, at the present time, afford to go hunting around for and hire an attorney. I appeal to you then as my representative to look into this matter and work to provide the citizens of this state with a medical board that serves the needs of the people first and foremost.

If you have any questions or require more information, please feel free to contact me.

Yours Sincerely,
William T. Sherman

Reference case numbers:
Stephen Burns, case #: 93-09-0043
Charles Weaver, case #: 93-09-0067

<div style="text-align:center">~~~*~~~</div>

July 26, 1994

Jeanne Kohl
300 West Harrison
Seattle, WA   98119-4081

Dear Representative Kohl,

As requested, here is a copy of the letter reporting my complaints to the Washington Medical Board.

One of the infuriating things about my experiences in raising my complaint both with the University hospital itself and the board is that I have not been allowed to discuss my case with any one person-to-person. They have allowed me to write them letters, which I have, but never once, despite the seriousness of my allegations, did either the hospital or the board allow me the opportunity to speak with some one personally. No one, it seems, wants to take any responsibility or be made accountable. In each instance, I have been told to differ to their authority without there being any obligation on their part to provide answers or explanations. In sum, the only consolation I have been provided with in each case is the assurance that "I can trust them," i.e. in their judgment. Well, I would much rather have some specific answers than a condescending pat on the back and being told that they "looked into it."

Although, I am only sending the copy of my letter to the board, the medical records of my case are available, only I do not have as ready access to them at the moment as the letter. If you would like copies of the medical records written by Dr. Burns and Weaver, please let me know and I will have copies made and sent to you. The Board itself, of course, has copies of these same records and you have my full permission to obtain these from them.

Thank you for looking into this matter.

Yours Sincerely,
William T. Sherman

~~~*~~~

August 2, 1994

Detective Roberts
University Police
1117 N. E. Boat Street
Seattle, WA 98105

Dear Det. Roberts,

Enclosed is the letter regarding allegations of molestation against staff of the University Medical Center that occurred in Oct. of 1992.

As I mentioned to you on the phone, I bring this to your attention not to gain your personal sympathy but to ask that you see that the law has not been violated. Having spoken with you at some length over the phone, you already have some idea of what my allegations pertain to. I would like you to please contact Dr. Stephen Burns and DR. Charles Weaver, both to my knowledge, still working for the University of Washington Medical Center and, as best you can determine the following:

1) Did Dr. Burns or someone on his staff place an EKG tape on my scrotum during the time I was unconscious in the Emergency Room?

2) Who was the staff person at the hospital (on my third visit) who "felt me out?"

3) Does this person latter admit to "feeling me out?" If so, how was this related either as a diagnosis or remedy for what was ailing me?

4) Why did the staff disallow me a urinalysis or other chemical test to determine whether or not I had been poisoned? Gen a report of poisoning one would think this would be a common sense kind of request to make.

5) Why did the medical staff attempt to discredit me with Officer Underwood of the Seattle Police Department when I attempt to make a report of having been poisoned?

Given the seriousness of what I charge I hope you will at least inquire, by asking these questions of Dr. Burns, Dr. Weaver and the staff, and thereby determine or help determine whether there is any foundation to them.

Sincerely,
William Thomas Sherman

~~~*~~~

November 3, 1994

Representative Jeanne Kohl
300 West Harrison, 5th Floor
Seattle, WA 98119-4081

Dear Representative Kohl,

I would like to reply to Leana Osterman's letter with the following:

Before doing so, let me first submit for later reference here the questions raised in my original complaint.

1. RE the EKG tape found in private area: Of all my allegations I realize this, though it might be the most serious, is the most difficult to prove. Yet due to the gravity of it I did not feel it could afford to be overlooked. Also, mentioning it perhaps in one way or other might help to explain the events which followed. This same conclusion as well applies similarly to the matter about the faulty prescription.

2. Two of the three physicians who filled out the report on me proceeded to adjudge my case psychological in nature without having making any serious or sincere effort whatsoever to determine whether a foreign substance was in my system at the time. Why was I on each occasion presumed in need of psychiatric care without even having been given a urinalysis test? Why, except until the last minute of the last visit were these refused me when I requested them? After all, if I was poisoned as I alleged what how would this conflict with the symptoms I am described by the doctor to have exhibited? If there was any possible doubt why from the beginning was I denied the benefit of it? Wouldn't the hospitals version make more sense if they had given me a proper blood and urinalysis test, and then presumed to judge whether my ailment was somehow psychosomatic in origin?

3. Is systematically feeling out a patient's *entire* body, sans groin, a proper procedure for treating some undergoing the effects of a street drug? While I admit

I am not in a position to answer that question, as a matter of common sense I don't see how feeling out a persons entire body has anything to do with remedying a complaint of poisoning.

4. Why did the staff feel they were in a position to interfere with and deprive me of my civil rights by preventing me from making a report to a Seattle Police officer?

Now to my response:

First, you will notice that none of the questions I raised earlier has been answered. Ms. Osterman brushes everything off without explanation other than to say that my allegations "have no rational basis." Once again the message is to leave everything to the doctor's they can do no wrong and we must trust their infallible judgment, even when possible allegations of criminal wrong doing are involved. Doctors, even though they empowered with the greatest trust over a person imaginable, because of their great prestige, are not, like every one else, required to provide full or rational explanations, even with regard to matters that are easy for a layman to understand. (*Note*. I am enclosing a recent article published in the *Seattle Times* which gives a demonstration of the kind of thing that can happen when matters involving a crime committed against an incapacitated patient by a medical staff person are left to the hospital itself to oversee.) As far as my case raises questions of general public concern, where on earth did this notion come from that medical personnel and doctors are not capable of committing serious crimes? One would think that this seems to be the assumption seeing that we allow them such extensive powers of self-regulation and governance. Please just for a moment stop and consider the power medical people, particularly those in a hospital, have over people in their care who might, for one reason or another, be incapacitated. If any crime is committed, where their are no witnesses other than the victim, then who is the first to deal with the problem - apparently the hospital itself. Now why, given human nature, should we necessarily assume that a hospital would necessarily be more concerned about the public interest or that of a single, violated crime victim, rather than it's own reputation and potentially being subjected to litigation.

As to my contacting the police: Yes, at the recommendation of an attorney I have, as you know, done so. They (the Seattle Police Dept.) said they would investigate but have yet to inform me what action (if any) they have taken. While it is not clear to me as yet how the police is going to address my complaint, it should be borne in mind that a Seattle police officer ran out on me at the hospital when I tried to make a report (see my original report), something which by the department's own admission was wrong of the officer to do. Sgt. Mark Kuehn of the Seattle police initially told me that he would not investigate. When I asked him merely to write a letter stating why he would not investigate, he at first said he would write me such a letter. Two weeks later when I called him back to find out when I was going to receive this letter, he said he had changed his mind and on second thought was going to have my complaint investigated. What is being done at this point, I am still in the process of endeavoring to find out.

When I called her, I asked Ms. Osterman whether feeling me out was a proper procedure. Her first reaction was to say that the only proof of my being felt out was myself. Notice that she did not defend what the staff person did, rather her first response was to deny that it happened. Now my question was not how did she respond to my charge, but what purpose did this medical procedure serve. Clearly, her initial answered implied that she considered what I described as an act of wrong doing, otherwise she would not have felt it necessary to respond by characterizing my question as an allegation when in fact all I posed was a simple objective question of whether, medically speaking, such a "procedure" was ever warranted in a case such as my own, and if so why. When I pointed this out to her she then said that such a "procedure" would be appropriate but would not comment on whether it should have been used in treating me.

How did the doctor's decide that I did not need a urinalysis? Ms. Osterman said they "looked" at me and decided that it wasn't necessary. Ms. Osterman's argument, based on what she said to me on the phone, in other words, is that I am crazy and that for this reason I should not be listened to. Now even if I were crazy, which needles to say I assure you I am not, does that mean that it would be impossible for a crazy person to be poisoned? Now granted we might suspect a crazy person complaining of being poisoned might suffering from might be a delusion, but does that mean crazy people could never be poisoned? The answer of course is no, such that there was no reason, given the seriousness of what I complained of, to deny me a simple chemical test which I was more that happy to pay the expense for. The grounds for Dr. Burns (my first visit) of saying that I was crazy was based on my statement that I thought that *if* I had been intentionally poisoned, which by the way I did not automatically assume, that my father *might* have been responsible. Well, Dr. Burns then assumed that such a thing must necessarily be beyond the realm of all possibility and therefore I must be crazy. While I will grant that such an allegation might ordinarily be odd and unusual, my merely rising it as a conjecture as far as what had caused my problem does not seem to me sufficient grounds to assume that I was crazy, such that he could refuse me a simple urinalysis or blood test to determine whether, accidentally or because of someone's intention, I had received a noxious substance into my system.

Ms.Osterman says main argument over the phone to me that what happened took place so long ago (two years) that there's no need to bother with it. Given the potential gravity of what I allege, I beg to differ.

In conclusion, why all the secrecy? Leaving aside my specific allegations why does the hospital refuse to explain the medical questions I raised, such as what is the purpose of feeling out a patient as I described. Why, given the seriousness of what I have alleged, was it necessary to wait so long to get even the terse, inadequate kind of response Ms. Osterman now provides us with? Clearly, there is much that still needs to be explained. Yet apparently we are required as a general rule to give medical people the benefit of the doubt, assume that they are beyond ordinary mortals, and are incapable of serious wrong doing because of the philanthropic nature of their calling.

Once again, why, whether I am crazy or no, was I refused a chemical test? This and other questions have yet to be satisfactorily addressed, let alone answered.

Will Dr. Burns and others involved be willing to subject themselves to a polygraph? This would make this whole matter very simple to determine and it is hard to see why the hospital refuses this. If money is a problem, I am more than happy to bear the expense of such., including paying the staff people for their time should they be able to pass such a test.

As to obtaining hospital records or information pertaining to my three visits with Dr. Burns, Mullins and Weaver, you have my full permission.

Also....
RE Washington State Medical Board: Why does it give the hospital the benefit of the doubt, without having bothered to discuss the case with me - at all? As I related to you earlier, not once did the Board trouble itself to discuss with me personally my questions or allegations. Instead, as with Ms. Osterman I am expected to merely take their word for it without rational explanation offered to the legitimate questions I raised in my initial complaint. Please then keep in mind that my complaint to you concerns the questionable handling of my case by the State Medical Board as well as the University of Washington Medical center.

Finally, while clearly I have a personal interest in this case, I respectfully hope you realize that its implications, regarding the how people in medicine police themselves, have a much more important relevance to the public at large. The enclosed copy of the Seattle Times news clipping I would think bears this out.

Thank you for your time and attention to this.

              Yours Sincerely,
              William Thomas Sherman

~~~*~~~

b. Miscellaneous

Below is something I wrote back about the mid nineties:

"IS THERE SOME WAY OF PROVING MY ALLEGATIONS?

"Yes, many of my allegations can be proved if someone who was honest would simply go and investigate and to attempt to confirm them.

"Here are some things which might be looked into to help verify my claims.

"a) For the possibility of mind control/torture technology see APPENDIX:

"b) Give any and everyone possible, that is in someway involved, and have or ask them to take a honestly administered lie detector test.

"c) Some kind of CAT Scan or MRI test to identify what is doing my head damage, specifically some kind of technology. Of course, someone will be sarcastic and say I ought to have my head examined, but this is just a reminder of the cruelty of such a technology and how it can be used to discredit its victims.

"I have had two CAT scans already and I was told nothing special showed up on them. Personally I believe some manner of machinations brought this result, either the doctors were crooked, or else there was other goings on such as my being given someone else's scan. "How could that happen?" one might reasonably ask? If we assume that by means of this device the said perpetrators can read my thoughts, they know in advance where I am going and can arrange it that I get a manipulated reception. These physicians who arranged to have the scans made then, hypothetically speaking may have been cajoled, bribed or intimidated to prevent their honestly helping me. Alternatively, the doctors who had the scans taken might have been acting in complete good faith, but that behind the scenes there was some switching or doctoring of the test results by other staff. This obviously will come across as very far fetched to some, yet if what I allege is true than such obstruction would be absolutely necessary as part of the criminal's plans. On top of all this, in most of my personal contact with them, I will be candid in saying that the demeanor of these doctors was hardly one of forthrightness and sincerity.

"Now, it has been brought to my attention by other victims and literature in the subject that this mind control/torture business can be carried out without an implant. Not being much of a technical minded person, how this is so I cannot quite fathom. It may after all be so in my case. Nevertheless, I view this possibility as unlikely and am thoroughly convinced (though admittedly I can't say 100% sure) that what I am dealing with is some kind of implant.

"d) On a human level if AN HONEST PERSON were to interview of the people I allege have participated in this scheme they would find themselves talking to people who are evasive, abrupt and dissembling. This is the kind of thing that cannot be used as hard objective evidence, yet I mention this for the reason that if an honest person were to interview in some degree of depth, such a one would not have too much difficulty seeing what I am talking about.

e) Earlier mentioned scars on my abdomen and inner thigh can be inspected for those who want proof on this point.

"f) Due to the complexity of the narrative there was one series of incidences I left out for brevity's sake. Before the events of 1992 I was seeing a girl every so often, named 'Cheryl Bowers.' Now without going into the whole story, let me sum up by saying this girl was somehow involved in setting me up. What is unique about her role was that she has a twin, something I did not discover till after I had come back from Los Angeles. Now she denied she had a twin sister, and she and her mother about 1993 moved, reportedly, to California. Now the proof in question on this point is this: to determine whether or not this gal had a twin, or else had cousins who were twins. I realize thissounds a bit convoluted, yet it is a relatively simple way to help establish my story. If it is proven that Cheryl Bowers has a twin (or cousins who are twins), this will leave her to explain why she (they) deceived me in the course of an on and off relationship that lasted about 3 years. If it is established that she has a twin, I would follow this up by inquiring why their long term deception of me: my reasoning being (based on evidence drawn from my contact with them) that these girls were originally involved in setting me up. There is a whole story behind my relationship to these girls, and I could write at length on the subject, including specific reasons why I believe they are connected with this business. However for the sake of keeping things as simple as possible, I have refrained from going into detail on this aspect of my story.

"Despite then the regrettable, yet practical necessity of omitting much detail in this matter regarding 'Cheryl Bowers,' what I am raising is really not as complicated as it might on the surface sound. To recap: 1) locate Cheryl Bowers, 2) find out if she has a twin (or cousins who are twins), 3) If she has a twin, ask her to explain why she decieved me on this point for three years - and much of the rest will begin to fall into place, or at least it will be a significant breakthrough.

"I have no information where 'Cheryl Bowers' moved to, however 'they' are apparently with their mother Carmel Bowers, (a former employee of, I believe JC Penney's in Bellevue), who was last residing at: 10728 NE 26, Bellevue, WA 98004

* Below are various names, addresses and phone numbers of people I dealt with in the course of this or whom I contacted:

Dr. Robert Aigner, MD , neurologist
Dr. John Chapman, MD
Dr. Dean Ishiki, MD

Ethical Committee
Los Angeles County Medical Association
1925 Wilshire Blvd.
Los Angeles, CA   90057

Special Crimes Division - Medico-Legal Section
320 West Temple St.
Room 780
Hall of Records
Los Angeles, CA   90012
974-7346

Los Angeles Police Dept.
213-485-2563
213-485-4063

Commanding Officer of Detectives
Los Angeles Police Department, N.E. Division
33353 San Fernando Road
Los Angeles, CA 90065

Dept. of Health Services
600 S. Commonwealth Ave. #800
Los Angeles, CA 90008
Attn. Licensing and Certification

Calif. State Licensing: 213-351-8200
8:00 to 5:00

Dept. of Health Services
600 S. Commonwealth Ave. #800
Los Angeles, CA 90008
Attn. Licensing and Certification

Calif.Medical Association
1201 K. St.
Sacarmento, CA 95814
916-444-5532

Calif. Med. Association
221 Main ST.
San Francisco, CA   94105
415-541-0990

Calif. Medical Association
P.O. Box 7690

San Francisco, CA 94120-7690

American Medical Associoation
515 N. State St.
Chicago, Ill. 60610
312-464-4818

Ethical Committee
Los Angeles County Medical Association
1925 Wilshire Blvd.
Los Angeles, CA 90057

Special Crimes Division - Medico-Legal Section
320 West Temple St.
Room 780
Hall of Records
Los Angeles, CA 90012
974-7346

Los Angeles Police Dept.
213-485-2563
213-485-4063
Commanding officer of Detectives
33353 San Fernando Road
Los Angeles, CA 90065

Amnesty International
322 8th Ave.
New York, NY 10001
1-800-AMNESTY

District 36
Senator Ray Moore, 284-8088
431 Cherberg Office Bldg.
Olympia, WA 98504

Helen Sommers, 283-6388, 1-206-786-7814
House Representatives, #203
John L'OBrien
Olympia, WA 98505
aide: Pauline Rice: 1-206-786-7814
P. O. Box 40671
Olympia, WA 98504

Jeanne Kohl, 285-1869, 281-5493, 1-206-786-7860
House of Representatives, #402

John L'OBrien Bldg.
Olympia, WA 98504
or
300 W. Harrison
Seattle, 98119-4081

State of Washington
Dept. of Health
P.O. Box 47852
Olympia, WA 98504-7852
Any further correspondence about this matter should be referred to as 93-077.
Hazel, 206-705-6612, or 1-800-633-6828
Gail V. Hughes, Manager
Office of Licensing Administration

Evergreen Legal Services - 464-5911
Fremont Legal Services - 548-8361
Lawyer Referral King County Bar Association - 623-2551
Volunteer Legal Services - 623-0281
John Alexander: 448-7172

Dept. of Health Facilities and Services Licensing
P.O. Box 47852
Olympia, WA 98504-7852
Attn: Hazel
Medical Disciplinary Board
Washington Board of Medical Examiners
1300 S.E. Quince St.
P.O. Box 47866
Olympia, WA 98504-7866
Attn. Betty
206-586-3335 Glen.
206-586-4574

Stephen Burns, case #: 93-09-0043
Charles Weaver, case #: 93-09-0067
Fax: 206-586-4573

American Medical Associoation
515 N. State St.
Chicago, Ill. 60610
312-464-4818

Adjutant General
State of Washington
Attn. Records

Camp Murray, Tacoma 98430-5186

University Police
Det. Roberts
543-9331

Asst. Chief of Police Roger Serra
1117 N. E. Boat St.
Seattle, WA 98105

Governors office: 360-753-6780
Lee Harris

Ray Moore's letter sent to:
Bruce Miyahara
Dept. of Health
P. O. Box 40002
Olympia, WA 98504-0002

Marty Brown, Secretary of the Senate
206-786-7550

Sgt. Mark Kuehn - 684-5590
(contacted in late Sept. 94)

Case # - 94-438-299

SAU - 296-9470
Gina (older one), Geena
Norm Maleng
516 3d Ave.
Seattle, WA
98104

Sgt. Campson or Sgt. Harper - 684-5583

Charles Hamilton: 623-6619
2003 Western Ave. Ste. 600
Seattle, WA 98121

Cheryl Bowers
(206) 488 - 6132
16015 124 N. E.
Woodinville, WA 98072

Carmel Bowers

10728 NE 26
Bellevue, WA 98004
822-0373

Janny Becker, Case Management Services: 322-5258

Sydney Thompson
213-250-4431
100 W. Edgeware
Los Angeles, CA 90026

Incident Officer P. Fox, Seattle PD, cse #96-499-535

Anthony Slide
4118 Rhodes Ave.
Studio City, CA 91604
(818) 769-4453

Michael Dunn (E! Entertainment network) 323-954-2682, MDunn@Eentertainment.com
Jenna Girard: 323-692-6482, 323-654-7655, JGirard@Eentertainment.com

Jack Thiem
(206) 778 - 5169, 543 - 8510
22815 Lakeview Dr., G - 315
Montlake Terrace, WA 98043

Norman Carl Rabin (Mind Control victim)
31 Cedar Drive East
Plainview, N.Y. 11803
516-349-0560

Mekka Robinson
Probation Counselor, Probation Services Division
Room 1400, Public Safety Bldg.
600 3rd Ave./MS 02-02-23
Seattle, WA 98104-1852
206-615-1966
mekka.robinson@ci.seattle.wa.us
Case:378914
684-7837
Room 1490, Dexter Horton Bldg., 710 2nd Avenue/MS 13-14-01
(case originally assigned to Ameo Butler)

Bob Powers, Western State Hospital

Admitted myself to clinic for immediate care: Oct. 16, 1992 at
5th Avenue Hospital
10560 5th Ave. NE, Seattle, 98125

Charles Hamilton
2003 Western Ave. Ste. 600
Seattle, WA 98121

The following two pairs of people lived upstairs above me at separate times at 3014 NW 75th St.
Shannon Hill
Suzy Coolidge

Thea De Young
Jeff Chrisafelli

Old addresses of mine:
William Thomas Sherman

6322 Woodlawn N.
Seattle, WA 98103
(206) 523-1464

P.O. Box 26225
Los Angeles, CA 90026
or
1377 Lucile Ave.
Los Angeles, CA 90026
213 - 660-0827

3014 NW 75th
Seattle, WA 98117
(206) 784-1132

presently:
1604 NW 70th St.
Seattle, WA 98117

~~~~~******~~~~~

ADDITIONAL REMARKS AND MATERIAL FROM THE POSTINGS AT MY WEBSITE AT GUNJONES.COM

On Friday, 13 August 2004, I had in-person delivered copies of my Narrative (plus all my writings on disk), with explanatory cover letter, to the Seattle offices of the Mayor, City Councilman in charge of Public Safety, Chief of Police, local head of the FBI, the two U.S. Senators, my U.S. Congressional representative, my State representative, and the Catholic Archbishop of Seattle. I did not and have not receive back a single response or reply. I complained about this to the police, and at one point Sgt. Liz Eddy, of the Seattle Crisis Intervention unit came to see me, and heard my story.

I didn't hear anything back from her and some months later on I contacted her office (by phone) again. On Oct 6, 2005, a woman came to see me pretending to be Sgt. Eddy. I know this woman to be an imposter and not the Sgt. Eddy I had previously met with. Now who do I contact to report this to? Another imposter?

It is not a little beyond extraordinary to me, indeed utterly preposterous knowing what I do, that I should be so alone after all these many years, when I have the facts, law, and argument vastly, if not completely and entirely, on my side. Really, these people are so guilty it is beyond hideous and unbelievable.

You have people such as lawyers, priests, professors who will not talk to me. Their behavior is not untypically shifty, evasive, gratuitously hostile. While some have been nice enough to receive me politely and decently – and for this who have been that way I am genuinely thankful – even so no one will call or visit.

And where they don't have an obvious legitimate excuse for indifference otherwise, what are they protecting? Evidently they are being blackmailed and or else are under the influence. Bribed is a possibility but to give them the benefit of the doubt I will assume it is one of the others. I cannot think they are intimidated by someone, rather they are somehow got to think I am someone to be avoided, and treated rudely. Yet it is those they are, knowingly or unknowingly, acting in behalf of that are the truly guilty ones. It is routine of Hell people to accuse and blame when they are the most guilty, in fact the most inhuman of monsters. For this reason, such as these professors, lawyers, or priests then would be led to think ill of me. But how and for what reason exactly, I can't understand. They do have a certain consensus. But to speak to them individually some come across as mad – honestly, without exaggeration or any desire on my part to impute them.

~~~****~~~

On Friday, 13 August 2004, I had in-person delivered copies of my Narrative (plus all my writings on disk), with explanatory cover letter, to the Seattle offices of the Mayor, City Councilman in charge of Public Safety, Chief of Police, local head of the FBI, the two U.S. Senators, my U.S. Congressional representative, my State representative, and the Catholic Archbishop of Seattle. I did not and have not receive back a single response or reply. I complained about this to the police, and at one point Sgt. Liz Eddy, of the Seattle Crisis Intervention unit came to see me, and heard my story.

I didn't hear anything back from her and some months later on I contacted her office (by phone) again. On Oct 6, 2005, a woman came to see me pretending to be Sgt. Eddy. I know this woman to be an imposter and not the Sgt. Eddy I had previously met with. Now who do I contact to report this to? Another imposter?

It is not a little beyond extraordinary to me, indeed utterly preposterous knowing what I do, that I should be so alone after all these many years, when I have the facts, law, and argument vastly, if not completely and entirely, on my side. Really, these people are so guilty it is beyond hideous and unbelievable.

~~~****~~~

For the benefit of those who do not already know and understand how this racket to obtain and hold power using spirit people works, this is essentially how it is. What they do is pretend to be or present themselves as (in effect) either Heaven or Hell depending on whom they want to persuade. If you reject "Heaven" then you can get in with the big money Hell has; if you accept "Heaven" you can't have any money (really) and will have to "go to Heaven" instead (with some possible latitude and variation permissible in between depending on how they want to use the person.) The reality is of course, it is the same group of spirit people; and the purpose is to hold and retain control over the public wealth and people's lives.

I can prove what I am saying with the following witnesses all of whom have refused me help:

* Father Michael Sweeney, former pastor of Blessed Sacrament church in Seattle
* Charles Hamilton, Seattle attorney
* Laurence Bonjour, Prof. of Philosophy, Univ. of Washington
* Cass Weller, Prof. of Philosophy, Univ. of Washington
* Charles Schulman, former English Prof., Univ. of Washington

And there are more I could name as well. Ask these people what they actually know about me, and what their reason then is for refusing me assistance. It is not my purpose to embarrass these individuals or put them on the spot; and if I could I would refrain from bringing their names up in this context in the first place; but their own credibility can throw light on my own. Go talk to these people; compare their various stories and explanations with each other and mine; see who is being honest and who is being rational; and you will, applying some plain rational analysis, see my claims vindicated. This said, again, I by no means intend to vilify or disparage these individuals in raising this point; for one thing they may act or hold the attitude as they do because they are being blackmailed or else under pressure from or under the influence of others whom they cannot deal with – and thus are prevented from speaking the truth as I assume they otherwise would. Yet enough is enough. Let's have the truth out about all this. It is not I who am preventing an open hearing; on the contrary I am the one being refused it!

~*~

...When I went to Harborview hospital last year (i.e., in 2004) to get my case of yellow jaundice, urinary tract infection, obstructed bowel movements, and liver poisoning treated, my assigned physician, Dr. Marcus Groffman, made an off hand comical quip to some staff people nearby about the "cone of silence" – on the basis of which remark I think you get the idea what I am dealing with...

~~~~~******~~~~~~

Appeal of Nov.2008 and Re-Introduction to Sherman "Narrative" (8 Nov. 2008.)

BRIEF SUMMARY AND ABSTRACT OF MOST OF WHAT FOLLOWS:

I am looking for honest, rational, and responsible persons to impartially examine, look into, assess and adjudicate some felony criminal related matters in which I am and have been a victim. Moreover, it is my contention that the reasons and grounds for my inability to receive any assistance in this up to this time – that is for 16 years – is and can be construed itself a serious crime, or at least shows a high degree of negligence and incompetence on the part of the government, police and legal community. This letter then is intended as an introduction and brief summary and not a full and proper exposition of my story and plight. A detailed account of my claims and experiences are to be found on my "Narrative" written a number of years ago, and which is contained on the CD that accompanies this letter (these same materials and more can be found at my website www.gunjones.com). If I could I would provide ready printed versions to you; however, I did something like that one time at a cost of something like $250.00 (I'm living on Social Security disability by the way) and did not receive back a *single* response (more on this particular incident in what follows below.) Consequently, I hope you will pardon me for not providing at this stage ready paper printed versions of the pertinent material.

Although I would not expect or insist that anyone necessarily have any sympathy for me personally, my case, assuming what I claim is true, has grave implications with respect to the welfare of others and indeed, in the estimation of some, could be reasonably deemed a public safety matter of the highest priority.

While I grant my story and claims are highly strange and unusual, this of itself, I would most doggedly insist, does not prove they cannot be true or have a discernable and verifiable basis in truth. And even if *some* of what I claim cannot, for one reason or other, be established, it does not follow that *all* that I allege of a serious nature cannot be proven. I mention this because some seem to take the view that if I cannot prove all of what I claim then it must be that I can prove *none* of it. This, naturally, I maintain, is a grossly false sort of logic, as well as being no less extremely unconscionably unfair to a petitioner claiming to have been a victim of violent crime and vicious abuse for a period spanning years.

Some seem to take the attitude that felony crime can be purchased – that is, if the price is right and the criminals take extra special care, say by means of blackmail, bribery, disinformation, and or other forms of strong arming and duress, to obstruct and prevent an investigation or prosecution that this legally constitutes *de facto* acquittal, and that consequently they are not answerable for their malfeasance, no matter how heinous and incomprehensible the offenses. Such a malicious and cynical attitude and conclusion, I hope, you will agree is in complete violation of the spirit of the Law, and should not on any grounds be tolerated or condoned, and that to tolerate or condone such only serves to attack the law at its very root and foundation.

Who are these people dogging and harassing me? The answer to this, I regret to say is not easy. Yet neither is it utterly baffling, let alone insoluble would but some time and investigation put into who and what I am talking about. Yet so powerful are these criminals that I am vying against that they can and have been successful in preventing inquiries and investigations into my claims. It is my contention then that by step-by-step logic using the scientific method it is applied it is possible to establish most of not all of pertinent claims of fact which I allege, including who specifically are the offenders. Indeed, the honest facts support what I say to such an extent that the substance of my claims can be proven irrefutably, at least to any honest, disinterested, and objective person who makes a reasonable effort to do so.

~~**~~

A wish one often and routinely finds expressed in Voter's Pamphlets and voiced by candidates for court judgeships is the desire that equal access to justice and the legal system be available to all, and in a manner unhindered by unfair prejudice and arbitrary discrimination.

Well, I am a case in point of someone very much in need – and I would argue deserving as well – of legal assistance. In fact, so true and so much is this so that to the degree I have failed to secure any for many years now does itself, it could be very reasonably be argued, constitute a crime and the willful and reckless obstruction of justice – a claim any half decent and fair person, who would they but makes themselves intelligently familiar with my case, would, I am positive, readily agree with.

Ordinarily, at least if we are an adult, we don't expert others to necessarily care about our own troubles and woes; unless perhaps they are family, a special friend, and or else unless we are willing and able, in some way, to purchase or hire their assistance. This sort of attitude and expectation when it comes to life's travails is perfectly normal. However, when even an isolated individual's plight is such as to qualify as horrifying and outrageous, and has gone on for an extended duration no less, then even the least decent citizen of just a little intelligence must be forced to consider that such a state of things may have serious implications for the community at large; not least of which when the said individual's plight is being caused by criminals who are both very violent, malicious, and highly organized and sophisticated.

Such I would plea is my own circumstance. In these past 16 years or so I have come *very painfully* to learn that a person can be put through the most extreme kinds of violence and abuse, and if they are alone and materially poor enough no one will listen to, let alone succor, them – all the less so if those they are up against are rich, powerful, criminals adroit at deceiving and sabotaging relationships and communications. Even if then you feel no need for sympathy for me personally, I very humbly and earnestly entreat you to at least consider what the predicament and victimization that I allege does or may mean for others and the community at large, including perhaps even loved ones of your own.

If "equal legal access to all" then *is* an important value we cherish, this of itself would warrant my receiving ample and speedy aid given that my nightmarish and truly agonizing and unspeakable ordeal, and that goes back at least 16 years and continues even to the present. Yet not only have I been refused assistance but have been denied help, characteristically under the flimsiest and childish kinds of excuses and pretexts, even when the reality and threat of violence were still present and imminent.

Where the positions are held in dire earnest, there are two sides to every question, and yet I who almost two decades have been crying bloody murder have difficulty even getting people to give me the time of day (to use a figure of speech.) Very sadly, for many there seems to be an unwritten law if one is alone no one need listen to them. Regardless of what the laws and legal system formally state or think, this is how it is for some people, and I have been put in a situation where because I have been intentionally and systematically made to be alone that this then opens the gate for the same people to physically (and in other ways) attack, assail and abuse me, sometimes in ways that are so cruel as to be imaginable to most people.

Yet even granting all that I assert above, explaining the wherefore of my past and present state of things is admittedly not without its challenges and difficulties. Some of what I allege some will and perhaps understandably reject out of hand as being preposterous. Yet even if, for the sake of argument (and to be brief), not all of my allegations are allowed, there still remains, I would argue, more than enough evidence to prove and establish that at least some serious crimes have taken place and that this itself and at the very least demands and requires the law's looking into the matters and charges I raise.

Taken as a whole, my circumstances and predicament could be considered rather involved and complex, and, admittedly, it might not be possible to address, get at get at and prosecute every matter with possibly serious legal implications. For one thing, there are and have been so many crimes committed and going on for so very long, it may not even be even possible to just list and record them, at least without a substantial expenditure of time and energy. However, if individual events are examined and weighed in isolation, the task of prosecuting the crimes I allege is *far less* demanding and problematical.

Now there are two subjects I raise in the course of my "Narrative" that some will think utterly fantastical, namely the question of a) spirit people (as in ghosts and the so-called paranormal), and b) brain torture radios (à la purported torture technology such as in an earlier time was famous for being used in former East Bloc countries.) Without here getting into the merit of these as concerns (I will and do so at other locations in the presentation of my case), let me at least put the irksome matter this way. Even if, as some will unthinkingly assume, I am crazy or delusional, does that mean that nothing that I say can be true? If the one, does it then necessarily follow that I could not possibly be a victim of malicious and violent crime? Or that my basic rights need not be considered or respected? Most alarmingly and disturbingly, in the course of my ordeal, some have answered "yes" to these questions. Needless to add, I would like to think there are others who do and would agree with me that such responses are extreme, unjust and grossly reckless and irresponsible.

Again, a major part of my story does involve the occult and people mixed up with the occult, or so I allege, and it is in no small part exactly this that has made these offenders so very dangerous and difficult to investigate or prosecute. What a nightmare it has proven then when in attempting to present my case attorneys and others require that I explain what I am talking about in 25 words or less! That it be further necessary for me to then go on and explain and account for something as outlandish as "brain torture radios" and you can see what an arduous and thankless task is mine. And yet I am and have not been dissuaded by all this from seeking aid and redress knowing as I do that the actual facts and evidence are, if availed of, truly on my side.

Aside from the purported brain torture radios, when I speak of *serious* crimes, it might well be asked – to what exactly am I referring?

First, let me say the amount of rank violence and other crimes I have been subject to now since 1992 is so numerous and so great that it would not even be possible for me to recollect all the occasions on which such took place, even if I were disposed to do so. However, and this said, the following is a *general* list of the kind and some of the things I do and have undergone; for their elaboration and specific instances otherwise, see my "Narrative."

* Poisoning, including such as involved use of a hypodermic needle
* Mutilation by way of scars and other excrescences placed upon my body
* Intentional infliction of severe and aggravated emotional distress
* My pets, some cats, abducted from me and or murdered
* Vandalism (including hacking of my computer and websites), breaking and entering, and document and file tampering on my computer
* My communications, whether e-mail, regular mail, phone obstructed and interfered with, and which has on numerous occasions hindered my obtaining legal or other necessary assistance

And also – though I list this separately for understandable reasons:
* Brain torture radios

* Witchcraft type assaults in which spirit person of varies kinds are and have been used to inflict harm and injury to me.

Is there not in any one of these, let alone all or some of them together, something amiss or some wrong taking place that it is in the purview, indeed the obligation, of the law to look into and, shown to be true as alleged, attempt to correct?

One attorney for years refused to help me on the grounds that I cannot explicitly by name identify who it is that has victimized and subjected me to these purported crimes. To this objection, my response has been and is that even a very slight bit of detective work and investigation of the facts would and will provide and uncover leads and evidence of substance necessary to get at this question. But true it is I concede, if there is no impartial investigation into my claims, I granted am at a loss to provide you with a specific offender. Yet let me if I might express the matter this way. If I can provide an investigator with at least tentative or reasonable grounds for some of the crimes I allege taking place than that, I maintain, is sufficient grounds and justification for either a formal or informal inquiry into my claims and allegations.

At this juncture then let me enumerate a general list of *some* proofs or possible proofs on my which my claim can be sustained. A number of these make reference to events recounted in my "Narrative" and to which I refer you for further explanation.

* There are (or should be) medical and police records connected to many of the incidences of my being attacked, and which records in some instances, and at very minimum, directly or indirectly show or imply something fishy going on.

* My contains an extensive listing of witnesses and possible witnesses who can, or can help to, corroborate my own claims of what took place. There are a number of persons, such as Bruce Long, a historical crime scholar, or Father Michael Sweeney, of the Dominican School of Philosophy of Berkeley, California, who I was at one time on very friendly and cordial terms with, but who refuse to speak with me now. Find out what such a person's reason for why they can't or won't speak with me or else, as applies for the given person, why they can no longer be reached (at least not by me anyway.)

* Try contacting a number of the professionals who rejected my petition, and compare with each other their reasons for doing so; accompanied by questions pertaining to what they actually know about me in the first place (as to have felt it all right to have dismissed my pleas.)

* Relating to my reckless driving offense of 2000, some claimed I hit them (in their car), and as result they collected insurance damages from Geico. This charge of being hit and collecting of insurance I know to be a complete fraud, and

there is not a doubt in my mind these persons who so collected are in some way connected with the group that has been making my life miserable and causing me all these problems these past 16 years.

* Back on 13 August 2004, copies of my "Narrative," with a suitable cover letter, were personally delivered by me to the Seattle offices of the following persons and public officials listed below. And yet I did not receive back a single reply from any of them, including not even a formal letter saying they were not in a position to be of any help. The question that arises then is 'why?' That some couldn't or wouldn't reply would be one thing. But that none could or did seems highly irregular.

* Patrick J. Adams, Special Agent in Charge Federal Bureau of Investigation Seattle Field Office 1110 Third Avenue Seattle, Washington 98101-2904
* Archbishop Alexander Brunett Catholic Archdiocese of Seattle 910 Marion Street Seattle WA 98104
* The Honorable Maria Cantwell 915 Second Avenue, Suite 3206 Seattle, WA 98174
* Councilman Jim Compton Committee of Public Safety 600 4th Avenue Fl 2 PO Box 34025 Seattle, WA 98124-4025
* Charles Hamilton, attorney 2003 Western Ave. Ste. 600 Seattle, WA 98121
* Chief R. Gil Kerlikowske Police Headquarters, 610 Fifth Avenue, Seattle, WA 98124-4986
* Senator Jeanne Kohl-Welles 157 Roy Street, Seattle, WA 98109-4111
* The Honorable Jim McDermott 1809 7th Avenue, Suite 1212 Seattle, WA 98101-1399
* The Honorable Patty Murray Jackson Federal Building, Suite 2988 915 2nd Avenue Seattle, WA 98174
* Mayor Gregory J. Nickels 600 4th Avenue Floor 7 PO Box 94726 Seattle, WA 98124-4726

* If I claim being assaulted by brain torture radios their must at least be some serious basis to my own belief of such; given that on at least two separate occasions back in the early 1990s I went to the medical risk, expense and trouble of having CAT scans done of my head.

* Why is it that there is no one, to my knowledge, able or willing or interested in at least looking into my astonishing claims and charges? One would think after the countless letters of appeal for assistance I have previously sent out, and after years of having my own website on the internet pleading my cause that there would have been at least a single person to at least ask and inquire into my claims if only out of curiosity if nothing else.

And there is other evidence but which requires a further knowledge and understanding of my case in order for it to be introduced.

In conclusion and in the meantime for now – as hard as it is evidently for some to admit and own up to, sinister evil, including certain extreme types of gangsterism, is real and does, in one form or other, exist. The proof of this, if none other, lies in the fact that and if for no other reason, than that there are those who do or will avail themselves of rank evil to secure, maintain, and increase their power and importance. This we see, hear and read about all the time in the news. That some should take things to a monstrous extreme as would constitute rank evil should then come as no real surprise to anyone. And yet who do you know of is willing and prepared to confront and deal with real evil? Not so many, obviously, as one might wish was the case. Those who are responsible for the crimes I allege are such as who do and have availed themselves of the most abject vileness, baseless malice, and cruelty. By my taking a stand against them, I am apparently one of the few that has the courage, intellect and intestinal fortitude to have tried to do so. And yet I am far from being the only one who is and or has been victimized by these monsters. If it is then possible for me to procure assistance in my cause, not only myself but who knows how many other similar sufferers stand to benefit, and regain that level of basic protection from big shot criminals and hoodlum violence that should be seen, if it isn't seen already, as among a citizen's most fundamental of God given rights and entitlements.

~~~****~~~

339

www.ingramcontent.com/pod-product-compliance
Lightning Source LLC
Chambersburg PA
CBHW052341220526
45465CB00003BA/900